Basic Concepts of Anatomy and Physiology

A Self-Study Program

Second Edition

Revising Author:
Patricia Schafer, R.N., M.S.
Assistant Professor, School of Nursing,
University of Pittsburgh

Programmed by the Education and
Training Department, Industrial Division,
Bolt, Beranek and Newman, Inc.
Cambridge, Massachusetts
Under the direction of:
W. B. Dean
G. E. Farrar, Jr., M.D.
A. J. Zoldos
All of Wyeth Laboratories

Basic Concepts of Anatomy and Physiology

A Self-Study Program

Second Edition

J. B. Lippincott Company Philadelphia Toronto

2nd Edition

For information address J. B. Lippincott Company, East Washington Square, Philadelphia, Pennsylvania 19105. Printed in the United States of America.
6 5 4 3 2 1
Library of Congress Cataloging in Publication Data
Main entry under title:

Basic concepts of anatomy and physiology.

1. Human physiology—Programmed instruction.
2. Anatomy, Human—Programmed instruction.
I. Schafer, Patricia. [DNLM: 1. Physiology—Programmed texts.
2. Anatomy—Programmed texts. QS 18 B311]
QP39.B26 1982 612'.007'7 81-20917
ISBN 0-397-54378-6 AACR2

Contents

Preface

Programmed textbooks are unique in that they present information to be learned in small, easily digested pieces and, in most cases, require an immediate response from the reader. Correct answers are rewarded by the satisfaction of accomplishment and errors are corrected at once. Thus the learning process becomes nearly painless and is given the added dimension of immediate feedback. Much accumulated experience has demonstrated that with a well-programmed text, knowledge may be acquired with less effort and in less time than that spent with a standard textbook. Programmed learning comes close to the one-student-one teacher relationship that educators and psychologists agree is the ideal situation for learning.

This revision retains the tone and format of the original book, but the information has been extensively updated and reorganized to emphasize the logical patterning inherent in scientific material. Two new sections—Protection from Disease and The Special Senses—are included to create a more complete presentation of core anatomic and physiologic knowledge.

The book will be useful to a wide range of persons whose programs of study require a basic knowledge of anatomy and physiology, such as those enrolled in allied health curricula or in general college or high school human science courses. The book may be used as the major text for such courses, or it may serve as a supplementary source, providing the student with an alternate style of information presentation. Also, persons whose occupations require that their knowledge of these subjects be refreshed, sharpened, or updated will find this text a flexible resource. It may be studied from beginning to end or examined for particular information. Cross-referencing facilitates the latter process. Originally prepared for training the personnel of a major pharmaceutical manufacturer, this thoroughly revised program assumes no prior contact with the subject matter, but it leads the reader into a fairly sophisticated knowledge of the basic concepts of anatomy and physiology.

Patricia Schafer

1 Basic Biologic Concepts

1–1. Human anatomy and physiology are part of BIOLOGY (bi—OL—o—gy), the science of all living things.

To understand how a living substance—human, animal, plant, or whatever—is arranged and how it functions, you must study the science of ____________________.

biology

1–2. ANATOMY (a—NAT—o—mi) is a subtopic of biology that describes how the body is put together so that it assumes its total form. How body parts are put together to assume a definite form is the study of ____________________.

anatomy

1–3. The human body and its parts have definite form and structure.

Structure is how the body parts are related to each other. This relationship may refer to either the part's appearance or position in the body.

Form is the three-dimensional appearance (not the outline) of the body or one of its ____________________.

parts

1–4. When you describe the body of one of its parts and you include all three dimensions, this is a description of the ____________________ of the body or the parts.

form

1–5. If you tell someone what a body part looks like and where it is located in the body, you are describing the ________________ of the part.

structure

1–6. Don't become confused. Form is not an outline but the three-________________ appearance of a box.

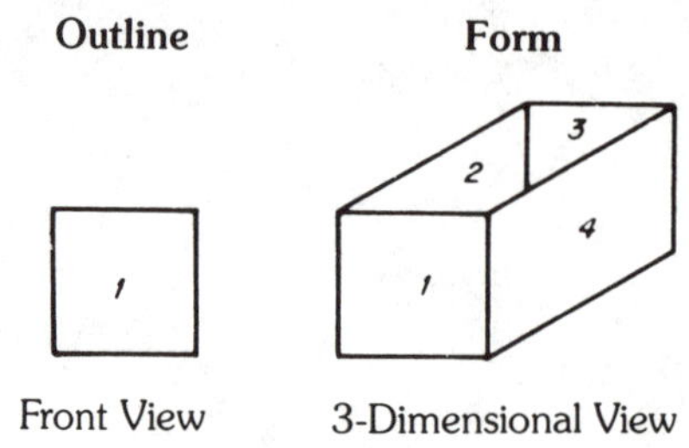

dimensional

1–7. A house has definite form and all of its parts should also have definite form. When you describe the position and appearance of the windows of a house, you are referring to the ________________ of the windows.

structure

1–8. The section of biology dealing with human form and structure is called ________________.

anatomy

1–9. Before the discovery of the microscope, the human body's ________________ and structure, or anatomy, was examined by using the eye as the principal investigating tool.

form

1–10. The anatomist was able to make only a <u>gross</u> or whole examination. This scientist was able to give a gross description of the ________________ body or one of its parts.

whole

1–11. When a doctor examines a human body or any of its related parts, he is making a ________________ examination.

gross

1–12. However, there are many minute parts of the body that cannot be seen by the eye alone. The discovery of the microscope enlarged the study of anatomy. A doctor could make not only a gross examination but also a ________________ examination.

microscopic

1–13. Microscopic anatomy, called <u>HISTOLOGY</u> (his—TOL—o—ji), is that discipline of science that does not depend upon the unaided eye but rather upon the ________________ to examine a body part.

microscope

1–14. When a microscopic examination of a body part is made, it is called microscopic anatomy or ________________.

histology	**1–15.** You might note by a gross examination that a woman was pregnant. If you wished to examine some of the affected parts in greater detail, you would use as a guide the study of ________________.
histology	**1–16.** If you describe the anatomy of a clock but not how it functions, you know but one-half of the story. The anatomy of the clock is important, but of equal importance is how the clock and its parts work or ________________.
function	**1–17.** Function simply means how something works. The study of how the body or its parts function is called PHYSIOLOGY (fiz—i—OL—o—ji). A study of how the clock functions would be called clock ________________.
physiology	**1–18.** The study of the function of the body can either mean how the body as a whole functions or how each individual organ functions. If you study how the stomach functions, you are studying the ________________ of the stomach.
physiology	**1–19.** The study of body form and structure is anatomy, and the study of how the body or its organs function is ________________.
physiology	**1–20.** A mechanic is asked by his boss to assemble a group of parts and describe the finished product. a. This is a study of the "physiology" of the machine. (Move to frame 21.) b. This is a study of the "anatomy" of the machine. (Move to frame 22.)
a.	**1–21.** You said the assembling of the machine and the description of the finished product is a study of the machine's "physiology." Not so! Physiology relates to how a machine—and the body is a machine—and its parts function, not to their form or structure. (Return to frame 20.)
b.	**1–22.** You said the assembling of the parts of the machine and the description of the finished product constitute a study of the machine's anatomy. Right! Anatomy is a study of form (organization of the entire machine) and structure (relationship of the parts). Physiology would be a study of how the machine functions (works).
	1–23. Most of us are able to recognize normal body form. When you study anatomy, you are guided by established guides or norms, which indicate the normal form and ________________ of the body.
structure	**1–24.** An anatomist is a scientist who establishes guides or ________________ for the study of anatomy.

norms	**1–25.** A good auto mechanic is one who understands not only how an automobile is put together but also how the engine ________________ normally.
functions	**1–26.** To assist the mechanic, the manufacturer of the automobile establishes the norms for the car's function. Likewise, the ________________ establishes the norms for the functioning of the human body.
physiologist	**1–27.** A physician, because of the norms established by the anatomist and physiologist, is able to distinguish between ________________ and abnormal body conditions.
normal	**1–28.** The study of abnormal body conditions is a separate medical science called PATHOLOGY (pa—THOL—o—ji). A pathologist is a scientist who studies abnormal ________________ and physiology.
anatomy	**1–29.** The study of disease and sickness is called ________________.
pathology	**1–30.** To understand how a disease influences the body, the ________________ must determine the: 1. cause of the disease 2. progress of the disease 3. effect of the disease on the body
pathologist	**1–31.** An auto mechanic is like the pathologist because he must know the cause of the disorder, how the disorder will progress, and the ________________ the abnormality will have on the operation of the automobile.
effect	**1–32.** A physician is trained to treat the sick. He must, therefore, have a knowledge of the (1) ________________, (2) ________________, and (3) ________________ of disease on the body.
1. cause 2. progress 3. effect	**1–33.** A scientist who is interested in the science of all living things would be called a: a. pathologist b. biologist c. anatomist (Select either a, b, or c and move to frame 34.)
b.	**1–34.** When you visit an automobile showroom and spend some time admiring a new car on display, you are in effect admiring the: a. anatomy of the car (move to frame 36) b. physiology of the car (move to frame 35)

b.

1–35. You said you were admiring the physiology of the car on display. Not so!
Physiology refers to how the car functions or works, not to appearance.
(Return to frame 34.)

---

a.

1–36. You said you were admiring the anatomy of the car on display. Right.
Physiology refers to function, and anatomy refers to the
(1) ________________ of the entire car and how its parts are related, or its
(2) ________________.

---

1. **form**
2. **structure**

1–37. If you examine an animal with your eyes, this study would be:

a. gross anatomy (move to frame 38)
b. histology (move to frame 39)

---

a.

1–38. You said the study would be gross anatomy, and with such an abbreviated description of what was being done to the animal, your answer was the only logical one.

Histology could not be considered because we did not include the use of a microscope in the description of the examination.

(Move to frame 40.)

---

b.

1–39. You said the study would be histology. But wait, was a microscope included in the description of the examination?

Histology is the study of microscopic anatomy.

(Return to frame 38.)

1–40. While walking down a busy street, you saw an attractive person walk by. What were you admiring?

a. her (or his) function
b. gross form and structure
c. histologic form and structure

(Select either a, b, or c and move to frame 41.)

---

b.

1–41. This is a tough one to defend, but let's allow logic to guide us. First, I am sure you did not want (intend) to look at her (or him) under a microscope, so histology is out. Second, you probably were not interested (or were you) in how her (or his) organs worked, so function is out.

(No response required—move to frame 42.)

1–42. You were undoubtedly impressed with the form of the entire body. There is also a good possibility that you were impressed by the structure, or how the parts were related to one another. Because your examination was done by using only the unaided eye, this was a study of the person's ________________ anatomy.

---

gross

1–43. Before you attempt to repair a machine, it is best to know how it functions normally. The study of machine function is called machine ________________.

physiology

1–44. If you understand the normal anatomy and physiology of a machine and its parts, you can predict how it will ________ if a part is broken.

function

1–45. Physicians examine a sick patient to locate any abnormalities of either the anatomy or physiology. If the doctor discovers disease, he must determine the cause, the progress, and the ________ of the disease on the body.

effect

1–46. The study of the cause, progress, and effect of the disease on the body is called ________.

pathology

1–47. You now have finished the first section of this unit. Before you go on you may want to take a short break.

1–48. All natural sciences are interrelated—you should expect biology, chemistry, and physics to overlap in essential areas.

An understanding of chemistry will help you better understand human physiology.

(No response required—move to frame 49.)

1–49. General chemistry has many subtopics among which is the chemistry of life, or BIOCHEMISTRY (bi—o—KEM—is—tree). Biology is the study of all living things and BIOchemistry is the chemistry of ________ things, or life chemistry.

living

1–50. You are constantly being influenced by chemical compounds and elements. For example, if you did not have ample water (H_2O), you would die. It is, therefore, important that you understand how the chemistry of living things, ________, is organized.

biochemistry

1–51. If you were asked to name some chemicals, I am sure you could easily do so. You might say sodium, gold, oxygen, and any number of common chemicals. These units of chemistry are called chemical elements. Hydrogen is such a unit or ________ element.

chemical

1–52. Some philosophers have stated that man is distinguished from all other living creatures by his use of symbols. This may be true when one considers that we have a word to describe our every action.

Chemistry, too, has symbols for each element, and together these symbols form a chemical language. For example, Fe is the symbol for the ________ ________ iron.

chemical element

1–53. There are 105 known chemical elements. Each element is a single substance and has a separate chemical ________ indicating what element it is.

symbol

1–54. A chemist must be able to use these symbols, as you do letters of the alphabet. You are not expected to master the chemist's language, but you should remember that the simplest chemical substance is an element and each element is represented by a ____________.

symbol

1–55. Fe - P - Na

These symbols represent the simplest form of the chemical ____________, iron, phosphorus, and sodium.

elements

1–56. Scientists wondered why some elements combined and others would not. The research of many scientists finally resulted in the discovery of the building blocks of elements—the ATOMS (at—oms). Elements are, therefore, made up of one or more of these building blocks called ____________.

atoms

1–57. When the chemical element gold is broken down into its smallest parts, the smallest part that still physically resembles the element gold and reacts as the gold element normally does with other elements is an ____________ of gold.

atom

1–58. There are 105 known chemical elements so there must also be 105 different building blocks or ____________, one for each of the chemical elements.

atoms

1–59. You would, if mining metals, soon be able to recognize the difference between gold and lead. It is not hard to conceive, therefore, that the difference between these elements (would/would not; choose one) ____________ be because of the fundamental difference in their building blocks, the atoms.

would

1–60. To summarize, you have discovered that

a. the smallest part of an element that can exist is an (1) ____________.

b. there are as many different kinds of atoms as there are (2) ____________.

1. atom
2. elements

1–61. The word HAT can be broken down into the simple symbols (letters) H—A—T. The chemical word H_2O (water) can be broken down into three symbols, H—O—H, to represent two (1) ____________ of hydrogen and one (2) ____________ of oxygen.

1. atoms
2. atom

1–62. As H and O atoms compound to make water, a druggist mixes two or more drugs together to ____________ a medicine.

compound

1–63. In its simplest terms a chemical compound is a combination of two or more different elements.

H + O + H = HOH or H_2O

atom atom atom = compound

When H + O + H are combined, the ____________ water is formed.

compound

1–64. Indicate which are elements and which are compounds (e for element, c for compound).

a. Fe ____________ c. P_2O_5 ____________

b. MnO_2 ____________ d. Na ____________

a. e
b. c
c. c
d. e

1–65. If you combine an atom of sodium (Na) with an atom of chlorine (Cl), you will produce the chemical ____________ table salt (NaCl).
Each combined form (NaCl) is also called a MOLECULE (MOL—e—kul) of salt.

compound

1–66. The chemical compound water, HOH, is composed of three atoms (H—O—H). HOH is a molecule of water. NaCl is the compound table salt—it is also a ____________ of salt.

molecule

1–67. The smallest part of an element is an atom.

Because compounds are made by combining the atoms of elements, the smallest part of a compound would be a ____________.

molecule

1–68. This symbol (Fe) represents an (1) ____________ of the element iron.

This formula (MnO_2) tells you that atoms of the elements Mn and O_2 have combined to form the compound MnO_2, the smallest part of which is a (2) ____________.

1. atom
2. molecule

1–69. During the early 1900s, it was thought that the atom was the smallest particle of matter that could exist. However, the development of the atom bomb was accomplished when the atom itself was split into ____________.

particles

1–70. But let's define what is meant by a PARTICLE (PAR—ti—cle) that is smaller than an atom. An atom is composed of three particles, the electron, the proton, and the neutron. (No response required—move to next frame.)

Hydrogen Atom

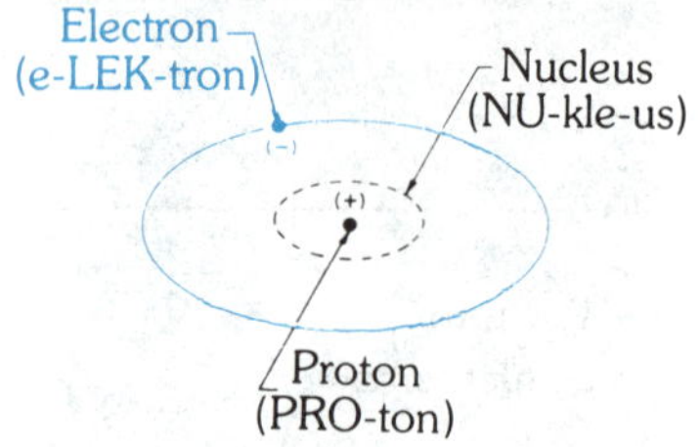

1–71. The hydrogen atom is a simple atom having only two of these small parts or particles, an ____________ and a proton.

Hydrogen Atom

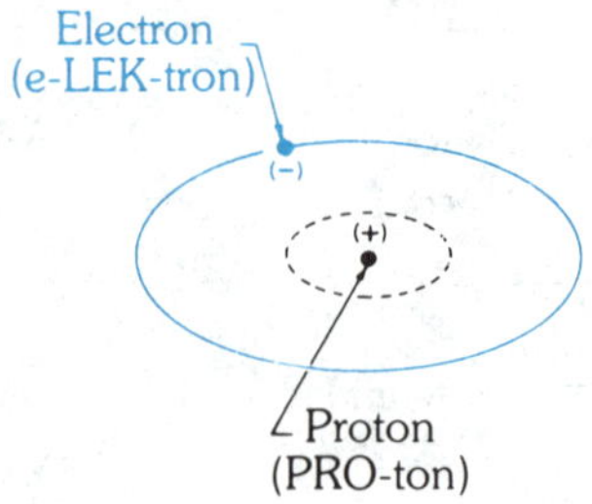

electron

1–72. The compound water (HOH) can be broken down into the atoms (H—O—H). These atoms are composed of small parts or ______________.

---

particles

1–73. The atom is the smallest part of an element that contains all the properties (how the element reacts and what it looks like) of the element.

The particles of the atom are not small bits of the ______________ but are parts of the atom itself.

---

element

1–74. This atom has three different particles: neutrons, (1) ______________, and (2) ______________.

Helium Atom

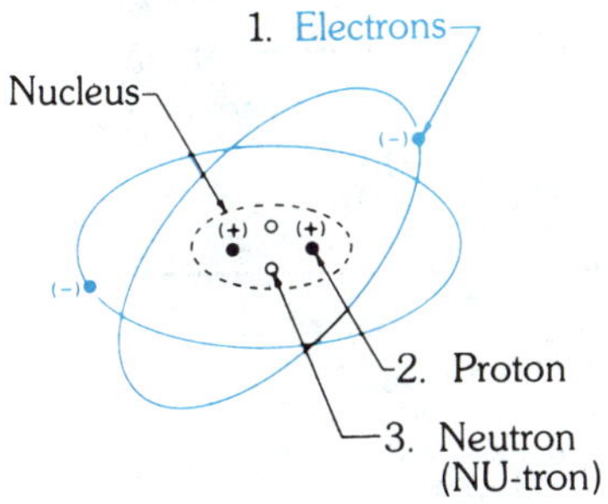

---

1. electrons
2. protons

1–75. All atoms (except hydrogen) are composed of three basic particles: the electron, the proton, and the ______________.

---

neutron

1–76. The atomic particles, except the neutron, carry an electric charge; the proton, a positive (+) charge, and the electron, a negative (−) charge. For example, a lithium atom has three protons or three (1) ______________ charges and three electrons or three (2) ______________ charges.

---

1. positive or (+)
2. negative or (−)

1–77. A beryllium atom has four (1) ______________ or four (+) charges and four (2) ______________ or four negative charges.

---

1. protons (+)
2. electrons (−)

1–78. Atoms are physically different because each atom's nucleus contains a different number of neutrons and ______________.

---

protons

1–79. Atoms have (1) ______________ rotating about the nucleus that is composed of positive (2) ______________ and neutral neutrons.

---

1. electrons
2. protons

1–80. Although the electron that is whizzing about the nucleus does not weigh as much as a proton (an electron weighs 1/1847 of one proton), it does influence how an atom will mix with its neighbors. So the chemical activity of an atom is dependent upon the rotating ______________.

electrons

1–81. Elements differ in physical characteristics and also differ in how they react chemically.

Chemical reactions are controlled by the (1) __________ action, and the physical characteristics are controlled by the (2) __________ and neutrons in the nucleus.

---

1. electrons'
2. protons

1–82. A lithium atom that contains three (+) positive charges (protons) and three (−) negative charges (electrons) is electrically neutral (carries no charge) because the (+) charges equally balance the __________ charges.

(−) or negative

1–83. If an atom has the same number of (−) and (+) charges, it (will/will not) __________ have an electric charge.

---

will not

1–84. Atoms with opposite charges are attracted to each other. A (−) atom (1) __________ a (+) atom.

Atoms with no charge or the same charge are not attracted to each other.

Negatively charged atoms (2) __________ (may/may not) combine with positively charged atoms.

---

1. attracts
2. may

1–85. When a neutral atom loses a (−) charge it has one more (+) charge and it then becomes a __________ charged atom.

(+) positively

1–86. The element oxygen has eight (+) charges and eight (−) charges and is, therefore, neutral.

If, however, the oxygen element gains two electrons, it will have a total of ten negative charges and will become __________ charged.

(−) or negatively

1–87. A charged atom is called an ION (i—on). An atom carrying an excess of electrons (more electrons than it has protons in the nucleus) is an example of an __________.

---

ion

1–88. It is important to remember that an atom can become charged by gaining or losing only its electrons. The protons are held within the __________ and are not able, under normal conditions, to leave the atom.

---

nucleus

1–89. Count the number of electrons and protons.

This atom is an __________ or positively charged atom.

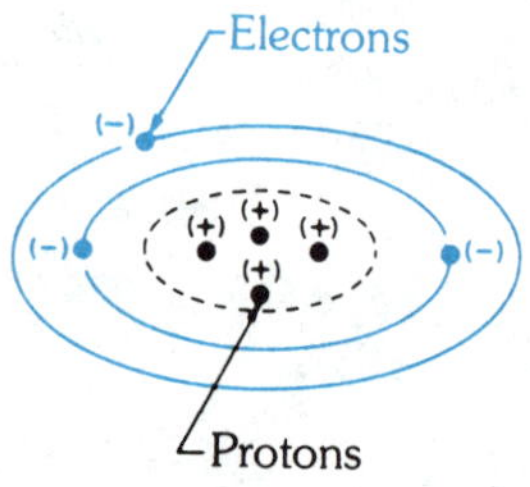

ion

1–90. This atom has three electrons and four protons. It is, therefore, an ion with a ________________ charge.

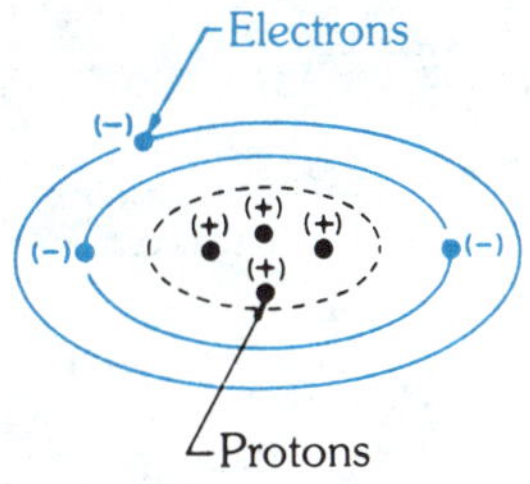

(+) positive

1–91. You can define an ion as an atom that has gained or lost one or more of its ________________.

electrons

1–92. A negative ion may attract a ________________ ion and form a chemical compound.

positive

1–93. Chemists have termed ions as either (−) charged ANIONS (AN—i—on—s) or as (+) charged CATIONS (KAT—i—on—s).

An atom that has lost one or more electrons is a CATION because it now carries a ________________ charge.

(+)

1–94. Atoms that gain one or more electrons are called ________________ because they now have a (−) charge.

anions

1–95. An ANION is attracted toward a ________________, a (+) charged atom.

cation

1–96. If you place an electrically (−) pole in a solution, it will attract the (+) ________________.

cations

1–97. Negative ions or (1) ________________ will be attracted to the (+) pole. Cations will be attracted to the (2) ________________ pole.

1. anions
2. (−)

1–98. This is a ________________ because it has a charge of one more proton than the total number of electrons.

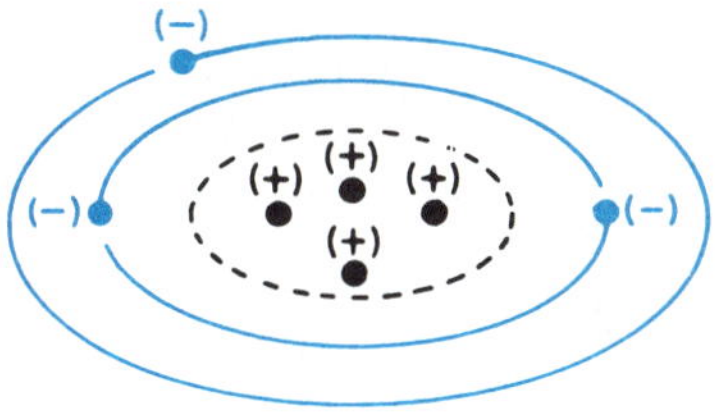

cation

1–99. An atom with six electrons and four protons would be an anion with a charge of (−2) because it has ________________ more electrons than the total number of protons.

2 (two)	**1–100.** You must master this chemical language now because later in the program when we refer to the fluid environment of the human body we will discuss how these ions combine to form chemical __________________.
compounds	**1–101.** If you could return to the "Dawn of Life," you would find all forms of life surrounded by an environment of sea water. As the surface of the earth changed, the fluid environment of living things also changed. Some living things remained in the fluid environment, others enclosed this fluid within their bodies. (Move to the next frame.)
	1–102. Today, human blood serum closely resembles the seawater of the early periods of the earth's history. Man no longer lives in a fluid environment but carries this environment __________________ his body.
within	**1–103.** This fluid environment within the body contains ions. When ions are placed in water, an electrical current can be passed through the solution. Such a solution is called an ELECTROLYTE (e—LEC—tro—lite) solution and is so because it contains __________________.
ions	**1–104.** Dry table salt will not easily allow electricity to pass through it (conduct electricity). However, when table salt (NaCl) is dissolved in water, it dissociates (breaks down) into Na^+ and Cl^- ions. The solution can now be called an __________________ solution.
electrolyte	**1–105.** Water is a poor electrolyte. However, if a chemical is dissolved in the water and it breaks down (dissociates) into __________________ the water becomes a good conductor.
ions	**1–106.** Compounds that break down into ions in solution make the solution a good conductor of electricity or an __________________ solution.
electrolyte	**1–107.** The blood serum, one of the fluid environments of the body, permits compounds to dissociate into ions and is, consequently, an __________________ solution.
electrolyte	**1–108.** The ions that dissociate in your blood serum are called ELECTROLYTES (e—LEC—tro—lites). When table salt is put into a solution and breaks down into Na^+ and Cl^- ions, these should be called the __________________ of the solution.
electrolytes	**1–109.** When your body contains too few or too many electrolytes, sickness or death can follow. If you take a saltwater fish out of its environment and place it in fresh water, it will generally __________________.

die	**1–110.** The fish was originally in the environment of an electrolytic solution with a (higher/lower) ________________ concentration of salt ions.
higher	**1–111.** When the ________________ balance of the fish's environment was changed, sickness and death followed.
electrolytic	**1–112.** It is extremely important that the concentration of electrolytes in solution and the ratio of solution to the available electrolytes be maintained at a normal level. If this balance is upset it becomes difficult to maintain good health. (Move to next frame.)
	1–113. You have been through a rather concentrated section of basic chemistry. A break is in order.
	1–114. The smallest part of an element containing all the properties of the element is a. an atom (move to frame 115) b. a molecule (move to frame 116)
a.	**1–115.** You said an atom was the smallest part of an element containing all the properties of the element. Right! An element is made up of one type of atom only. (Move to frame 117.)
b.	**1–116.** You said a molecule was the smallest part of an element containing all the properties of the element. Not so! A molecule is the smallest part of a compound. (Move to frame 115.)
	1–117. Elements are pure substances that cannot, by normal chemical means, be broken down into parts smaller than an ________________.
atom	**1–118.** H_2SO_4 The above figure is an exampe of a a. symbol b. element c. compound (Select either a, b, or c and move to the next frame.)
c.	**1–119.** Symbols are the letters chemists use to denote elements—sodium (Na)—iron (Fe). A compound or molecule is made by combining two or more ________________.

elements	**1–120.** The number of electrons that revolve about the nucleus of an atom influence how an element will ________________ with other chemical elements.
react	**1–121.** Protons and neutrons that are confined within the nucleus of an atom determine the ________________ appearance of the atom.
physical	**1–122.** Atoms with eight electrons and six protons are called: a. cations (move to frame 124) b. anions (move to frame 123)
b.	**1–123.** You said atoms with eight electrons and six protons are called anions. Right! 8 electrons = −8 6 protons = +6 = −2 charge This atom will go to the (+) charged pole. (Move to frame 125.)
a.	**1–124.** You said atoms with eight electrons and six protons are called cations. Not so! (Review frames 1–93 & 1–94; then return to frame 122.)
	1–125. When a compound is dissociated in a solution it forms ions called ________________.
electrolytes	**1–126.** If common table salt (NaCl) is put into solution, it dissociates into sodium (Na^+) or (1) ________________ (+ions) and chlorine (Cl^-) or (2) ________________.
1. cations 2. anions	**1–127.** An electrolytic solution such as blood serum is one that contains electrolytes and will allow ________________ to pass through it.
electricity	**1–128.** If the electrolytic solution of your body becomes unbalanced (too many or too few electrolytes in solution), it is difficult to maintain good ________________. Certain types of electrolytes in body fluids give the solution the chemical characteristic of acidity or basicity (alkalinity).
health	**1–129.** The acid–________________ balance for each type of body fluid is described according to the concentration of hydrogen ions (H^+) in the fluid.
base	**1–130.** The concentration of H^+ is expressed in units called *p*H units. *p*H is a measure of the acidity of a solution. H^+, therefore, causes a solution to have the chemical characteristics of (an acid/a base).

an acid	**1–131.** Pure water is considered to be neutral and has a pH of 7. Acids have a pH less that 7 and bases have a pH greater than 7. A solution with a pH of 3 would be (an acid/a base).

neutral
← increasingly acidic ↑ increasingly basic →

0	1	2	3	4	5	6	7	8	9	10	11	12	13	14

an acid	**1–132.** A solution with a pH of 1 is ________ (more/less) acidic than a solution having a pH of 5.
more	**1–133.** List the following pH values in order of increasing acidity. pH 7, pH 9, pH 2, pH 12
pH12, pH9, pH 7, pH 2	**1–134.** A pH of 7 represents a solution that is (1) ________. A pH of 1 represents a solution of marked (2) ________. A pH of 14 represents a solution of marked (3) ________.
1. neutral 2. acidic 3. basic	**1–135.** Two kinds of balance that must be maintained in body fluids are the numbers of (1) ________ and the concentration of acids and (2) ________. Each special body fluid has a particular acid-base balance, expressed in units called (3) ________.
1. electrolytes 2. bases 3. pH	**1–136.** Chemical compounds that are NOT produced by living things are called INORGANIC (in—or—GAN—ik). These compounds constitute the ASH of the body. The ash of the body contains INORGANIC chemical ________.
compounds	**1–137.** Chemical compounds that ARE made by living substances are called ORGANIC (or—GAN—ik). Ninety-six percent of your body is composed of ORGANIC compounds; the remaining 4% are not made by your body and are, therefore, ash or ________ compounds.
inorganic	**1–138.** Plants manufacture simple sugars. These compounds, because they are produced by living things, are ________ compounds.
organic	**1–139.** Compounds of calcium, sodium, potassium, and chlorine that are not manufactured by living things are ________ compounds.
inorganic	**1–140.** The three main organic compounds of the body are CARBOHYDRATES (kar—bo—HI—drates; sugars and starches), LIPIDS (LI—pids; fats), and PROTEINS (PRO—te—ins). The protein of meat is an ________ compound.

organic

1–141. Carbohydrates, lipids, and proteins are all organic chemicals because they are made by ______ things.

living

1–142. The three main organic chemical compounds used by the body as the major sources of food are ______, ______, and ______.

lipids (fats); carbohydrates (sugars); proteins; (any order)

1–143. You may accurately describe man as a blob of "living stuff." A more scientific term for "living stuff" is PROTOPLASM (PRO—toe—plazm).

This "living stuff" or ______ is composed of both organic and inorganic compounds.

protoplasm

1–144. All living things, plants and animals, are composed of this living material called ______.

protoplasm

1–145. Blood cells, skin, and bone (are/are not) ______ all forms of protoplasm.

are

1–146. Protoplasm is a colorless, viscous material that is in a semi-liquid state because of the particles suspended in it.

Try to imagine protoplasm as being like raspberry jam. The raspberry seeds would be the largest particles that are ______ in the protoplasm.

suspended

1–147. Protoplasm is composed of (1) ______ compounds containing calcium, sodium, potassium, and chlorine, and (2) ______ compounds containing lipids, proteins, and carbohydrates.

1. inorganic

2. organic

1–148. So far, we have talked about living things, but what does it really mean "to be alive?"

The basic characteristics of a living substance are things you do every day, such as moving, eating, and responding to stimuli.

(No response required; move to frame 149.)

1–149. Without being prompted or prodded, protoplasm can move from one area to another. This property of protoplasmic movement is called MOTILITY (mo—TIL—i—ti). MOTILITY is one of the life functions that is characteristic of all ______ things.

living

1–150. A piece of equipment may be mobile or easily moved. Such equipment depends upon an outside force for its movement.

Protoplasmic motion or ______ can take place without the assistance of an outside force.

motility	**1–151.** Some protoplasm moves by a flowing action similar to jam moving down the side of the jar. At other times protoplasm depends upon small hairlike oars for its ________________.
motility	**1–152.** Have you noticed that many times you are not hungry until you look at the clock? The clock is the STIMULUS (STIM—u—lus) that causes you to act and in this case to think of ________________ or eating.
food	**1–153.** The process of responding to a stimulus is termed IRRITABILITY (ir—i—ta—BIL—i—ti). Protoplasm exhibits this function of irritability or responding to a ________________.
stimulus	**1–154.** Environmental change is an example of a stimulus that can cause an ________________ reaction in a living thing.
irritability	**1–155.** When protoplasm responds to a stimulus, the protoplasmic reaction is called ________________.
irritability	**1–156.** Protoplasm, when stimulated, sets up a wavelike motion that is similar to a stone's being dropped into a pool of water. This wavelike motion carries the ________________ to every section of the protoplasm.
stimulus	**1–157.** There are many stimuli that may affect protoplasm and cause it to respond. Heat, electricity, or mechanical stimuli may induce the protoplasmic response called ________________.
irritability	**1–158.** Protoplasm is truly dynamic—it never rests. Changes within the protoplasm are occurring continually. Some of the protoplasmic material is being built up and other portions broken down. Protoplasm is in a constant state of ________________.
change	**1–159.** This process of construction and destruction within protoplasm and the removal of waste is called METABOLISM (me—TAB—o—lis—m). Some of the metabolic activity is ________________, whereas another phase is devoted to the destruction of the protoplasm.
constructive	**1–160.** When protoplasm is in the constructive phase, it is termed ANABOLISM (a—NAB—o—lizm). New protoplasm is made during the anabolic phase of ________________.

metabolism	**1–161.** Metabolism has a destructive phase. When protoplasm is being destroyed, it is called CATABOLISM (ca—TAB—o—lizm). Protoplasm is being broken down during the: a. anabolic phase. (Move to frame 163.) b. catabolic phase. (Move to frame 162.)
b.	**1–162.** You said catabolism is the phase of metabolism in which protoplasm is being broken down. Right! It is during catabolism that protoplasm is broken down to release energy for body functions such as irritability. (Move to frame 164.)
a.	**1–163.** You said anabolism is the phase of metabolism in which protoplasm is being broken down. Not so! (Move to frame 164.)
	1–164. During anabolism, protoplasm is being built up to balance the action of ____________, or destruction of protoplasm.
catabolism	**1–165.** The greatest protoplasmic growth would take place during: a. catabolism (move to frame 167) b. anabolism (move to frame 166)
b.	**1–166.** You said that the greatest growth would take place during anabolism. Right! During the process of anabolism, protoplasm is used and new protoplasm made. Remember, anabolism means to build up—catabolism, to tear down. (Move to frame 168.)
a.	**1–167.** You said that the greatest growth would take place during catabolism. This is not so! (Return to frame 165.)
	1–168. It is not difficult to imagine what would happen to the body if the process of catabolism were more active than the process of anabolism. The body (would/would not) ____________ be in a declining or unhealthy state.
would	**1–169.** As protoplasm is broken down during catabolism to produce energy, the waste formed must be eliminated. This process of elimination of waste, or EXCRETION (eks—KRE—shun), is another metabolic activity of all ____________ things.
living	**1–170.** Waste is sifted out and eliminated from the body by the process called ____________.

excretion

1–171. The activities that all living substances must exhibit are
a. spontaneous movement, or (1) ____________________
b. response to a stimulus, or (2) ____________________

---

1. motility
2. irritability

1–172. c. metabolism, or the (1) ____________________ and (2) ____________________ of protoplasmic material, and the selection and elimination of waste or (3) ____________________.

---

1. anabolism
2. catabolism
3. excretion

1–173. There is a fourth life function—that of perpetuating protoplasm by growth and division to make more protoplasm. This function of producing more of ones kind is called ____________________.

---

reproduction

1–174. The life functions enable living substances to maintain internal stability. HOMEOSTASIS (HO—me—o—STAY—sis) is the term that describes this internal ____________________.

---

stability

1–175. ____________________, or internal stability, is maintained by the abilities of the body to respond to internal and external changes.

---

homeostasis

1–176. The life functions are regulated by special mechanisms that respond to changes in the (1) ____________________ and (2) ____________________ environments.

---

1. internal
2. external

1–177. Special mechanisms that help the body maintain ____________________ are called REGULATORY MECHANISMS.

---

homeostasis

1–178. Just as the mechanic can "read messages" from the auto that certain parts are not functioning properly, the regulatory ____________________ of the body "read" chemical messages that some internal or external stimulus has occurred. These mechanisms, like the mechanic, must respond and adjust some part or parts so that the whole machine can run properly again.

(Information about particular regulatory mechanisms will be found as you proceed through the following chapters.)

---

mechanisms

1–179. The ability of the body to respond to stimuli, make adjustments, and restore balance between internal and external environments is called ADAPTATION.

(No response required.)

1–180. For adaptation to occur, the body must be able to respond to environmental ____________________. Internal and external changes that cause the regulatory mechanisms to act to restore balance are called STRESSORS.

stimuli	**1–181.** Stressor is a term coined by Selye in the 1950s to describe any factor that produces stress; stress is the sum total of all biologic reactions to internal or external ________________.
changes	**1–182.** When a stressor occurs, the body tries to adapt to it by calling on regulatory ________________ to "read" the stressor message and make adjustments in the physiologic activities of the body.
mechanisms	**1–183.** The goal of the stress-adaptation cycle is to help the body maintain internal stability, a state of ________________.
homeostasis	**1–184.** Maintaining the acid-base balance of body fluids is an important type of internal ________________, or homeostasis.
stability	**1–185.** Buffer systems are one type of regulatory mechanism that acts to maintain acid-________________ balance.
base	**1–186.** Buffer systems are chemical pairs of acids and ________________ that are available in the fluid portions of the body.
bases	**1–187.** During the processes of metabolism many chemical substances are produced or broken down. To maintain a normal *p*H range in body fluids during metabolic activities, chemical ________________ systems release acidic or basic compounds.
buffer	**1–188.** This is just one of the regulatory mechanisms the body has available to maintain acid-base balance. Other methods will be discussed in later chapters. No response required.

Now Let's Review

	1–189. Compounds that comprise 4% of the body and are not made by the body are ________________ chemical compounds, or the ash of the body.
inorganic	**1–190.** Carbohydrates, lipids, and proteins are all examples of ________________ chemical compounds.
organic	**1–191.** All living things are composed of the "living stuff" called ________________.

protoplasm

1–192. Protoplasm is a combination of water, inorganic and organic compounds. You could describe its physical appearance as being a colorless, viscous material in a ________ state.

semi-liquid

1–193. The total sum of the anabolic and catabolic activity plus excretory action is called ________.

metabolism

1–194. You can classify a substance as being alive if it performs the following four life functions:

1. motility
2. irritability
3. metabolism
4. ________

reproduction

1–195. The life functions help living things to maintain ________.

homeostasis

1–196. When an internal or external change occurs, the human body can adjust itself to those changes by the action of ________ ________.

regulatory mechanisms

1–197. The ability of the body to react to stressors and adjust is called the stress-________ cycle.

adaptation

1–198. When any part of the stress-adaptation cycle breaks down, the body will not maintain internal stability, or ________, and disease will occur.

homeostasis

1–199. The biologic language that you have studied in unit one will be used in every unit of this course. For example, unit two will contain information about cells which are specialized forms of protoplasm. Cells have a definite anatomy, and their physiology is dependent upon the actions of ions, organic, and inorganic compounds. Like all living substances, cells must carry out the four life functions.

You should take a break of an hour or more before beginning Chapter 2—The Human Cell.

2 The Human Cell

2–1. The human body is composed of many highly specialized protoplasmic masses called CELLS. Your arms, legs, and other body parts are composed of these "building-block" units called ____________.

cells

2–2. The human body has a definite anatomy because of the three-dimensional ____________ and structure of the cell.

form

2–3. Because a human cell cannot be seen by the unaided eye, people tend to be confused about the size of body cells.

Scientists who study cellular anatomy use an optical instrument, the ____________, to identify minute cell structures.

microscope

2–4. Human cells are not measured in inches but rather in microns. The symbol for a micron is the Greek letter μ (pronounced "mu"). In measure, a micron is equal to 1/1000 of a millimeter. Because one millimeter is about 1/25 of an inch, a ____________ is equivalent to approximately 1/25,000 of an inch.

micron or μ

2–5. A nerve cell measuring 60 μ or 60 (1) ______________ in length is equal to approximately:
(2) a. 60/25,000 of an inch
b. 60/250,000 of an inch
c. 30/1250 of an inch

1. microns
2. a.

2–6. The average size of a human body cell is 20 microns, which can also be written using the micron symbol as 20 ______________.

μ

2–7. It is difficult to conceive what a micron really means, or to visualize accurately how small many microscopic organisms are. However, most single-celled bacilli (a form of bacteria) measure about one micron. This means that it would take more than 1500 of these bacilli laid side by side to span 1/16 of an inch!

(No response required—move to frame 8.)

2–8. You have probably noted that a single isolated soap bubble is spherical. When a single soap bubble is "sandwiched" in among many soap bubbles, it is no longer spherical but compressed into an ______________ shape.

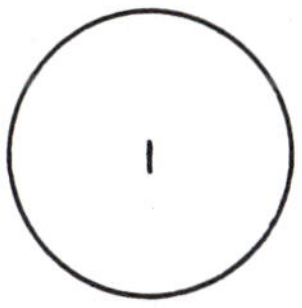

Spherical

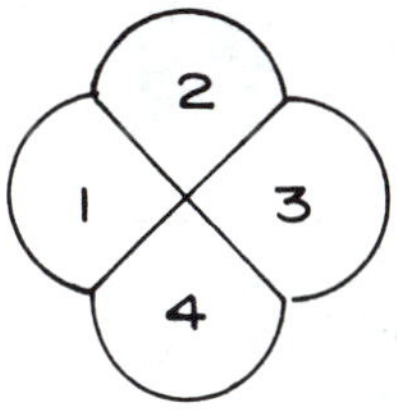

Irregular

irregular

2–9. A single isolated body cell can be compared to an isolated soap bubble because it has a ______________ ______________ and is three-dimensional.

spherical form

2–10. The form of an isolated cell may not always be spherical. For example, the forms of those specialized cells that have ______________ body functions to perform are not spherical.

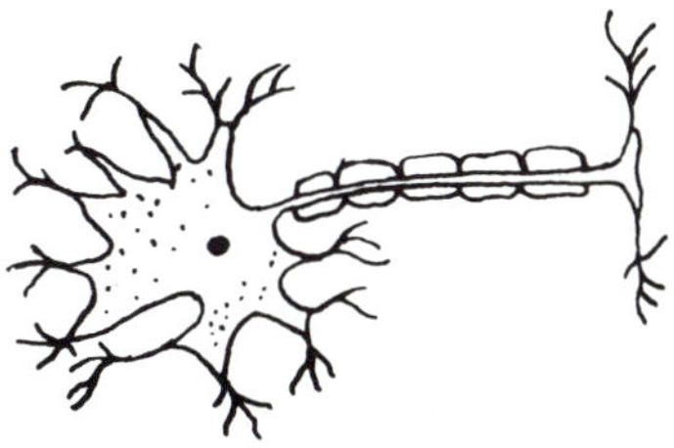

Nerve Cell

Muscle Cells

specialized

2–11. Because of printing limitations, the cells illustrated in this program are drawn in two dimensions. You must remember however, that your body cells have form and structure and are all ________________.

Anatomy of a Body Cell

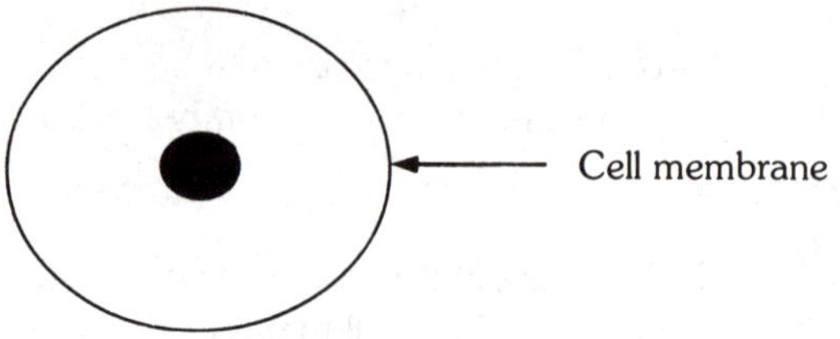

Cell

three-dimensional

2–12.
Every body cell, whether it has a spherical or an elongated form, is surrounded by a protective cover, the cell ________________.

membrane

2–13. The cell membrane controls the transfer of body substances into or out of a cell. For example, a cell membrane allows nutrients to pass into a cell and waste to pass out of the cell. Those substances that normally stay within the cell are kept inside because of ________________ ________________ action.

cell membrane

2–14. A membrane that permits only selected substances to pass through it is called a SEMIPERMEABLE (PUR—me—a—bl) membrane.
All body cells have semi ________________ membranes.

permeable

2–15. Nerve cells and muscle cells, like other body cells, have a ________________ cell membrane.

Cell

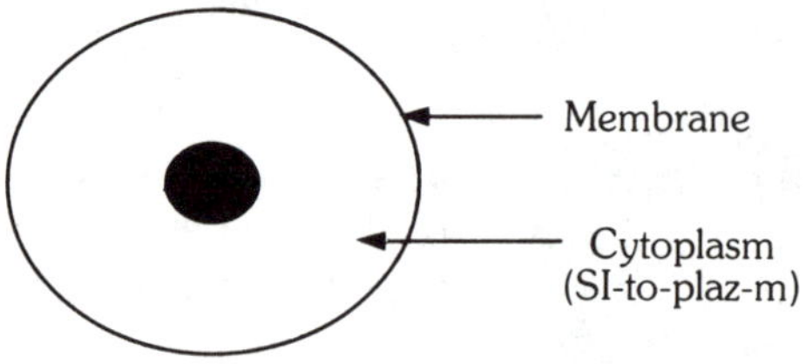

semipermeable

2–16.
Within the cell membrane is the "machine shop" of the cell, called the ________________.

cytoplasm

2–17. The cell functions of motility and irritability are carried out by this cell "machine shop," the ________________.

cytoplasm	**2–18.** You know that the cell is a mass of protoplasm, or "living stuff." The individual parts of a cell are composed of the same material as the whole, but because these parts perform specialized functions, they are said to be a specialized form of ________________.
protoplasm	**2–19.** The cell cytoplasm is an example of ________________ protoplasm. Further specialization is seen in the particles suspended in cytoplasm.
specialized	**2–20.** Because the cytoplasm is a specialized form of protoplasm, it has organic and inorganic substances ________________ in it.
suspended	**2–21.** In addition to suspended particles, cytoplasm contains special structures called ORGANELLES. Each type of organelle has a special ________________, and a particular function, or metabolic activity.
structure	**2–22.** MITOCHONDRIA are special structures in cells that are responsible for energy production. Mitochondria are one type of ________________.
organelle	**2–23.** LYSOSOMES are special structures in cells that break down large particles and digest waste products. Lysosomes are another type of ________________.
organelle	**2–24.** Organelles that are active in energy production are called ________________.
mitochondria	**2–25.** Organelles that break down large products and digest waste products are called ________________.
lysosomes	**2–26.** The metabolic activities carried out by organelles require the action of particular proteins called ENZYMES. Special proteins that act to regulate metabolic activities are called ________________.
enzymes	**2–27.** Production of protein occurs in another special cellular structure, or ________________, called a RIBOSOME.
organelle	**2–28.** Some of the organelles that are important in the metabolic activities carried out by cells are (1) ________________, (2) ________________, and (3) ________________.
1. mitochondria **2. lysosomes** **3. ribosomes**	**2–29.** In addition to protein, two other nutrient compounds are (1) ________________ and (2) ________________.

1. lipids 2. carbohydrates	**2–30.** Just as ribosomes are active in the production of protein, other cellular structures, or ________________, participate in carbohydrate and lipid production.
organelles	**2–31.** The ENDOPLASMIC RETICULUM (EN—do—plaz—mic re—TIC—u—lum) is an organelle shaped like long channels that extend through the cytoplasm. Two forms of endoplasmic reticulum are present: 1. rough endoplasmic reticulum, which is grainy in appearance due to the presence of ribosomes on its surface. 2. smooth endoplasmic reticulum, which contains no ________________ on its surface.
ribosomes	**2–32.** The function of rough endoplasmic reticulum is to participate in the production of ________________ that will be exported from the cell.
protein	**2–33.** Because smooth endoplasmic reticulum contains no ribosomes, it is not active in ________________ synthesis but helps in the production of LIPIDS.
protein	**2–34.** Another organelle that works with rough endoplasmic reticulum to ________________ protein from the cell is the GOLGI (GOL—gee) APPARATUS.
export	**2–35.** The Golgi apparatus produces large carbohydrate molecules that combine with protein formed by the ________________ ________________ ________________. These carbohydrate-protein compounds are released by the cell to its surroundings.
rough endoplasmic reticulum	**2–36.** The organelles listed below are active in producing particular organic compounds. Identify each compound. 1. rough endoplasmic reticulum a. ________________ 2. smooth endoplasmic reticulum b. ________________ 3. Golgi apparatus c. ________________
a. protein b. lipid c. carbohydrate	**2–37.** You recall from Chapter 1 that one of the life functions, ________________, the construction and destruction of nutrient material, is carried out in the protoplasm of cells.
metabolism	**2–38.** When organelles are active in producing organic substances and constructing new material, the metabolic process is called ________________.
anabolism	**2–39.** Anabolism uses energy to build up new substances. Breaking down substances, the process of ________________, releases energy.

catabolism	**2–40.** For the body to have energy available to help it adapt to environmental changes, and carry out life functions, there must be a balance between the processes of (1) ________ and (2) ________.
1. anabolism 2. catabolism	**2–41.** The organelles active in energy production are the ________.
mitochondria	**2–42.** In the mitochondria, substances are broken down and the ________ produced is stored in phosphate compounds called ADENOSINE (a—DEN—o—sin) TRIPHOSPHATE, or ATP.
energy	**2–43.** When energy is needed by organelles in the cytoplasm to build up substances such as protein, ATP releases its stored ________ in the form of a phosphate ion. In this process, ATP is converted to ADP, or ADENOSINE DIPHOSPHATE.
energy	**2–44.** Anytime the body is called on to do work, such as breathing, walking, or building new cell substances, energy must be released from ________ and be made available to the cellular organelles in the active area.
ATP	**2–45.** Before you go on, review what you have learned so far. An isolated body cell is roughly ________ in form.
spherical	**2–46.** The size of an average human body cell is approximately 20 ________, or 20/25,000 of an inch.
microns	**2–47.** Every human body cell is surrounded by a cell ________.
membrane	**2–48.** The "machine shop" within the cell membrane is a type of specialized protoplasm called ________.
cytoplasm	**2–49.** The cytoplasm carries out the life functions of (1) ________ and (2) ________.
1. motility 2. irritability (either order)	**2–50.** The cytoplasm contains special structures called ________, which have particular functions in the cell's metabolic activity.

organelles

2–51. Near the center of the cell is an egg-shaped body, the NUCLEUS (NU—cle—us). The word nucleus comes from a Latin word meaning the kernal of a nut.

In the same way that the kernel of a nut is within the nut, the ____________ of a cell is also located inside the cell proper.

Cell

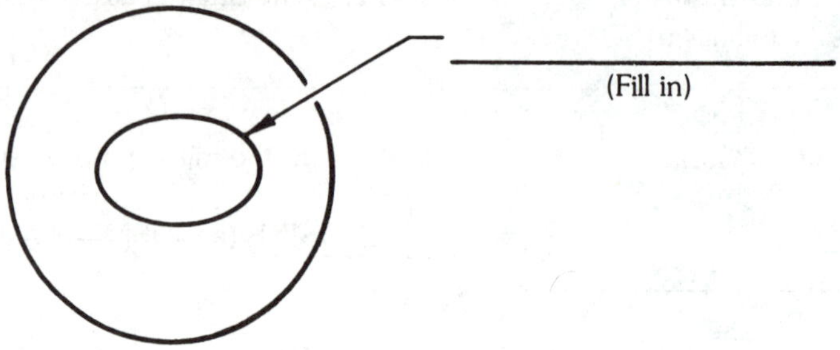

nucleus

2–52. The arrow in this drawing points to the ____________ of the cell.

Body Cell

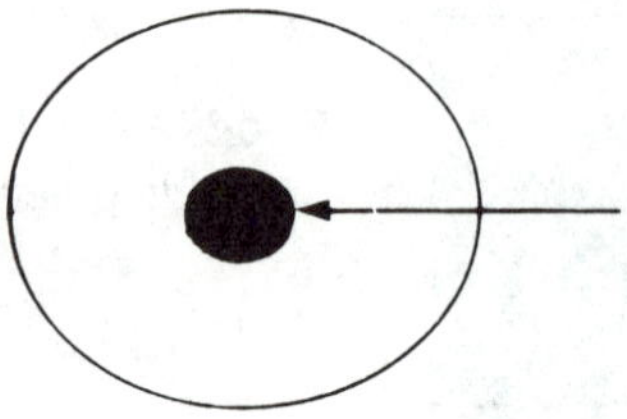

nucleus

2–53. A BINUCLEATE (bi—NU—kle—ate) cell is one with two nuclei (nuclei is the plural of nucleus).

A MULTINUCLEATED (multi—NU—kle—ated) cell is a cell with many ____________.

nuclei

2–54. Although most cells have only one nucleus, there are cells with two nuclei, or binucleate cells, and cells with many ____________, or multinucleate cells.

nuclei

2–55. The nucleus acts as a "control center" for the vital activities of the cell such as growth, reproduction, and the development of new cells. If the body requires new cells to repair damage, the ____________ will initiate this action.

nucleus

2–56. The living substance within the nucleus, called NUCLEOPLASM (NU—kle—o—plazm), is somewhat more dense than the remaining protoplasm of the cell. The cell protoplasm is composed of specialized types called (1) ____________ and (2) ____________.

1. **nucleoplasm**
2. **cytoplasm**

2–57. A pliable membrane, the NUCLEAR MEMBRANE, surrounds the ____________, or living nuclear substance.

nucleoplasm

2–58. NUCLEOPROTEINS are acidic materials contained in the nucleoplasm. Nucleoproteins are compounds of nucleic acids and ________________.

proteins

2–59. Proteins and nucleic acids combine to form the ________________.

nucleoproteins

2–60. Here is a chart to help you picture the composition of the nucleoproteins:

NUCLEOPROTEINS

PROTEINS

NUCLEIC ACIDS

1. RNA
2. DNA

2–61. The nucleic acids are divided into two types, called RNA AND DNA. The full names of these acids are ribonucleic acid and deoxyribonucleic acid, respectively, but for the sake of simplicity they are usually referred to by their abbreviations, ________________ and ________________.

RNA, DNA (either order)

2–62. RNA and DNA are both classified as ________________ acids.

nucleic

2–63. Research done by Avery in 1944 indicates that DNA plays a role in the transmission of heritable variations. The characteristics you inherit from your parents appear to be influenced by the nucleic acid ________________.

DNA

2–64. RNA is found in the nucleoplasm in small, rounded bodies called nucleoli. In the NUCLEOLUS, RNA and protein combine to form the organelles involved in protein synthesis, the ________________.

ribosomes

2–65. Once formed in the ________________, the ribosomes can move to the cytoplasm, where some become attached to the endoplasmic reticulum.

nucleolus

2–66. Since they are composed of RNA and protein, ribosomes are ________________.

nucleoproteins

2–67. The primary RNA activity occurs in the cytoplasm of the cell and involves the formation of specific cellular proteins. However, this ________________ formation function, though carried out by RNA, is controlled by DNA.

protein

2–68. DNA is found in all nuclei, and it is confined to the dark heavy threadlike substances of the nucleus called CHROMOSOMES (KRO—mo—soms). Human body cells have 46 of these threadlike ________________, arranged in 23 pairs.

chromosomes

2–69.

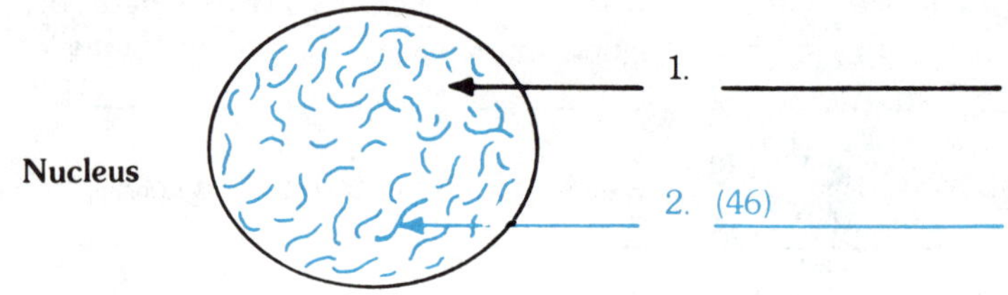

1. nucleoplasm
2. chromosomes

2–70. The chromosomes within the nucleoplasm are concerned with the cell's HEREDITY (he—RED—i—ty)—characteristics such as size, shape, and number of nuclei in the cell. When a cell reproduces—by dividing in two—the ________________, which are concerned with the hereditary characteristics, are split.

chromosomes

2–71. Each chromosome consists of thousands of GENES arranged in a particular pattern along its length. The hereditary messages are encoded in the particular structure and arrangement of ________________ along each chromosome.

genes

2–72. Genes carry the messages for the cell's activity and provide information so that the activity can be regulated. Special proteins that regulate cell activity are called (1) ________________. Enzyme activity is guided by messages from (2) ________________.

1. enzymes
2. genes

2–73. Since it is DNA that is responsible for the (1) ________________ characteristics, the structure of each gene is a special arrangement of (2) ________________ particles.

1. hereditary
2. DNA

2–74. Let's review what you know about the nucleus.

The protoplasm of the nucleus is called the (1) ________________ and is surrounded by a (2) ________________ ________________.

1. nucleoplasm
2. nuclear membrane

2–75. The most important materials of the nucleoplasm are the ________________, composed of nucleic acids and proteins.

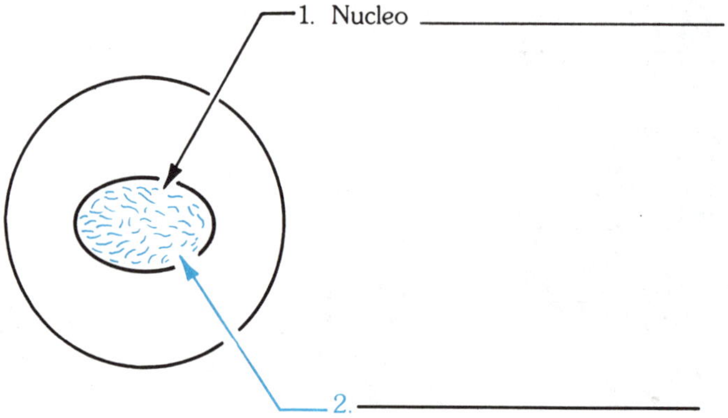

nucleoproteins **1. plasm** **2. nucleoproteins**	**2–76.** The nucleic acid that influences heritable variations is (1) ________________. The nucleic acid that functions primarily in the cytoplasm is (2) ________________.
1. DNA **2. RNA**	**2–77.** The cell's hereditary characteristics are carried by threadlike substances within the nucleus, called ________________.
chromosomes	**2–78.** The hereditary messages are coded by the specific arrangement of (1) ________________ particles for each of the (2) ________________ that are organized on each chromosome in a special sequence.
1. DNA **2. genes**	**2–79.** Cell theory states that: 1. All (a) ________________ things are composed of cells. 2. All cells are a mass of (b) ________________. 3. All cells exhibit the same four life functions.
a. living **b. protoplasm**	**2–80.** The life function of reproduction indicates that all new cells come only from previously existing ________________.
cells	**2–81.** When the body parts are repaired, or the body itself grows, the cell must exhibit the life function of ________________.
reproduction	**2–82.** The repair of skin after a superficial injury and the growth of hair and fingernails are all examples of cell ________________.
reproduction	**2–83.** Most cells reproduce when they are mature. An exception is the red blood cell, which loses its nucleus before maturity, and must, therefore, be produced through another process involving immature cells. But <u>sometime</u> during their life histories, ________________ body cells are able to reproduce.
all	**2–84.** Each cell reproduces by dividing roughly in half to form two new cells. The normal process by which cells ________________ to make new cells is called <u>MITOSIS</u> (my—TOE—sis).
divide	**2–85.** Mitosis proceeds by a series of definite, orderly phases. At the end of the series the one old cell has been divided into two entirely new cells, and the process of ________________ is completed.

mitosis	**2–86.** When cells are not exactly engaged in reproduction, they are in a state called INTERPHASE. They can be compared to the engine of your car while it is idling at a stoplight: it is performing all the functions of running except moving you and the car forward. When cells are in interphase, they are performing all the normal functions of life except carrying out the process of ________________.
mitosis	**2–87.** When cells are not reproducing but are otherwise carrying out all the normal life functions, they are said to be in the ________________.
interphase	**2–88.** Several hours before mitosis begins, the chromosomes double in number by forming an exact copy of each paired strand. Chromosome duplication occurs during the cell state called ________________.
interphase	**2–89.** Mitosis actually begins with the PROPHASE (PRO—phase). The 46 chromosomes, with a copy of each strand attached to its partner, become thick and more conspicuous as the cell enters this first phase, the ________________.

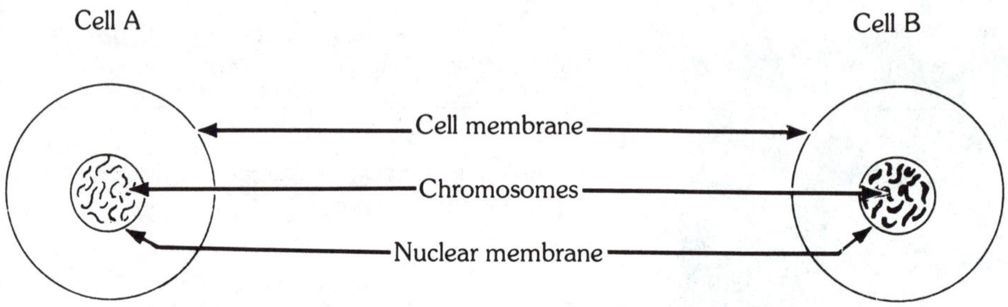

prophase	**2–90.** Cell A is in interphase. The thickening of the chromosomes in cell B indicates that it has entered the ________________.
prophase	**2–91.** As the prophase continues, the individual chromosomes split lengthwise into exact pairs, and the nuclear membrane disappears. The 46 original chromosomes are ________________ in number, arranged in 46 pairs.

Prophase

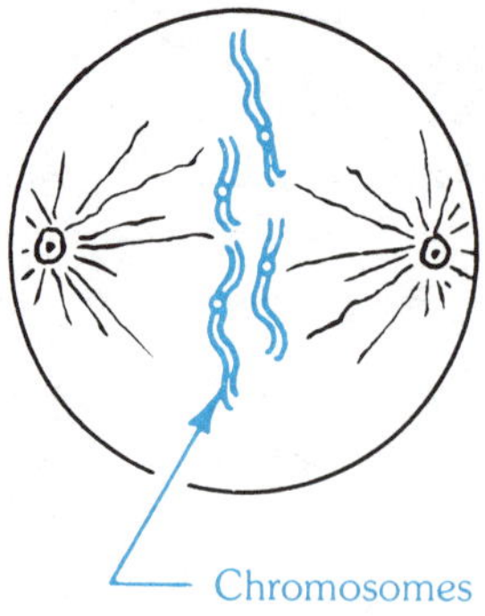

Cell A | Cell B

Prophase

92 (or double)

2–92.
Cell A has just entered the prophase. In cell B the chromosomes have (1) ____________, and the (2) ____________ has disappeared.

---

1. split
2. nuclear membrane

2–93. After the prophase is completed, the <u>second</u> stage of mitosis, the METAPHASE (MET—a—phase), begins. During the metaphase, each chromosome strand and its copy become attached to a delicate system of fiberlike substances that have arisen between the opposite poles of the cell. The chromosomes are suspended midway between these 2 ____________ of the cell.

Metaphase

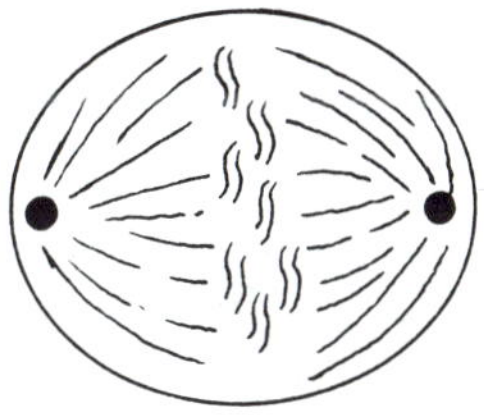

poles

2–94.
This drawing shows the chromosomes attached to the fiberlike substances extending from the poles of the cell during the ____________.

---

metaphase

2–95. The stage when the nuclear membrane disappears and the 46 chromosomes split into pairs is termed the late ____________.

---

prophase

2–96. The stage after the prophase, when the chromosomes are attached to the fiberlike substances extending from the poles of the cell, is called the ____________.

---

metaphase

2–97. The <u>third</u> phase of mitosis is the <u>ANAPHASE</u> (AN—a—phase). During the anaphase, each chromosome strand and its copy finally separate completely and move toward the opposite poles of the cell, one to one pole and the other one to the other, until there is a full set of ____________ (number) chromosomes at each pole.

Anaphase

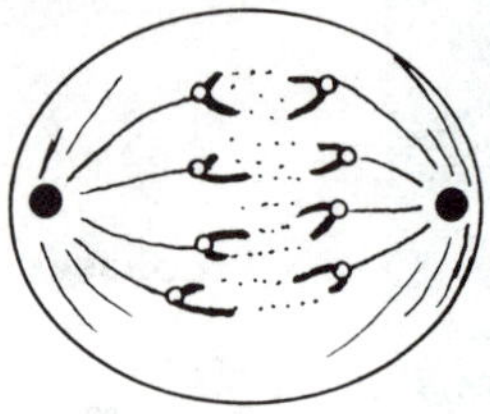

46

2-98.
This drawing shows the movement of the chromosomes toward the opposite poles of the cell during the ________________.

anaphase

2-99. After the anaphase is completed, the fourth step of mitosis, the TELOPHASE (TEL—o—phase), begins. The prefix TELO means end, so this phase is the ________________ step in the mitosis or reproduction of the cell.

Telophase-I

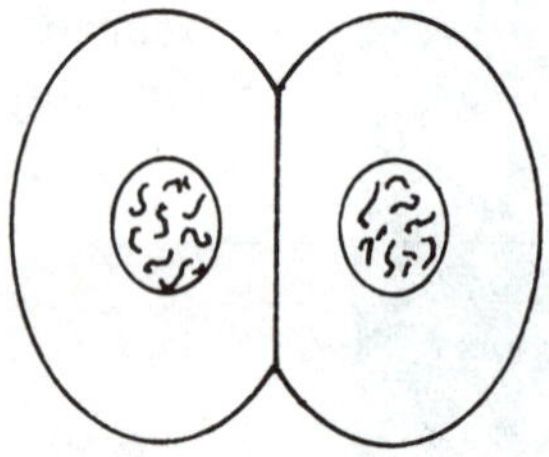

last

2-100.
The fourth and final stage of mitosis, called the ________________, is the two-cell stage.

telophase

2-101. As the telophase begins, the chromosomes return to their normal size and are surrounded again by the nuclear ________________.

membranes

2-102. The process of mitosis is completed with the conclusion of its final phase, the telophase, with the formation of ________________ separate cells.

two

2-103. Each of the two separate cells now contains ________________ (number) chromosomes.

Telophase-II

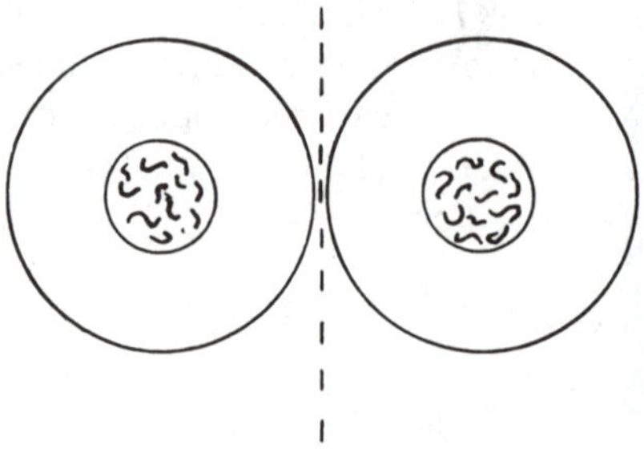

46

2–104. Let's review what you have learned about mitosis.
The objective of cell ________________ is to produce new cells.

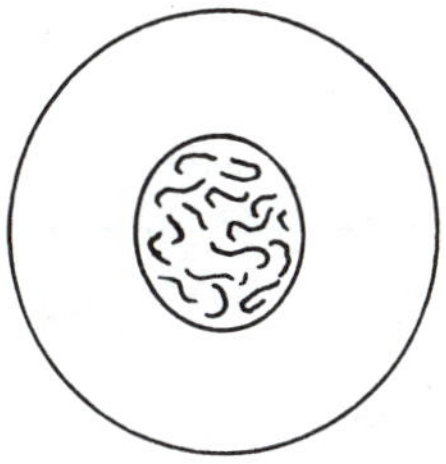

mitosis

2–105. This drawing shows the cell while in the ________________, in other words, performing all its normal metabolic functions except reproduction.

A

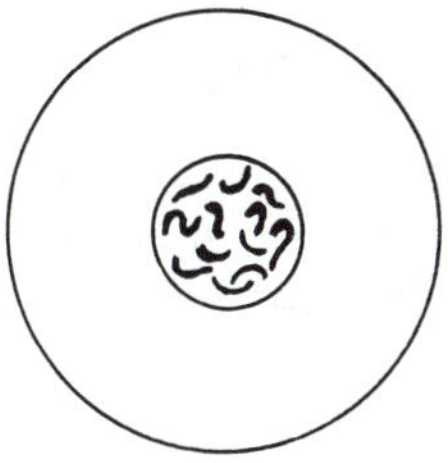

B

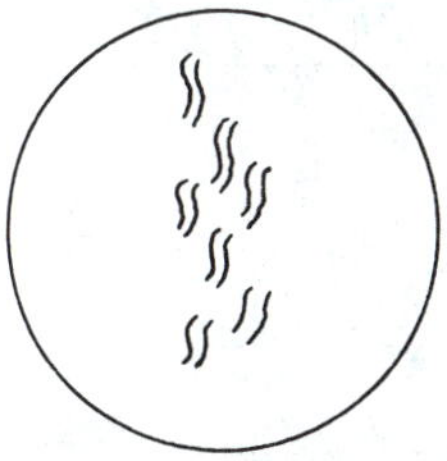

interphase

2–106.
The chromosomes thicken, the nuclear membrane disappears, and the chromosomes spread out around the nuclear area during the
________________.

A

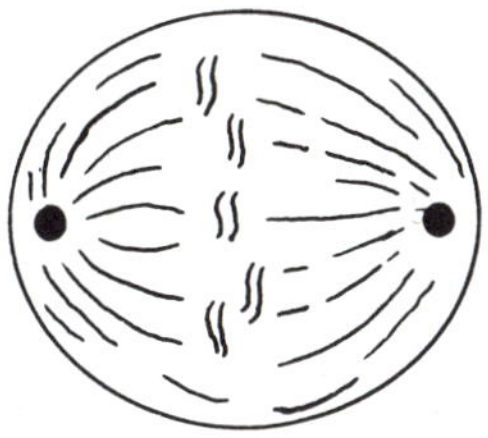

B

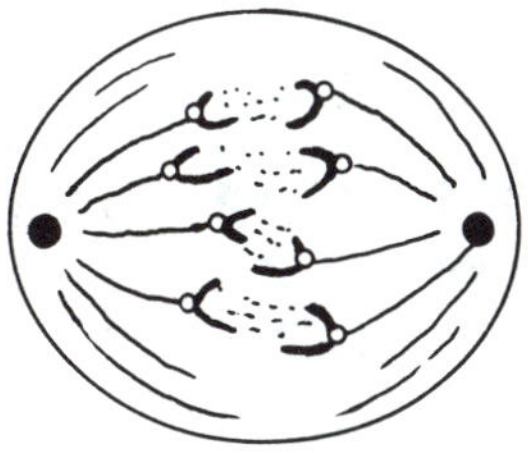

prophase

2–107.
Drawing (1) ________________ shows the metaphase. Drawing B illustrates the (2) ________________.

---

1. A
2. anaphase

2–108. The chromosomes become attached to fiberlike substances extending from the poles of the cell during the ________________.

---

metaphase

2–109. The anaphase is the stage during which the split chromosomes separate and move toward the ________________ of the cell.

poles

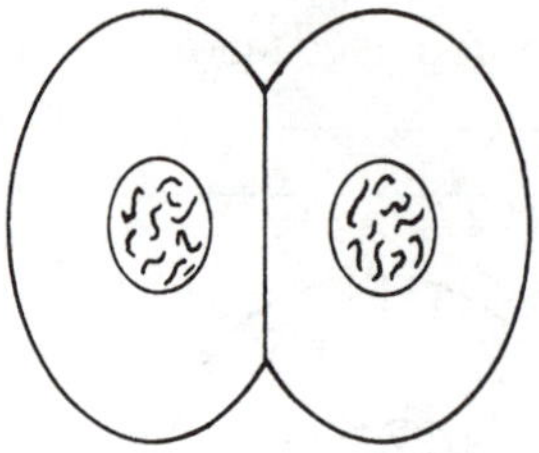

2–110. This is an illustration of the (1) ________________, or two-cell stage. Each new nucleus has (2) ________________ (number) chromosomes.

1. telophase
2. 46

2–111. The number of chromosomes is kept constant, 46 in each cell, because during the prophase each chromosome pair ________________ lengthwise.

splits

2–112. Each body part has a different rate of speed at which its cells divide. Some body cells complete all five phases in 20 to 40 minutes, but this figure is not constant because cells divide at ________________ rates.

different

2–113. When you were first conceived, the total cell count of this "you" was one. It's amazing to consider that this one cell divided, and then the two new cells divided, and so on over and over through the process of ________________ until the final result was a human adult.

mitosis

2–114. If cells continued to divide without pausing—if each new cell proceeded to divide again as soon as it was "born"—you can see that the new cells would become progressively smaller. It is the interphase that allows the cells to ________________ to their normal size.

grow

2–115. The cells in the various body parts not only divide at different rates but also have different rates of ________________.

growth

2–116. Some interesting facts related to growth are:

1. Your size, since conception, has increased 26 billion times.
2. If you had continued to grow at the same rate you did during the eighth month of your mother's pregnancy, your weight would be two trillion times the weight of the earth.

(No response required—Move to frame 117.)

2–117. Obviously, you don't weigh more than the earth, and the reason you don't is that various factors have influenced the rate and nature of your cell ________________.

growth

2–118. Some growth factors are inherited in the ________________ of the cell.

genes

2–119. Nutritives—foods—influence growth, and so do certain chemicals. The vitamins, hormones, and the nucleoprotein ________________ are examples of growth-influencing chemicals.

DNA

2–120. You have developed a working understanding of the cell, its composition, and how it reproduces and grows.

Take a break before you continue.

(No response required.)

2–121. When a cell is not actively reproducing, it is said to be in the ________________.

interphase

2–122. The cell takes this opportunity to grow; it may also alter the physical and chemical properties of the cellular protoplasm during the ________________.

interphase

2–123. The altering of the chemical properties of protoplasm allows the cell to specialize. SPECIALIZATION of a cell's development takes place because the ________________ and physical properties of its protoplasm are changed.

chemical

2–124. Some cells develop into skin cells, others into bone cells or cells of other body parts. These are all examples of cell ________________.

specialization

2–125. Once a cell begins to develop into a cell of a specific body part, it cannot change its course. When a cell's protoplasm is altered and the cell begins to develop into a specific stomach cell, it (will/will not) ________________ change and develop into a liver cell.

will not

2–126. If you recall, you developed from one cell that divided, and these new cells in turn divided, and so on until a many-celled being had developed. During the intervals between each division, certain cells became ________________ and developed into cells of specific body parts.

specialized

2–127. A cell that has the potential of developing into a cell of a variety of body parts is termed MULTIPOTENTIAL (mul—ti—po—TEN—shal). Cells that can be either skin cells or bone cells, depending on how their protoplasm is altered, are examples of ________________ cells.

multipotential

2–128. The body keeps some of these versatile ________________ cells in reserve at all times, in case of emergency.

multipotential

2–129. All body cells go through the phase of reproduction, and then most go through a period of chemical and physical change during which they become cells of specific body parts. This process of change is called ________________.

specialization

2–130. As a cell becomes more specialized, it has less and less need for reproduction. The closer a group of cells comes to developing into, for instance, a completed nose, the less likely the individual cells of the nose are to ________________.

reproduce

2–131. Human cells rest first, then divide by the process of ________________, and then rest again.

mitosis

2–132. After the cell divides, the chemical and physical properties of its (1) ________________ may alter, and the cell enters a period of (2) ________________.

1. protoplasm
2. specialization

2–133. All body cells go through a cycle of (1) ________________ and rest. Most cells then proceed to (2) ________________ as cells of a particular body part until that particular body part is developed.

1. mitosis
2. specialize

2–134. These specialized cells must be cemented together to give form and ________________ to a body part.

structure

2–135. Between the cells is a cementlike substance called INTERCELLULAR (in—ter—CEL—u—ler) CEMENT.

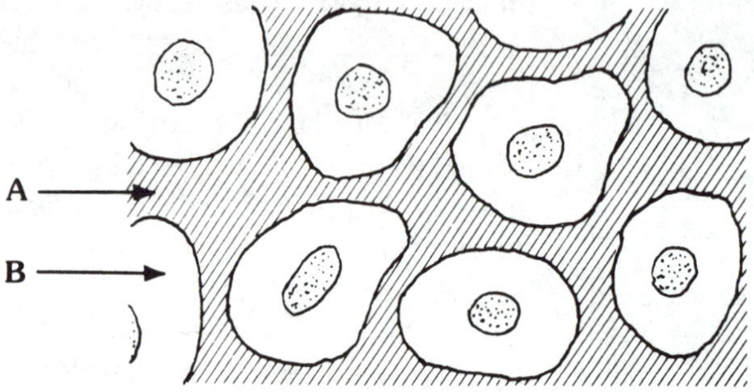

Arrow B points to the cytoplasm of the cell. Arrow A points to the intercellular ________________.

cement

2–136. A brick wall when being erected has mortar placed between the bricks to make the bricks adhere to each other.

Cells also have an adhesive substance, the ________________ cement, that causes the cells to retain their relationship to each other.

intercellular

2–137. The cement between the bricks is supplied by the bricklayer, not made by the bricks.

Cells, however, make their own adhesive, the ________________ ________________.

intercellular cement

2–138. The intercellular cement is called a cell product. This indicates that intercellular cement is not itself a cell, but rather a cell ________________.

product

2–139. The intercellular cement is only one of the many ________________ made by the cells.

products

2–140. The space between the cells occupied by the intercellular cement is referred to as INTERSTITIAL (in—ter—STISH—al) SPACE, "interstitial" meaning "between one thing and another." The fluid that is also found in this space would be called ______________ fluid.

interstitial

2–141. Each cell is bathed on at least one side by ______________ fluid.

interstitial

2–142. You should not confuse the interstitial fluid with the fluid inside the cell, the INTRAcellular fluid. Inter means between, as in "intercellular" and "interstitial," and intra means inside.

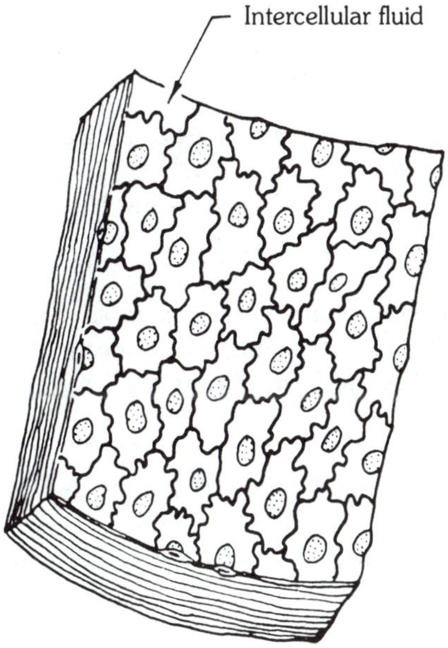

Cytoplasm contains:

a. intracellular fluid
(Move to frame 143.)

b. interstitial fluid
(Move to frame 144.)

a.

2–143. You said that cytoplasm contains intracellular fluid. Right!

The prefix "intra" means inside; so the fluid within the cytoplasm, which is located inside the cell, is intracellular fluid.

(Move to frame 145.)

b.

2–144. You said that cytoplasm contains interstitial fluid. Not so!

Cytoplasm is found inside the cell, not in the space outside the cell.

(Return to frame 142.)

2–145. Body fluids may also be located in body tubing called VESSELS (VES—ls). The Latin word for vessel is vascellum, and the adjective made from vascellum is VASCULAR (VAS—cu—lar). Because these particular body fluids are transported inside the vessels, they are called ______________ vascular fluids.

intra

2–146. Blood PLASMA (PLAZ—ma) would be an example of ______________ fluid.

intravascular

2–147. About 70% of the body weight of an adult human is water. This 70% can be broken down into:
50% inside the cell, or (1) ________________ fluid;
15% between the cells, or (2) ________________ fluid;
5% in vessels, or (3) ________________ fluid.

1. intracellular
2. interstitial
3. intravascular

2–148. A body fluid that is not intravascular can be either ________________ or interstitial.

intracellular

2–149. If a body fluid is not intracelluair, it is either ________________ or intravascular.

interstitial

2–150. Intercellular cement is an example of:
a. interstitial substance (Move to frame 151.)
b. intracellular substance (Move to frame 152.)

a.

2–151. Intercellular cement is an example of interstitial substance.
Interstitial refers to any substance between the cells.
(No answer required. Move to frame 153.)

b.

2–152. You said that the intercellular cement was an example of intracellular substance.
The prefix intra means "inside." If the cement were intracellular, this would indicate that it is inside the cell, which is not the case.
(Return to frame 151.)

2–153. Let's review what you've learned in this unit.
The small "building blocks" of the human body are specialized forms of protoplasm called ________________.

cells

2–154. Fill in the blanks:

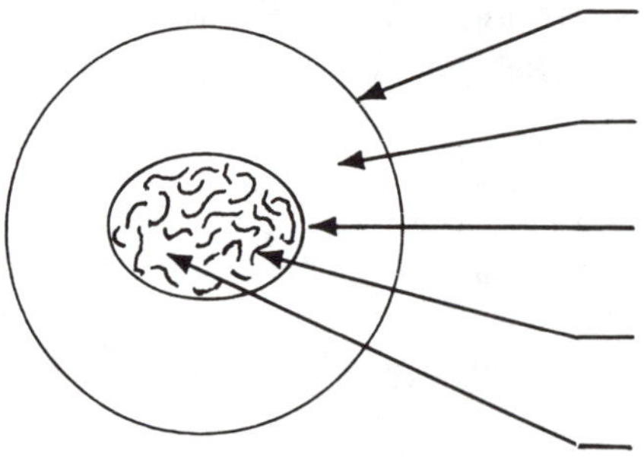

1. ________________
2. ________________
3. ________________
4. ________________
5. ________________

1. cell membrane 2. cytoplasm 3. nucleus 4. chromosomes 5. nucleoplasm	**2–155.** The cell nucleus influences cellular ________________.
reproduction	**2–156.** Cellular metabolism is carried out in the cell "work house," the ________________.
cytoplasm	**2–157.** Another name for the normal process of cell reproduction in the human body is ________________.
mitosis	**2–158.** Cell division, or mitosis, is divided into four stages or phases. The nuclear membrane disappears and the chromosomes split during the ________________.
prophase	**2–159.** During the metaphase the chromosomes become attached to delicate fiberlike substances that extend to the opposite cell ________________.
poles	**2–160.** During the anaphase the split chromosomes move toward the poles of the cell. There are now ________________ chromosomes on each side of the cell.
46	**2–161.** The last phase of division, when two distinct and separate cells are formed, is the ________________.
telophase	**2–162.** Cells divide, then pause to grow and________________ into different forms before reproducing again.
specialize	**2–163.** A multipotential cell is one that can grow into a cell of one of ________________ body parts.
many	**2–164.** Cells are held together by an interstitial substance called ________________ cement.
intercellular	

3 Human Tissues

3–1. Through the processes of mitosis and specialization, groups of cells gain particular structural arrangements and functions. Groups of cells with similar (1) ________________ and (2) ________________ are called TISSUES.

1. structure (form)
2. function

3–2. Tissues are groups of cells with similar anatomy, or (1) ________________, and similar physiology, or (2) ________________.

1. structure (form)
2. function

3–3. Muscle, nerve, bone, blood, and skin ________________ are composed of groups of similar cells.

tissues

3–4. Your finger is made of bone (1) ________________, muscle (2) ________________, nerve (3) ________________ and fat and skin tissues.

1. tissue
2. tissue
3. tissue

3–5. Each of the tissue types in your finger has its particular structure and all the cells of the tissue ________________ in unison.

function

3–6. If you were to study tissues, you would need to use a microscope. Such a study would be an examination of the ________________ of the tissue.

(Review frames 1–13 to 1–15 if you need help.)

histology

3–7. Histologically there are 4 basic tissues:

1. EPITHELIUM
2. CONNECTIVE
3. MUSCLE
4. NERVE

(No response required. Move to frame 8.)

3–8. The EPITHELIUM (ep—i—THE—li—um) is a sheetlike tissue with one surface attached to your body and the other surface exposed. Skin is an example of epithelial tissue because one surface is attached to your body and the other surface is ________________ to the air.

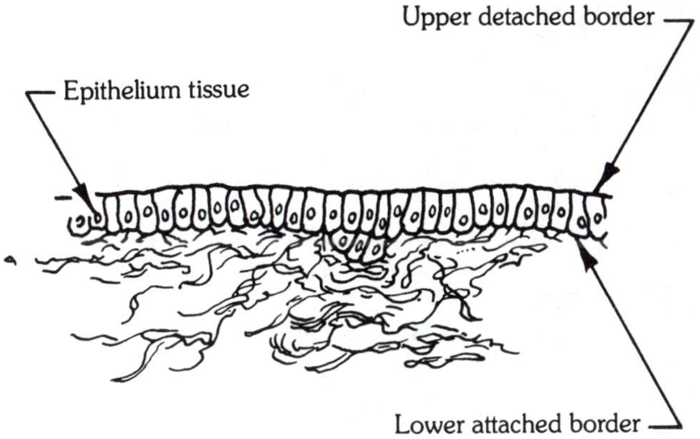

exposed

3–9. The cells that form epithelial tissue are so tightly packed that very little ________________ ________________ is required to attach the cells together in order to give the tissue form and structure.

intercellular cement

3–10. There are many hollow body spaces that are covered by this sheetlike ________________ tissue.

epithelial

3–11. Epithelial tissue protects the body and its parts in many ways. Skin, for example, prevents foreign substances from ________________ the body and causing infection.

entering

3–12. The body parts such as stomach, kidneys, and intestines are lined with epithelial tissue and are therefore ________________ by this tissue.

protected

3–13. Epithelial tissue forms GLANDS, groups of specialized cells used for the elimination of waste, the secretion of important fluids, and other jobs.

The life function of ________________ is carried out by the epithelial tissue.

(Review frame 1–162 to 1–166 if you need help.)

excretion	**3–14.** Epithelial tissue also serves to absorb substances needed by the body. Food, for instance, is broken down by enzymes in the stomach and small intestine, then is ________________ into cells lining the intestine.
absorbed	**3–15.** Epithelial tissue also assists in performing the life function of irritability or the ________________ of an organism to a stimulus.
response	**3–16.** If you have ever burned your tongue on food that was too hot, you have experienced an example of your tongue's epithelium performing the function of ________________.
irritability	**3–17.** Finally, epithelial tissues exhibit the life function of reproduction. If your skin is damaged, for instance, it is the epithelium that ________________ it.
repairs	**3–18.** To summarize, epithelial tissue: 1. exhibits the life functions of (1) ________________, (2) ________________, and (3) ________________. It (4) ________________ and secretes important substances and protects and (5) ________________ parts of your body.
1. excretion **2. irritability** **3. reproduction** **4. absorbs** **5. repairs**	**3–19.** You can always recognize epithelial tissue because 1. It is a sheetlike cover or (1) ________________ for a body part; 2. One of the tissue's surfaces is (2) ________________ to the body part, and the other surface is (3) ________________.
1. lining **2. attached** **3. exposed**	**3–20.** The epithelium is one of the 4 basic body tissues. The other 3 tissue types are connective, (1) ________________, and (2) ________________.
1. muscle **2. nerve**	**3–21.** CONNECTIVE TISSUES are grouped together because of functional similarities, how they act, rather than structural similarities. Although they may not look alike, ________________ tissues do function alike.
connective	**3–22.** You know that epithelial tissue protects and repairs damaged body areas. Connective tissue also functions as a ________________ and repair agency.
protective	**3–23.** When you think of connective tissue you should remember: Connective tissues are more similar in ________________ than they are in structure.

function

3–24. Connective tissues support, (1) ________________, and (2) ________________ the body.

1. repair
2. protect
(either order)

3–25. Epithelial tissue has small amounts of intercellular cement between the cells. Connective tissue contains much more intercellular substance than does epithelial tissue. In some types of connective tissue—bone, for example—the ________________ substances are more abundant than the cells.

intercellular

3–26. All cells are bathed on at least one side by interstitial fluid. The epithelial tissue cells receive support from the ________________ cement that occupies the interstitial space.

intercellular

3–27. The interstitial substance of connective tissue is called the tissue MATRIX (MA—triks). For example, the calcium that surrounds bone cells is part of the connective tissue ________________.

matrix

3–28. Connective Tissue MATRIX (interstitial substance)

FIBERS   GROUND SUBSTANCE

The connective tissue matrix is composed of FIBERS and ________________ ________________.

ground substance

3–29. Epithelial tissue cells are surrounded by (1) ________________ fluid. Supporting tissue cells are surrounded by a tissue matrix that is composed of (2) ________________ and (3) ________________ substance.

1. interstitial
2. fibers
3. ground

3–30. The ground substance is semi-fluid or solid, depending upon the type of supporting tissue. Bone would be an example of a structure whose ground substance is ________________.

solid

3–31. Ground substance is often amorphous, like gelatin; in other words, unlike the body, it has no definite ________________.

form

3–32. If you mixed some liquid gelatin and added to this a cup of fruit and a cup of shredded coconut, you would have, when the mixture set, an imitation of a ________________ tissue.

connective

3–33. Let's examine the gelatin analogy. The pieces of fruit embedded in the gelatin would be the:

a. cells of the tissue (move to frame 35)
b. ground substance of the tissue (move to frame 34)

b.	**3–34.** You said that the pieces of fruit would be the ground substance. Do you recall in frames 26–29 (review if necessary) that the ground substance surrounds the cells? Your choice was wrong. Move to frame 35 for more information.
a.	**3–35.** You said that pieces of fruit would be the cells. Right! The ground substance, or gelatin in this example, surrounds the cells (pieces of fruit) in a semi-liquid to solid state. (Move to frame 36.)
	3–36. You have accounted for the gelatin—the ground substance—and the fruit—the cells. The shreds of coconut would be the second component of a tissue matrix, the ____________.
fibers	**3–37.** The ground substance of the tissue matrix can be in either a semi-fluid or a ____________ state.
solid	**3–38.** One group of connective tissue cells is called FIBROBLASTS (FI—bro—blasts). The suffix, —blast, means to sprout or germinate. The word fibroblast indicates that the connective tissue cells sprout or form ____________.
fibers	**3–39.** It is interesting to note that the fibers formed by the connective tissue cells or ____________ are not attached to the cells.
fibroblasts	**3–40.** Fibers formed by the fibroblasts are cell ____________ similar to intercellular cement.
products	**3–41.** Fibroblasts contain a large amount of rough endoplasmic reticulum, indicating that these cells are important in the production of ____________ used outside the cell to form the intercellular substance.
protein	**3–42.** Let's review what you have learned. Connective tissues are groups of cells more alike in (1) ____________ but not necessarily in (2) ____________.
1. function 2. structure	**3–43.** Connective tissue is composed of a. cells and matrix (move to frame 44). b. cells and ground substance (move to frame 45).
a.	**3–44.** You said that connective tissue is cells and matrix. Right. (Move to frame 46.)

b.

3–45. You said that connective tissue is cells and ground substance. Not so. Review frames 28–37 and then return to frame 43.

3–46. Ground substance is produced by the ____________________ of connective tissue.

Loose Connective Tissue

fibroblasts

3–47. The fibers that are germinated or made to grow by the fibroblasts determine what type of tissue is present. For example, COLLAGENOUS (kol—AG—in—us) connective tissue is so called because the protein collagen is the chief ingredient of its ____________________ .

fibers

3–48. Collagenous connective tissue is made of wavy fibers.

This is a fiber of ____________________ connective tissue.

collagenous

3–49. These wavy fibers are composed, in turn, of wavy layers of FIBRILS (FI—brils). A fibril is simply a very small fiber.

Look back at the drawing in frame 48. There are three fibrils in this fiber of ____________________ connective tissue.

collagenous

3–50. The size of a collagenous fiber is determined by the number of ____________________ that it contains.

fibrils

3–51. Fibers have a cement that surrounds the fibrils somewhat like the insulation on electrical cables that holds strands of wire together.

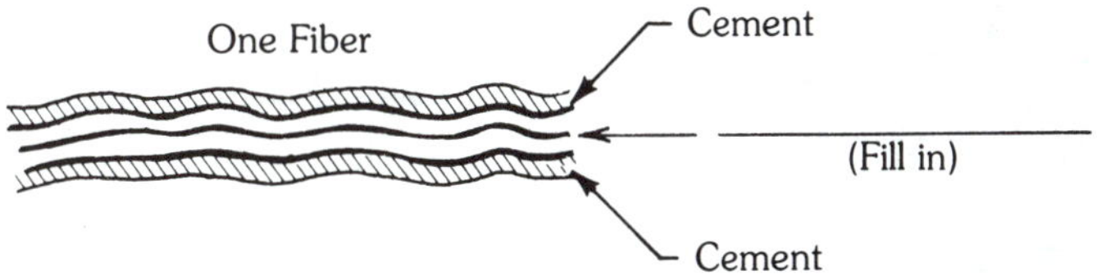

fibril

3–52. What might be the advantage of wavy collagenous fibers?

a. to provide the tissues with great strength and elasticity. (Move to frame 53.)
b. to prevent fibroblasts from leaving the tissue. (Move to frame 54.)

Answers	Frames
a.	**3–53.** You said the wavy collagenous fibers provide great strength and elasticity. Right. Then fibrils can be stretched until straightened and the fiber then becomes very resistant to any further change. If the fiber has not been stretched so much that it is damaged, it will return to its original position when released. (Move to frame 55.)
b.	**3–54.** You said that the wavy collagenous fibers prevent the fibroblasts from leaving the tissue. Not so. Go back and review frame 39 and then reread frame 52.
	3–55. Collagenous fibers are extremely tough. This quality of toughness of collagenous fibers makes this connective tissue resistant to ________________ (your own words).
damage or injury	**3–56.** When fat and other impurities have been removed from certain collagenous tissues, the fibers can be used as surgical gut to repair injured tissues. (No response required. Move to frame 57.)
	3–57. There are other connective tissue fibers whose most important component is the protein elastin, and these fibers are called ELASTIC fibers. The name elastic tells you that these ________________ are more flexible than collagenous fibers.
fibers	**3–58.** The most resilient connective tissue fibers are the ________________ fibers.
elastic	**3–59.** Both elastic and collagenous fibers are embedded in the amorphous ________________ ________________ of the tissue matrix.
ground substance	**3–60.** The consistency of the tissue matrix varies with ACTIVITY, AGE, and the general HEALTH of the body. The tissue matrix of a healthy child may vary considerably from that of a healthy ________________.
adult	**3–61.** Connective tissue has three types of intercellular fibers, elastic, RETICULAR (re—TIK—u—ler), and ________________.
collagenous	**3–62.** Reticular connective tissue is composed of a mesh of fine ________________ fibers.
reticular	**3–63.** In some locations, reticular connective tissue is called the BASEMENT MEMBRANE because it acts as a base or foundation for the epithelium. In other locations it provides the framework for various organs; the liver, for instance, is constructed around a meshwork of ________________ tissue.

reticular

3–64. Don't become confused.

Reticular tissue is so-called because it consists of reticular fibers, which may be present in many tissues other than reticular connective tissue.

(No response required. Move to frame 65.)

Adipose Connective Tissue

3–65. Another type of connective tissue containing reticular fibers is ADIPOSE (AD—i—pose) or simply fat ________________.

tissue

3–66. Fat storage and the production of fat cells are the main functions of ________________ tissue.

fat (adipose)

3–67. The state of nutrition and overall health determines how much fat or ________________ tissue is present in the body.

adipose

3–68. No one is literally a "fat-head" because the brain, like the eyes and the lungs, is unable to store ________________.

fat

3–69. Adipose cells are supported by a mesh of ________________ ________________ and collagenous fibers.

reticular fibers

3–70. Concrete is often reinforced by the addition of a wire mesh.

Adipose tissue is (1) ________________ by the network of reticular fibers and (2) ________________ fibers.

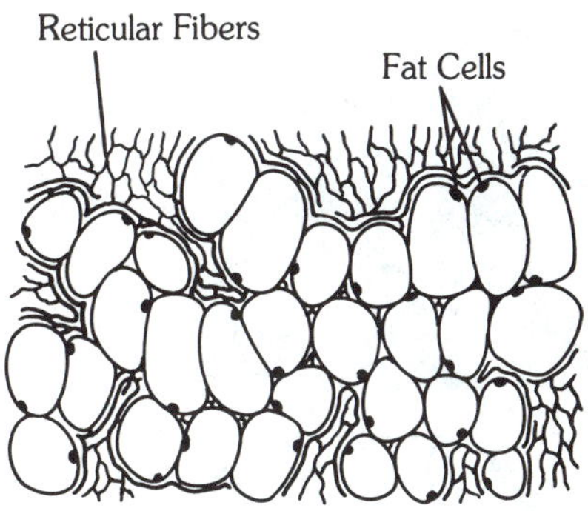

1. reinforced
2. collagenous

3–71. Adipose tissue is a good body packing agent and shock absorber. It acts also as ________________ against cold.

insulation

3–72. Pause for a moment and review what you've learned.

Connective tissue has three kinds of fibers:
the tough, resilient (1) ________________ fibers;
the even more resilient (2) ________________ fibers;
and the mesh-forming (3) ________________ fibers.

1. collagenous
2. elastic
3. reticular

3–73. A collagenous fiber is composed of wavy ________________.

fibrils

3–74. The type of connective tissue present is determined in part by the type of fiber in the tissue.

Collagenous tissue, for example, contains the protein collagen, which is the chief ingredient of its tough ________________.

fibers

3–75. Collagen is also an element of the somewhat finer fibers that form the meshlike (1) ________________ tissue, which is sometimes called the basement (2) ________________.

Dense Connective Tissue

1. reticular
2. membrane

3–76. So far you have read about connective tissues that have been loosely arranged. Connective tissue can also be dense, so dense that the matrix contains more ________________ than ground substance.

fibers

3–77. Flex your foot upward, and then reach down and feel the large cordlike structure that connects your calf muscle to your heel bone. You are feeling the ACHILLES (a—KIL—ez) TENDON (TEN—dun), a good example of ________________ connective tissue.

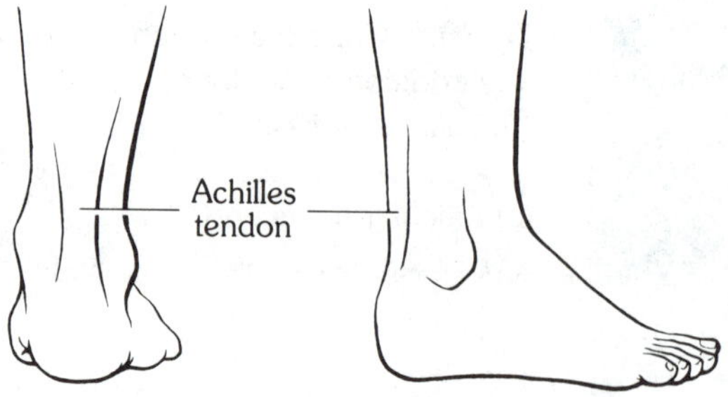

dense

3–78. The Achilles tendon is an example of connective tissue in which the ________________ are especially numerous and closely packed.

fibers

3–79. Tendons are composed predominantly of wavy collagenous fibers. If you recall, these fibers can be stretched to a point that ________________ the tissue.

damages
(or injures)

3–80. Tendons are stronger, in some respects, than bone. For example, the ________________ fibers of a ¼″ tendon can sustain a 1000-lb. load!

collagenous	**3–81.** Watch a human or a dog jump, and you will see that stretching tendons allow muscles to contract and when tendons recoil, muscles relax to their original position. You should remember that tendons attach (1) ________________ to (2) ________________.
1. muscles **2. bones**	**3–82.** Another dense connective tissue that may be cordlike is the LIGAMENT (LIG—a—ment). Ligaments are composed of both elastic and collagenous fibers. One function of ligaments is to connect bones together. Tendons attach bones to muscles. ________________, on the other hand, help connect bones to each other. (You will meet other connective tissue ligamentous structures with other functions later.)
ligaments	**3–83.** Feel the lower inch of your nose. Wiggle it about. Is it connected to the upper body structure? This material under the skin is a. connective tissue (move to frame 85) b. soft bone (move to frame 84)
b.	**3–84.** You said that the material was soft bone. Not bad! The material is called CARTILAGE (KAR—ti—lij). Cartilage is a dense connective tissue that is firm, like plastic, but not hard, like bone. In infancy, cartilage occupies some areas that later become bone. (Move to frame 86.)
a.	**3–85.** You said that the material was connective tissue. Very good! This connective tissue is called CARTILAGE (KAR—ti—lij). Cartilage is a dense connective tissue that is firm, like plastic. In infancy, cartilage occupies some areas that later become bone.
	3–86. The dense connective tissue in your ear is another example of ________________.
cartilage	**3–87.** As you walk, jump, or run, your bones do not rattle against one another because of the pads of ________________ between them.
cartilage	**3–88.** It is important for you to note that certain types of dense connective tissue have few or no blood vessels passing through their intercellular substance. If tissues such as cartilage, tendons, and ligaments are damaged, they will a. repair slowly (move to frame 90) b. repair fast (move to frame 89)
	3–89. You said that the damaged dense connective tissues will repair rapidly. Not so! Recall from Chapter 2 that nutrients and chemicals are important for the growth of cells. (Return to frame 88.)

3–90. You said that injured dense connective tissue will repair slowly. Right!

Because there are fewer blood vessels in this type of tissue, there is a lesser supply of nutrients and chemicals important for cell growth and tissue repair.

3–91. As you become older, the cartilage between the bones becomes, for reasons unknown, increasingly inflexible, and the joints become less and less mobile.

Cartilage can be converted to ________________ early or late in your life.

Bone Connective Tissue

---

bone

3–92. The next connective tissue you will study supports your body and protects such vital organs as your brain and your heart. This most dense supporting tissue is ________________.

---

bone

3–93. Bone, like all other supporting tissue, contains cells and a matrix of (1) ________________ ________________ and (2) ________________.

---

1. ground substance
2. fibers

3–94. There is one distinguishing characteristic of bone that makes it different from other connective tissue. The ground substance of bone contains a high content of inorganic calcium compounds that make the bone tissue ________________.

---

rigid or hard

3–95. Bone tissue is not a solid structure but has many internal canals. Among other functions, these canals allow ________________ to pass through the bone and supply nutrients to the bone cells.

---

fluids

3–96. The connective tissue cells you already learned were the fiber-producing ________________.

Bone cells are called OSTEOCYTES (OS—te—o—sites).

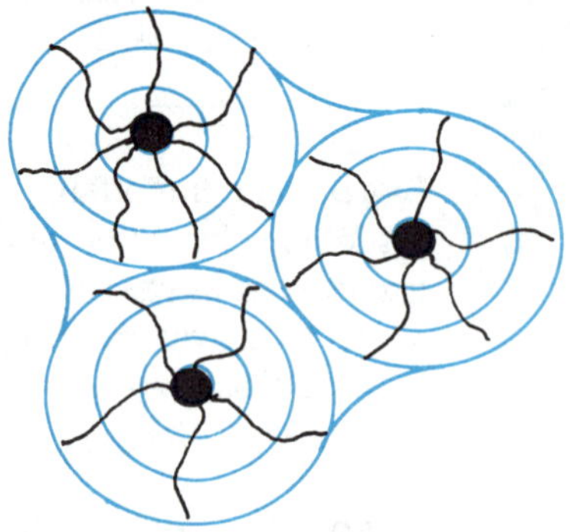

fibroblasts

3–97.

This cross section of part of a bone shows three bone cells or ________________.

osteocytes	**3–98.** The bone cells or ______________ are situated in small <u>cavities</u> in the bone.
osteocytes	**3–99.** Look at the illustration in frame 97. The cell and its parts are drawn in blue; the portion drawn in black is the bone matrix. Note that: 1. The osteocyte membranes spread out like roots. 2. The bony rings are similar to the growth rings of a tree (No response required. Move to frame 100.)
	3–100. Bone tissue is not solid. In addition to cavities for the osteocytes, it contains canals that allow the osteocytes to receive ______________.
nutrients	**3–101.** The canals pass <u>up</u> and <u>down</u> through the bone and <u>across</u> the rings of bone that contain the ______________ cavities.
osteocyte	**3–102.** In this illustration the matrix is white outlined in blue, and the ______________ are black. Small canals connect the osteocyte cavities to one another. Larger canals come into contact with the smaller ones and contain blood vessels to carry nutrients to and from the osteocytes. 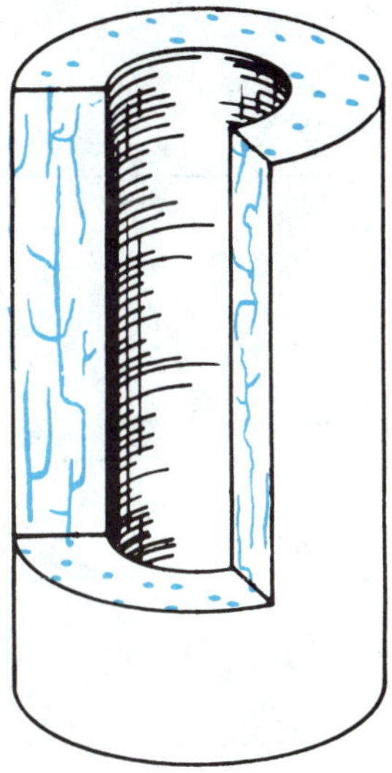
canals	**3–103.** Nutrients pass through the smaller canals themselves or are conducted by blood vessels within the larger ______________.
canals	**3–104.** But why is such an elaborate system of canals needed in bone tissue? (a) bone is so dense that the nutrients could not pass between cells easily. (Move to frame 106.) (b) the tissue length requires a canal system. (Move to frame 105.)
b.	**3–105.** You said that the tissue length requires an elaborate canal system. Not so. Collagenous and adipose tissue may be equally long; yet they need no such system. (Return to frame 104.)

a.	**3–106.** You said that the bone tissue was so dense that a canal system was required. Right! The nutrients from the blood vessels in larger canals could not easily penetrate the calcified intercellular substance to nourish the ________________. (Move to frame 107.)
osteocytes	**3–107.** The fibers in the matrix of one growth layer of the bone intertwine with the fibers of other layers to provide greater bone ________________.
strength	**3–108.** The next time you have a chicken or turkey leg, place the bone in a jar and cover it with a vinegar solution. In a few days, you will be able to tie this customarily brittle bone into a knot. Why? a. The vinegar made the bone more elastic. (Move to frame 110.) b. The vinegar dissolved the calcium compounds in the bone. (Move to frame 109.)
b.	**3–109.** You said that the vinegar dissolved and removed the calcium from the bone. Right! When the calcium compounds were dissolved, all that was left were the ________________ of the matrix. (Move to frame 111.)
a.	**3–110.** You said that the vinegar made the bone more elastic. Well, yes and no. Yes, the bone was more elastic. But the vinegar <u>did something</u> to the bone to make it more elastic. It dissolved and removed the calcium from the bone. (Move to frame 111.)
fibers	**3–111.** The calcium compounds that are stored in bone are in a constant state of flux. While some calcium is being deposited, other deposits are being removed for use elsewhere in the body. The portions being transported <u>out</u>, like the nutrients that are transported <u>in</u>, are carried by the ________________ ________________.
blood vessels	**3–112.** Bones also manufacture and store <u>MARROW</u> (MAR—o). There are two types of marrow, red marrow and yellow. Your bones contain both red and yellow ________________.
marrow	**3–113.** Marrow is responsible for the production of blood cells. If the marrow is diseased, a bone or ________________ disease could develop.
blood	**3–114.** Bone and bone marrow are separate entities. Marrow is a cell ________________ of bone tissue.

product

3–115. Bones have still another function in the body—they are good levers. Aside from housing marrow and providing a mooring place for muscles, the four functions of bones are that they:
1. protect the body.
2. (1) ________________ the body.
3. act as (2) ________________ allowing motion.
4. store and release (3) ________________.

Hemopoietic Connective Tissue

1. support
2. levers
3. calcium

3–116. Bone marrow is a cell product of bone tissue and is responsible for production of ________________ cells.

blood

3–117. The tissues responsible for the production of blood cells are called HEMOPOIETIC (HE—mo—po—etic) tissues and include ________________ marrow and LYMPHATIC (lim—FAT—ic) tissue.

bone

3–118. The blood cells that are produced in ________________ tissue carry out the functions of protection and repair of the body.

hemopoietic

3–119. Since blood is composed of specialized cells suspended in a fluid ground substance, blood, like cartilage and bone, is a ________________.

tissue

3–120. The fluid portion of blood is called PLASMA (PLAZ—ma) and the remainder consists of cells. The cells are (1) ________________, (2) ________________ and (3) ________________.

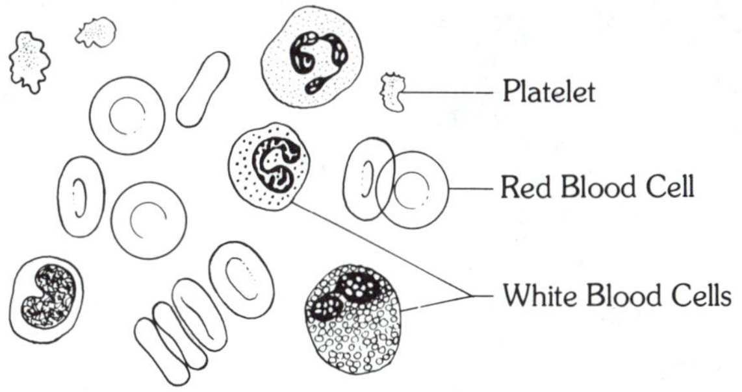

1. red blood cells
2. white blood cells
3. platelets

3–121. The red blood cells, white blood cells, and platelets are formed in the bone marrow portion of ________________ tissue.

hemopoietic

3–122. Lymphatic tissue is located in various parts of the body. The major function of lymphatic tissue is to produce specialized white blood cells that ________________ the body from infection.

protect	**3–123.** In some lymphatic tissue sites, LYMPH NODES, specialized white blood cells called MACROPHAGES (MAC—ro—fa—zhs) are attached to the reticular fibers of the ________________ ________________.
tissue matrix	**3–124.** Bacteria that have invaded the body are filtered through the lymph tissue so that the ________________ can attack the bacteria and prevent injury to the body.
macrophages	**3–125.** At the beginning of this unit you learned about loosely arranged connective tissues—the collagenous, reticular, and elastic tissues. Lets review what you have read about bone and cartilage, or ________________ connective tissue.
dense	**3–126.** What makes connective tissue either more or less dense is the proportion of (1) ________________ to (2) ________________ ________________ in the matrix. The higher the proportion of (3) ________________, the denser the tissue.
1. fibers **2. ground substance** **3. fibers**	**3–127.** A cordlike connective tissue that attaches bones to muscles is a ________________.
tendon	**3–128.** A connective tissue—also sometimes cordlike—that attaches bones to each other is a (1) ________________. A connective tissue that is firm and flexible, like plastic, is (2) ________________.
1. ligament **2. cartilage**	**3–129.** Because of their density, these tissues allow proportionately less blood to pass through them than the rest of the body and therefore heal more ________________.
slowly	**3–130.** The densest supporting tissue in the body is (1) ________________. It derives its hardness or rigidity from its high (2) ________________ content.
1. bone **2. calcium**	**3–131.** Bones are not solid but contain cavities for both the (1) ________________ and (2) ________________ that allow nutrients to enter.
1. osteocytes **2. canals**	**3–132.** Bones also manufacture and store (1) ________________, which is in turn responsible for the production of certain (2) ________________ cells.
1. marrow **2. blood**	**3–133.** Bone marrow and lymphatic tissue are called ________________ tissue.

hemopoietic	**3–134.** These tissues produce the blood cells, called (1) ________________, (2) ________________, and (3) ________________.
1. red blood cells **2. white blood cells** **3. platelets**	**3–135.** Specialized white blood cells, important for protection of the body from bacterial infection, are produced in(1) ________________ tissue and are called (2) ________________.
1. lymphatic **2. macrophages**	**3–136.** You may want to take a break before beginning the next tissue type.

Muscle Tissue

	3–137. The next type of tissue to be studied after epithelial and connective is the "meat" or "flesh" of the body, the ________________ tissue.
muscle	**3–138.** Muscle tissue differs from other tissues because a single muscle fiber is a muscle cell. Epithelial tissue has no fibers; however, each muscle cell is a ________________ ________________. Muscle cells are called SARCOLEMMA (SAR—ko—LEE—ma).
muscle fiber	**3–139.** The primary function of muscle tissue is to do work by either contracting or ________________.
expanding	**Muscle Cells** 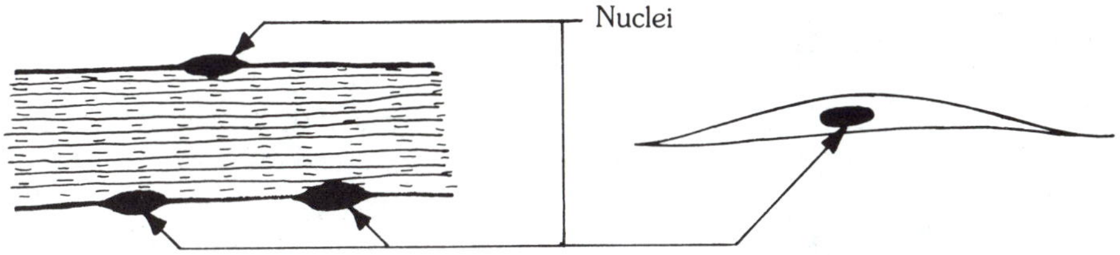 1. Striated (STRI—a—tid) 2. Smooth **3–140.** Sarcolemma are classified as being either striated, meaning striped, or ________________.
smooth	**3–141.** Look at the illustration in frame 140. The cross-striped cell is a (1) ________________ muscle cell. The unstriped cell is a (2) ________________ muscle cell.
1. striated **2. smooth**	**3–142.** "Striated" and "smooth" refer not to the function of a muscle cell, but to the muscle's ________________.

structure

3–143. Muscles can also be further classified as voluntary or involuntary. This classification refers to how the muscle cells ______________.

function

3–144. A muscle such as the large muscle (biceps) in your arm is striated, and because you have control over the actions of this muscle it is referred to as a striated (voluntary/involuntary) ______________ muscle.

voluntary

3–145. You do not have any control over the smooth muscles of your body. Therefore, there are no smooth ______________ muscles.

voluntary

3–146. All smooth muscles of the body are ______________.

involuntary

3–147. The striated muscles are all voluntary with one exception, the heart, which is cardiac muscle. Because you have no control over your heart, it is an ______________ muscle.

involuntary

3–148. Skeletal muscles, like those in your arms and legs, are striated. Your leg muscles are therefore striated ______________ muscles.

voluntary

3–149. Remember, there are (1) ______________ smooth voluntary muscles and but one involuntary striated muscle, the (2) ______________ muscle.

1. no
2. heart

3–150. Smooth muscle cells have simple cellular structures with tapering ends, a central ______________, and many small organelles.

nucleus

3–151. The organelles of smooth muscles are embedded in the cell's cytoplasm, which in sarcolemma is called the SARCOPLASM (SAR—ko—plazm). The organelles called MYOFIBRILS (MI—o—fi—brilz) are embedded in the ______________ of the muscle cell.

sarcoplasm

3–152. The contracting elements of the muscle are the small organelles called ______________.

myofibrils

3–153. Fill in the blanks.

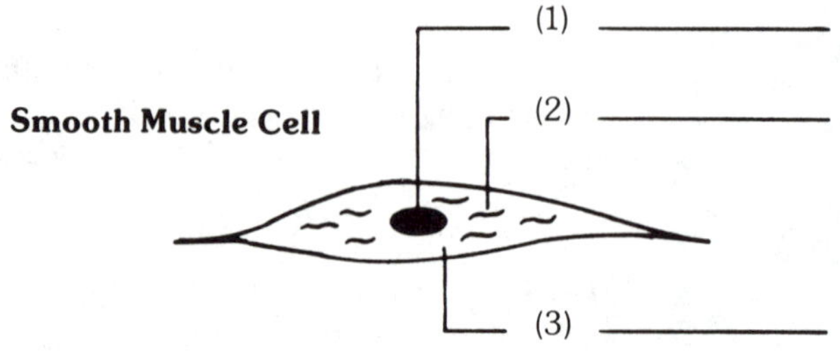

1. nucleus
2. myofibril
3. sarcoplasm

3–154. You recall that muscle tissue does not have separate cells and fibers. The muscle fiber is the cell and includes the nucleus, (1) ______________, and the myofibrils that are the (2) ______________ elements of the muscle.

Smooth Muscle Tissue

1. sarcoplasm
2. contracting

3–155.

Each muscle fiber is separate and is contained within a delicate sheet of specialized connective ______________.

---

tissue

3–156. The separate ______________ are then bound together into bundles by more connective tissue.

---

fibers

3–157. If a contraction impulse is transmitted to one fiber in the bundle, it in turn transmits the stimulus to every fiber in the bundle, and the myofibrils within each cell react by ______________.

---

contracting

3–158. The skeletal or striated muscle fibers also contain cytoplasm called (1) ______________ and the contracting elements,the (2) ______________.

Striated Muscle Fiber

(Longitudinal Section)

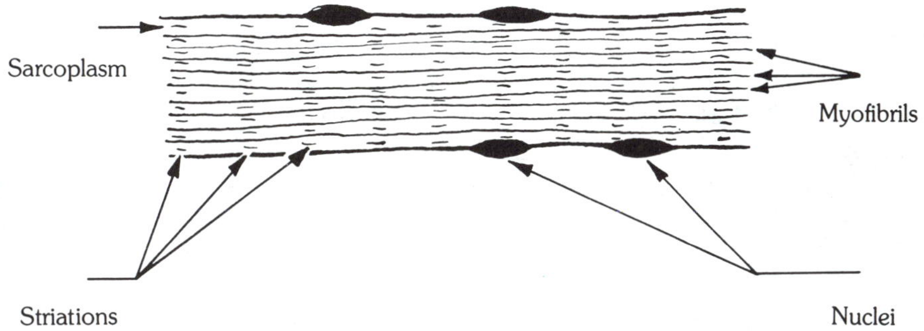

1. sarcoplasm
2. myofibrils

3–159.

The dark and light bands that give this muscle its striped or striated appearance are due to the structure and arrangement of the ______________.

---

myofibrils

3–160. The cells of striated muscles contain not just one nucleus but several, making them ______________-nucleated cells.

---

multi

3–161. Like smooth muscle fibers, the individual striated fibers are surrounded by (1) ______________ tissue. Groups of fibers are bound together to make (2) ______________ of fibers.

1. connective **2. bundles**	**3–162.** Striated muscles respond in much the same way that smooth muscles do. The impulse is transmitted from fiber to fiber so that the entire ________________ contracts at once.
muscle	**3–163.** When a muscle contracts, work is accomplished. The energy required to do this ________________ is stored within the muscle.
work	**3–164.** Muscle cells store phosphate compounds, ATP and ADP. When the muscle is called on to do work, some ATP molecules break down to ADP. Energy is released by the breaking of a phosphate ion bond and is used to accomplish ________________.
work	**3–165.** The energy supply for muscular contraction is produced in the mitochondria when the ________________ molecules break down to ADP.
ATP	**3–166.** The organelles in which ATP is broken down to ADP to produce energy are called ________________.
mitochondria	**3–167.** If mitochondria are present in proportionately large quantities, the muscle fiber is less susceptible to fatigue. The muscle of your heart—which is constantly at work—is composed of muscle fibers with (more/fewer) ________________ mitochondria than are available in cheek muscles.
more	**3–168.** Lesser numbers of mitochondria within the muscle cells makes the muscle (1) ________________ susceptible to fatigue. Greater numbers of muscle cell mitochondria enable the muscle to work for a (2) ________________ (longer/shorter) duration. Most muscles in the human body are composed of a combination of both slow-and fast-acting muscle fibers.
1. more **2. longer**	**3–169.** Let's review some of what you've been learning. 1. Connective tissue has distinct cells and fibers. 2. Epithelial tissue has no (1) ________________. 3. In muscle tissue, the individual muscle (2) ________________ are the (3) ________________.
1. fibers **2. cells** **3. fibers** **(2 and 3— either order)**	**3–170.** All muscles can be classified either as voluntary or involuntary. How many smooth voluntary muscles are there in your body? (none, few, many) ________________.
none	**3–171.** Heart muscle is striated; you would classify this as ________________ striated muscle.

involuntary	**3–172.** All striated muscles except ________________ muscle can also be referred to as skeletal muscles.
heart	**3–173.** Fill in the blanks on this illustration of a (1) ________________ muscle cell. 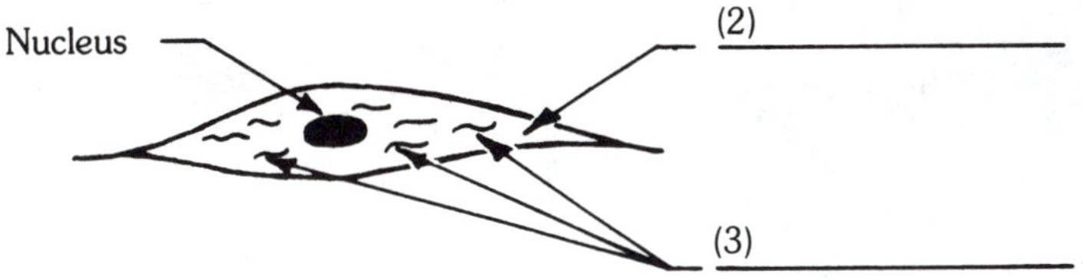
1. smooth 2. sarcoplasm 3. myofibrils	**3–174.** Muscle cells are enclosed in separate sheets of tissue and are combined into bundles that are also enclosed in specialized ________________ ________________.
connective tissue	**3–175.** The ability of a group of cells to respond to a stimulus, the life function of ________________, is one of the most fascinating of life functions.
irritability	**3–176.** When the body or one of its parts responds to a stimulus, it is the ________________ cells that are responsible for transporting the impulse from the receiving cells to other areas of the body.
nerve	**3–177.** Remember that nervous tissue is primarily a receiver and conductor. A muscle responds when the nerve tissue receives a stimulus and ________________ the impulse to that muscle.
conducts	**3–178.** All protoplasm exhibits the qualities of irritability and conductivity to some degree, but these qualities are most highly developed in ________________ ________________.
nerve cells or nerve tissue	**3–179.** A nerve cell is called a NEURON (NU—ron). The special functions of neurons are to receive ________________ and to conduct impulses.
stimuli	**3–180.** The structural and functional unit of nervous tissue is the ________________ or nerve cell.

neuron	**3–181.** A single nerve cell is composed of one cell body and several extensions called processes, the DENDRITES (DEN—drites) and the AXON (AX—on). Fill in the blanks in the illustration. Nerve cell processes are also called NERVE FIBERS. 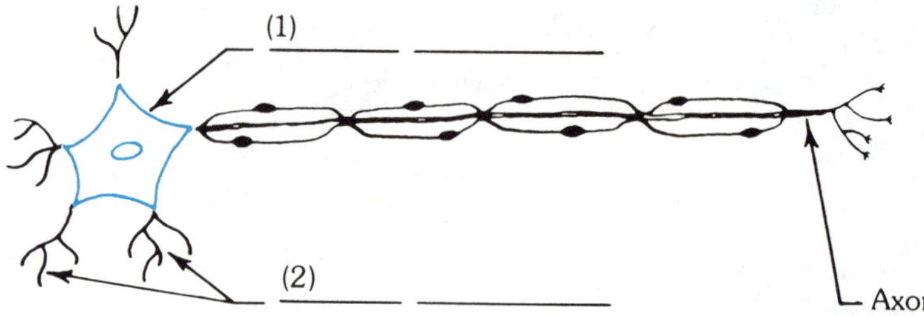
1. cell body 2. dendrites	**3–182.** The cell processes of a neuron are extensions of the cell ______.
body	**3–183.** When a nerve cell is stimulated, the impulse is picked up by the rootlike ______ spreading from the cell body. 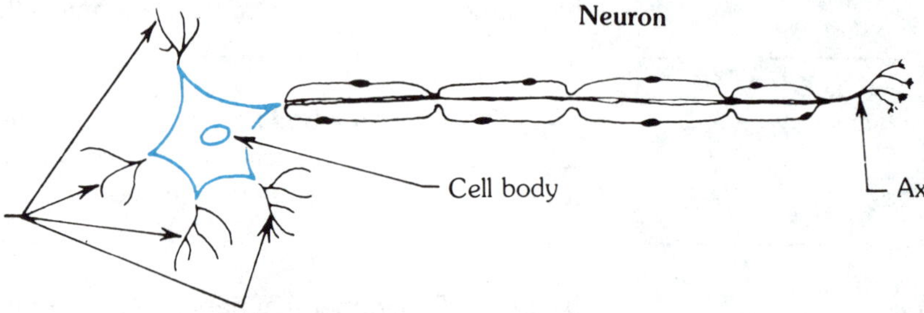
dendrites	**3–184.** The dendrites branch from the cell body and are in contact with axons and cell ______ of other neurons.
bodies	**3–185.** The process of the neuron that carries an impulse away from the cell body is the (1) ______. The nerve fibers that transmit the impulse toward the cell body are the (2) ______.
1. axon 2. dendrites	**3–186.** Neurons may have many dendrites; however, they have but one cell body and one ______.
axon	**3–187.** The impulses that travel through a neuron are classed as either AFFERENT (AF—er—ent) or EFFERENT (EF—er—ent), depending on the direction of the impulse. When you come into a theater, you are admitted by the ticket-taker; when you go out, you leave by the exit. The prefix ex means out, and the prefix ad means ______.

in or into

3–188. An AFFERENT (AF—er—ent) impulse is one that travels into or toward the cell body of a neuron. Afferent impulses would thus first travel by way of the ________________ of a neuron.

dendrites

3–189. An EFFERENT (EF—er—ent) impulse is one that travels away from the cell body of a neuron. Leaving the cell body, efferent impulses would travel by way of the ________________ of a neuron.

axon

3–190. Dendrites transmit (1) ________________ impulses.
Axons transmit (2) ________________ impulses.

1. afferent
2. efferent

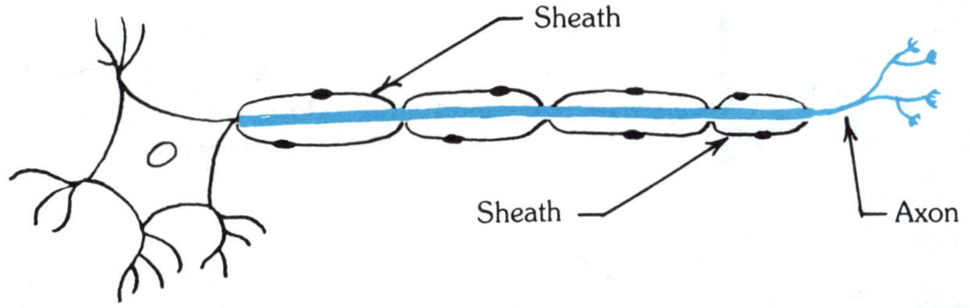

3–191.
The axon, drawn here in blue, is enclosed in protective covers called ________________.

sheaths

3–192. Depending upon the location and function of an axon there may be more than one ________________ covering it.

sheath

3–193. Two types of sheaths are identifiable, the MYELIN (MY—e—lin) and the NEUROLEMMA (nu—ro—LEM—a). An axon may have either one type of sheath or the other, or both. When both are present, the myelin is always the inner sheath, and the ________________ the outer sheath.

neurolemma

3–194. If an axon is enclosed in two sheaths, the neurolemmal sheath is always ________________ the myelin.

outside (of)

3–195. The neurolemmal sheath is formed by the cytoplasm of special cells, SCHWANN cells. The ________________ sheath is composed of protein-lipid compounds from these same cells and from other special cells.

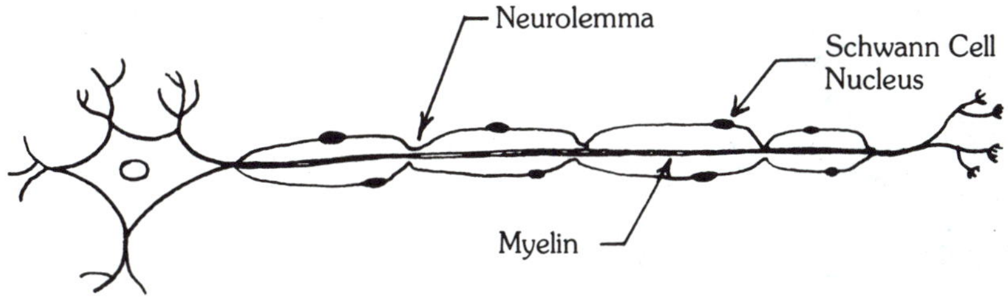

myelin	**3–196.** The myelin sheath is a protein-lipid substance. This (1) ________ sheath acts as an insulator to protect the (2) ________ ________.
1. myelin 2. nerve fibers or cell processes	**3–197.** When myelinated fibers are arranged in bundles they have a whitish color. In the brain, groups of neurons with ________ fibers are referred to as WHITE MATTER.
myelinated	**3–198.** The GRAY MATTER of the brain is composed of groups of nerve cell bodies whose ________ ________ are unmyelinated.
nerve fibers	**3–199.** The neurolemmal sheath also acts in protecting the nerve fibers. More importantly, the neurolemma plays an active role in cell repair. Fully developed nerve cells cannot reproduce, but the sheath-covered cell process, the ________, can be repaired.
axon	**3–200.** For example, if an axon is cut, the part that is separated from the cell body degenerates, but the sheath survives. If the separated neurolemmal sheath and the sheath that is still attached heal together again, a new ________ grows out into the reconnected sheath.
axon	**3–201.** You will discover later in this course that certain processes of the nervous system cannot be repaired. The action of the neurolemma referred to in frames 199–200 applies to many but not all processes.
	3–202. Let's review some of what you have learned about nervous tissue. A neuron is the (1) ________ and (2) ________ unit of nervous tissue.
1. structural 2. functional	**3–203.** Neurons have two main functions, to (1) ________ stimuli and (2) ________ impulses.
1. receive 2. conduct	**Neuron** 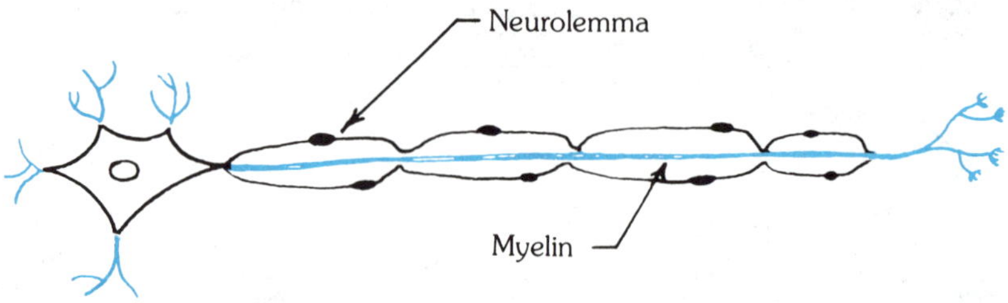 **3–204.** Label the parts drawn in blue.

1. dendrite 2. axon	**3–205.** An afferent impulse is conducted by a. an axon (Move to frame 206.) b. dendrites (Move to frame 207.)
a.	**3–206.** You said that an afferent impulse is conducted by an axon. Not so. Remember, afferent impulses are those traveling toward the cell body,and axons carry impulses away from the cell body. (Move to frame 208.)
b.	**3–207.** You said that an afferent impulse is conducted by dendrites. Right! (Now move to frame 208.)
	3–208. When cell axons are protected by two sheaths, the (1) __________ sheath is the outer insulation, and the (2) __________ sheath is the inner insulation.
1. neurolemma 2. myelin	**3–209.** If you were to describe the myelin, you would say that it was a (1) __________ substance. An example of myelin is the (2) __________ matter of the brain.
1. protein-lipid 2. white	**3–210.** Although some neurons can be repaired, mature neurons cannot __________.
reproduce	**3–211.** The repairing of an axon can take place if the detached __________ sheath heals and a new protective sheath is formed.
neurolemmal	**3–212.** You have discovered that neurons transmit impulses. You are called, and you react; you are physically pushed, and again you react. How does a stimulus get from your ear or skin to the central computer, your brain? (No response required. Move to frame 213.)
	3–213. When individual neurons are placed end to end, they form extensive neural pathways. When you are called, the receptor neurons in your ears send an impulse to other __________ that finally transmit the message to the brain.
neurons	**3–214.** It is important to remember that the impulse travels from the axon of one neuron either to the __________ or directly to the cell body of the next neuron.

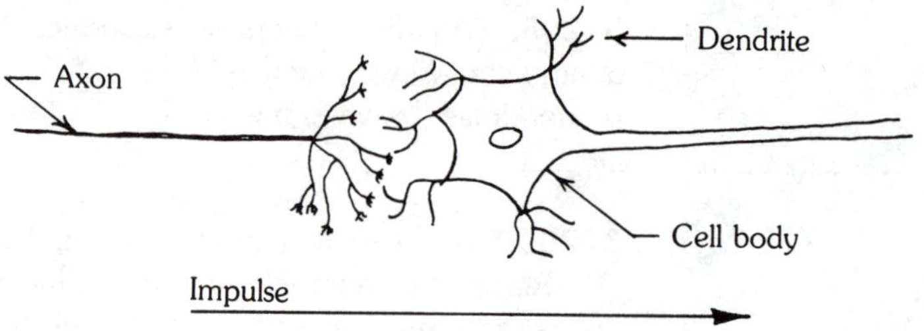

dendrites

3–215.
Note that there is never a structural connection between the ________ of one neuron and the dendrites of another neuron.

axon

3–216. The association between the axon of one neuron and the dendrites or cell body of another is called a SYNAPSE (si—NAPS). The synapse is a functional connection rather than a ________ connection.

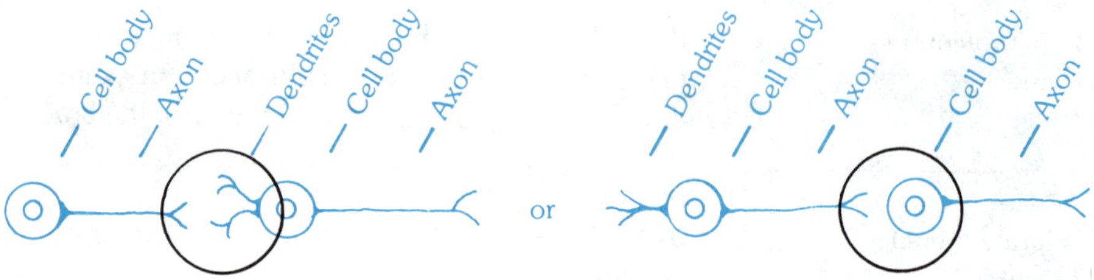

structural

3–217.
The blue circles indicate the points of impulse transmission called ________.

synapses

3–218. More than one axon may feed impulses into a ________ or functional connection.

synapse

3–219. Let's review the passage of an impulse as it travels to the brain.
A stimulus is received by the (1) ________ of a neuron. The impulse then travels through the cell body and along the axon until it comes to the (2) ________.

1. dendrite
2. synapse

3–220. The impulse then crosses the synapse to the dendrites or directly to the cell body of another neuron, and this process is continued until the impulse reaches the ________.

brain

3–221. All impulses do not travel to the brain. The brain was used here as an example because it is widely recognized as a central receiving station for impulses.
(No response required. Move to frame 222.)

	3–222. A synapse prevents an impulse from traveling back towards its source. Impulses can travel in only ______________ direction across a synapse.
one	**3–223.** Have you even been examined by a physician and had your knee-jerk REFLEX tested? The physician taps your knee slightly below the kneecap—this is the stimulus; an impulse is transmitted; and you ______________ by jerking your leg upward.
respond (or react)	**3–224.** The impulse has traveled—in one direction—from a receptor, your knee, first to the spine, and then to an effector organ, the muscle of your ______________.
leg	**3–225.** A neuron that receives a stimulus is called a SENSORY neuron. When the doctor hit you, the stimulus was picked up by a sensory neuron in your ______________.
knee	**3–226.** The sensory neuron routed the impulse to the spinal cord, which passed it on to a MOTOR neuron. The impulse then traveled by way of motor neurons to the ______________ of your leg, and you kicked involuntarily.
muscle	**3–227.** You now have all the elements of the reflex pathway: Stimulus received by (1) ______________ neuron; impulse transmitted to spine; finally from spine to muscle by way of (2) ______________ neuron.
1. sensory **2. motor**	**3–228.** An impulse can be compared to electricity that is produced in your automobile battery. The impulses that travel along a neuron are electrical in nature and are caused by the charged atoms or ______________ inside and outside a neuron.
ions	**3–229.** Another electrical analogy can be made about the sheath that covers the axon. The sheath is similar structurally to the ______________ that protects an electrical wire.
insulation	**3–230.** The myelin and neurolemmal sheaths are unlike electrical insulation because they permit ions to enter and leave the neuron, creating an ______________ charge that allows an impulse to flow along a neural pathway.
electrical	**3–231.** Neural pathways are composed of tissues because the nerve cells that make up the pathways, like those of all tissues, have a similar (1) ______________ and (2) ______________.

1. structure
2. function

3–232. Do you recall, however, that sometimes unlike tissues combine? Each muscle cell is separate and is surrounded by a delicate connective tissue. The separately contained cells are a combination of two ________________ tissues.

unlike

3–233. When unlike tissues combine, an ORGAN may be formed. For example, your heart is composed of muscle tissue, nervous tissue, and connective tissue. The heart is therefore an ________________.

organ

3–234. An organ is an organization of dissimilar ________________.

tissues

3–235. Your stomach is composed of connective, muscle, and nerve tissue. It is, like your heart, an ________________.

organ

3–236. Organs can combine together so that they make a functional unit. An example of this would be your digestive SYSTEM, which consists of your esophagus, stomach, intestines, and some accessory or helping organs.

(No response required. Move to frame 237.)

3–237. The brain and spinal cord are both (1) ________________ of the nervous (2) ________________.

1. organs
2. system

3–238. Chapter 4 will introduce you to yet another system composed of several organs, the INTEGUMENTARY (in—teg—u—MEN—ta—ri) SYSTEM. The skin is the integument of your body.

4 The Integumentary System

4–1. The word "integument" means "skin" or "covering". Your skin, nails, hair, and certain glands form the INTEGUMENTARY (in—teg—u—MENT—a—ree) SYSTEM of your body, the subject of this unit.
(No answer required.)

4–2. Let's begin by reviewing some things you have already learned.
Cells that have a similar anatomy and physiology group together to form
________________ .

tissue

4–3. An organization of dissimilar tissues is called a(n) ________________ .

organ

4–4. A group of organs that form a functional unit is a(n) ________________ .

system

4–5. Look at your skin. Obviously the tissue you are looking at has one surface attached to deeper tissues and one surface exposed. Although you may see blood vessels through the tissue, the surface tissue itself is avascular. Therefore you know that the surface of your skin is ________________ tissue.

epithelium (epithelial tissue)	**4–6.** If you could peel back the epithelial tissue on the surface of your skin, you would see another layer of tissue that is also part of your skin. This second layer of tissue is highly vascular. It serves to support the surface tissue and to connect it to underlying tissue. The second layer of your skin, then, is composed of (epithelial/connective) ________________ tissue.
connective	**4–7.** If you answered these five frames correctly, you are doing fine. If your answers were not correct, you might do well to pause for some review of <u>Chapters 2 and 3</u>. (No answer required. Go on to frame 8 or review, as your conscience dictates.)
	4–8. On the surface of your skin is a layer of (1) ________________ tissue. Beneath this surface is a layer of (2) ________________ tissue.
1. epithelial 2. connective	**4–9.** You have learned that the skin is composed of two different types of tissue. An organization of dissimilar tissues is an organ. The skin, therefore, is an ________________ .
organ	**4–10.** The surface layer of the skin is called the (1) ________________. The inner layer of the skin is called the (2) ________________ .

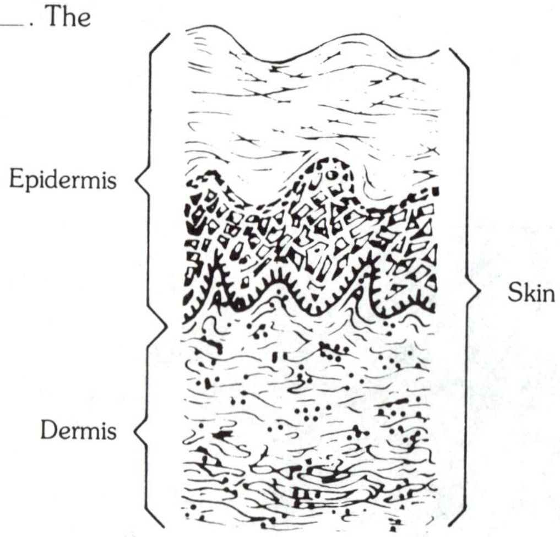

1. epidermis 2. dermis	**4–11.** A note on word origins: The dermis is sometimes called the "true skin." The word dermis comes from the <u>Greek</u> work for skin, and the Greek prefix "epi" means "on" or "over." The epidermis is, literally, an "over-skin." You can easily remember that the <u>epi</u>dermis consists of <u>epi</u>thelial tissue, because both words begin with the same prefix, "epi". (No answer required.)
	4–12. Another note on word origins: "Cuticle" (literally, "little skin") and "cutaneous" (pertaining to the skin) are both derived from the <u>Latin</u> word for "skin." The Latin prefix "sub" means "under." Thus, the <u>SUBCUTANEOUS</u> (sub—kew—TA—ne—us) <u>TISSUE</u> is the layer of tissue ________________ the skin.

under

4–13. Beneath the dermis of the skin is a fibro-fatty layer of tissue known as ______________ ______________.

subcutaneous tissue

4–14. Label the three parts of this drawing:

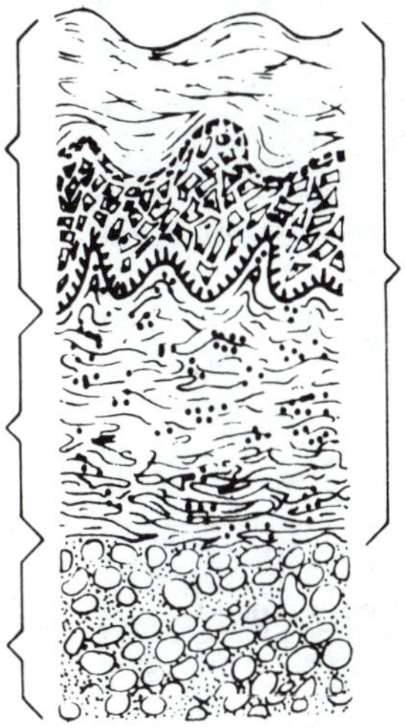

1. epidermis
2. dermis
3. subcutaneous tissue

4–15. The surface layer of the skin, the epidermis, consists of from two to four layers. On a thick-skinned surface, such as the palm of the hand or sole of the foot, the epidermis will have a maximum of (how many?)
(1) ______________ layers, whereas on even the thinnest-skinned surface, it will have a minimum of
(2) ______________ layers.

1. four
2. two

4–16. The cells of the epidermis begin to develop or <u>germinate</u> in the bottom layer, called the ______________ ______________.

stratum germinativum

4–17. As new cells are produced, the older ones move toward the top layer of the epidermis, which is called the ______________ ______________.

stratum corneum	**4–18.** As the cells of the epidermis grow, they move toward the (inner/outer) __________ layer.
outer	**4–19.** As the cells move from the stratum germinativum towards the stratum corneum, their protoplasm is gradually transformed into a horny substance, and the cells die. The stratum corneum then consists of a layer of (living/dead) __________ cells.
dead	**4–20.** The dead cells are often described as "horny" (semi-opaque). Since the Latin word for "horn" is "cornu" (as in "unicorn"), this horny substance probably inspired the name for the layer of cells in which it is found, the __________ __________.
stratum corneum	**4–21.** When epidermis consists of only two layers, those layers are the lower or germinating layer and the surface layer of dead cells. Label this drawing of two-layered epidermis.

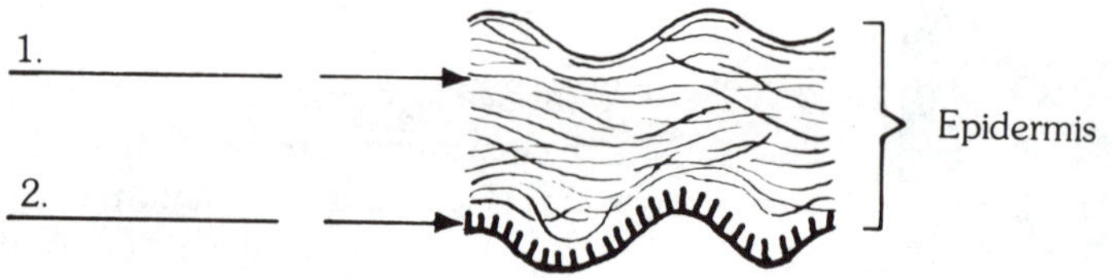

1. stratum corneum 2. stratum germinativum	**4–22.** The horny substance found in the cells of stratum corneum is called KERATIN (KER—a—ten). Keratin is a nonliving substance that replaces the __________ of the cells as they move toward the surface of the skin.
protoplasm	**4–23.** The horny substance found in the cells of the stratum corneum is called __________.
keratin	**4–24.** Label this drawing of the skin:

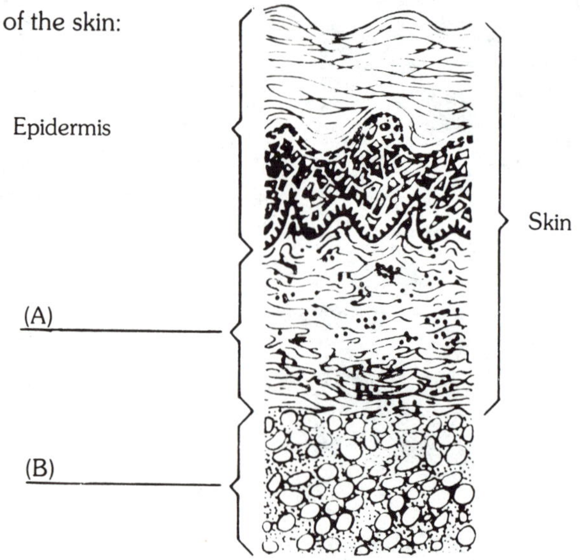

a. **dermis**
b. **subcutaneous tissue**

4–25. The dermis consists of two layers. The layer just below the epidermis has numerous nipplelike projections called PAPILLAE (pa—PIL—ee). Which number in this diagram labels the papillae?

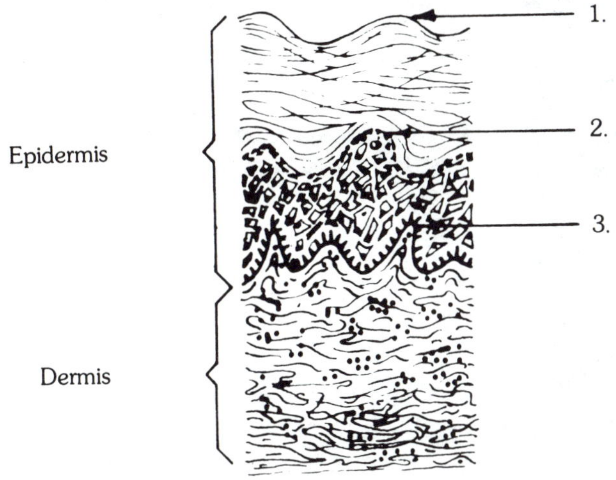

3.

4–26. The outer layer of the dermis is called the PAPILLARY (PAP—i—ler—ee) LAYER, and the inner layer, or basement membrane, is called the RETICULAR (re—TIC—u—ler) LAYER. Thus, the layer of the dermis that touches the epidermis is the (1) ____________ layer, and the layer that touches the subcutaneous tissue is the (2) ____________ layer.

1. **papillary**
2. **reticular**

4–27. The whorls on your fingers that create fingerprints result from the presence of projections called ____________ in the papillary layer of the dermis.

papillae

4–28. Voluntary movements of the face and neck are possible because there are skeletal muscles in the lower or ____________ layer of your dermis.

reticular

4–29. With movement, the skin can stretch in many directions and return to its previous shape. Such elasticity and resiliency are due to the presence of (1) ____________ and (2) ____________ fibers in the reticular layer.

1. **elastic**
2. **collagenous**

4–30. Label the following drawing:

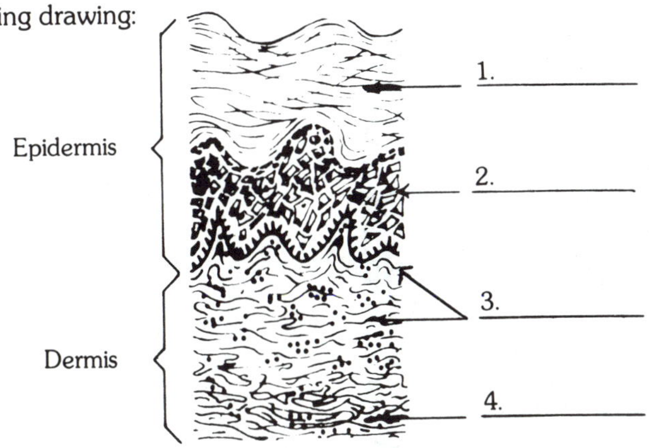

1. stratum corneum 2. stratum germinativum 3. papillary layer 4. reticular layer	**4–31.** Beneath the skin is a layer of tissue called subcutaneous tissue. Like the dermis, the subcutaneous tissue is a layer of ________________ tissue.
connective	**4–32.** The subcutaneous tissue serves as a storage place for fat cells or adipose cells. Thus, a fat or adipose person is one with an overabundance of (epidermis/subcutaneous tissue) ________________ .
subcutaneous tissue	**4–33.** The fat cells of the subcutaneous tissue are called ________________ cells.
adipose	**4–34.** Until now, we have described the skin as an organ composed of layers of tissues. These tissues, however, are not simple structures laid one upon the next like a pile of blankets. As you will see, a network of glands, blood vessels, and nerves is woven through the tissues and functions with them. (No answer required.)
	4–35. You will recall that a <u>vascular</u> tissue is a tissue that (contains/does not contain) ________________ blood vessels.
contains	**4–36.** An <u>avascular</u> or <u>nonvascular</u> tissue is one that does not contain ________________ ________________ .
blood vessels	**4–37.** You can see from the drawing below that the tissue of the epidermis is (vascular/avascular) ________________ 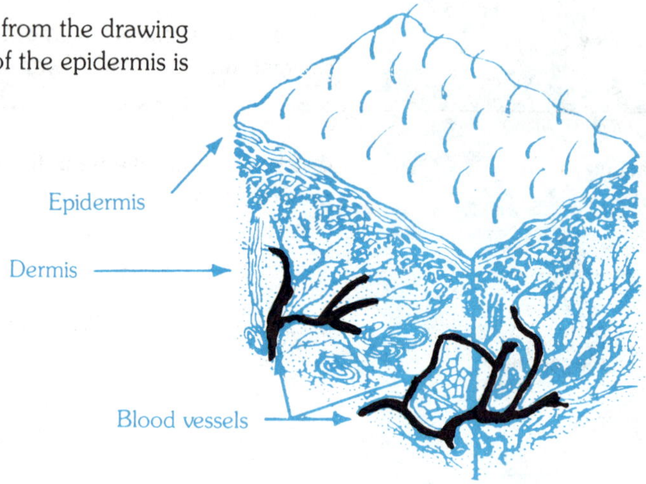
avascular	**4–38.** The tissue of the dermis is (vascular/avascular) ________________ .
vascular	**4–39.** Because the epidermis contains no blood vessels from which to obtain nourishment, it must receive its nutrients from the blood vessels closest to it; namely, the blood vessels in the upper or ________________ layer of the dermis.

papillary

4–40. Nutrients pass out of the blood vessels of the papillary layer and move up into the intercellular spaces of the ______________.

epidermis

4–41. In addition to blood vessels, the dermis contains another system of vessels known as lymphatic vessels. The lymphatic vessels remove tissue fluid from the tissue spaces. Thus the dermis receives its nutrients from the (1) ______________ vessels and excretes its excess tissue fluid into both the blood vessels and the (2) ______________ vessels.

1. blood
2. lymphatic

4–42. Both lymphatic vessels and blood vessels are found in great abundance in the (epidermis/dermis) ______________.

dermis

4–43. Tissue fluid is carried away from the tissue spaces of the dermis by the blood vessels and by the ______________.

lymphatic vessels

4–44. Differences in skin color exist between races, between individual members of the same race, and even between the various parts of one individual's body. Let us see what substances are responsible for these color differences.

(No response required.)

4–45. A substance called MELANIN (MEL—a—nin) is responsible for colorations in skin. Freckles and brown moles occur at spots where the epidermis contains a high concentration of ______________.

melanin

4–46. Brown and dark skinned persons have high concentrations of melanin distributed through the deep layer of the ______________.

epidermis

4–47. When a pin pricks your skin, or when something hot or cold touches your skin, you are aware of the sensation. Thus you can infer that your skin (contains/does not contain) ______________ numerous nerves.

contains

4–48. Your skin serves as a sense organ because there are ______________ located in the dermis.

nerves

4–49. The integumentary system contains numerous glands. Among these are the sweat glands, which produce ______________.

perspiration (sweat)

4–50. When the body is exposed to heat, it produces perspiration. The evaporation of the perspiration helps to regulate body temperature by making the skin (cooler/warmer) ______________.

cooler	**4–51.** The sweat glands of the integumentary system help to regulate body ________________.

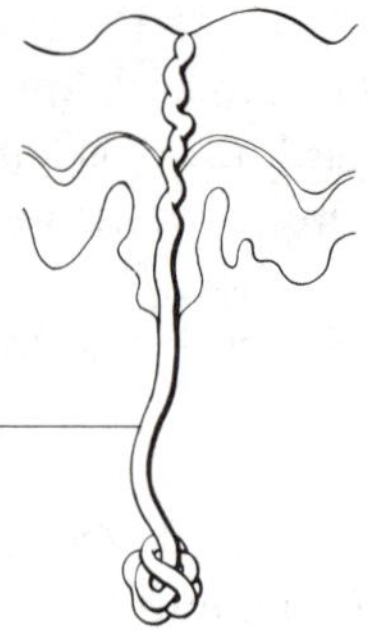

temperature	**4–52.** Both skin and sweat glands are members of the ________________ system.
integumentary	**4–53.** CERUMINOUS (ser—OO—min—us) GLANDS are modified sweat glands that produce CERUMEN (ser—OO—min) or ear wax. Ceruminous glands are located in the external ________________.
ear	**4–54.** Ear wax is the product of the ________________ glands in the ear.
ceruminous	**4–55.** MAMMARY (MA—ma—re) GLANDS are located in the mammae (singular: mamma) or breasts. The main function of these specialized glands is the production of ________________.
milk	**4–56.** A nursing infant is nourished on the milk produced by the ________________ glands of its mother.
mammary	**4–57.** SEBACEOUS (se—BA—shus) GLANDS lubricate the skin and hair by producing an oil called sebum. If too little sebum is produced by the sebaceous glands, a very (oily/dry) ________________ skin may result.
dry	**4–58.** The function of the sebaceous glands is to ________________ the skin and hair.
oil or lubricate	**4–59.** Certain glands help to control body temperature by secreting liquid from the skin. These are called ________________ glands.
sweat	**4–60.** Mammary glands produce (1) ________________. Sebaceous glands produce (2) ________________. Ceruminous glands produce (3) ________________.

1. milk
2. oil
3. ear wax

4–61. So far, you have studied <u>two</u> components of the integumentary system. These two components are (1) ________________ and (2) ________________.

1. skin
2. glands

4–62. Hair, like skin and glands, is a part of the ________________ system.

integumentary

4–63. The main body of a hair is the (1) ________________. The enlarged portion at the bottom of the hair is the (2) ________________. The opening in which hair is enclosed is the (3) ________________.

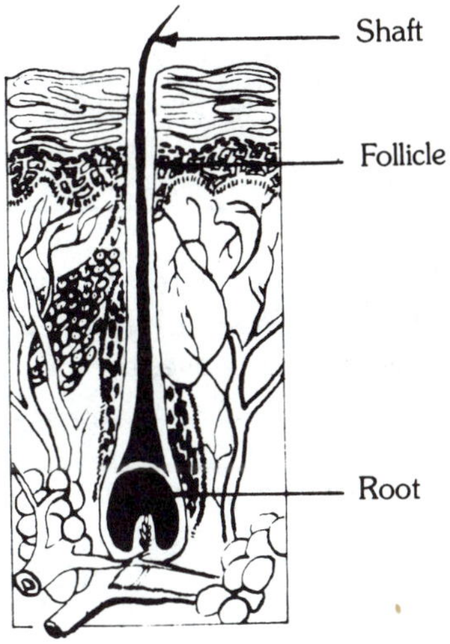

1. shaft
2. root
3. follicle

4–64. Label this drawing of a hair:

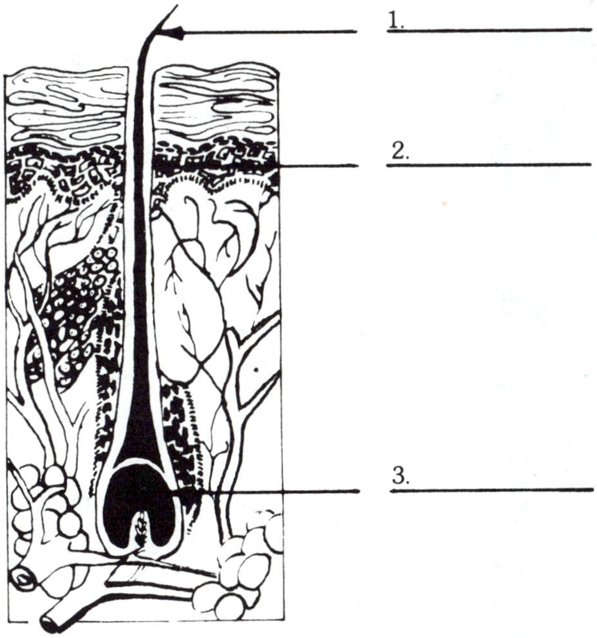

1. shaft
2. follicle
3. root

4–65. The three components of the integumentary system that you have studied are (1) ____________________, (2) ____________________, and (3) ____________________.

---

1. skin
2. glands
3. hair
   (any order)

4–66. Hair is a specialized form of skin; so, too, are fingernails and toenails. It is a common belief that nails are a form of bone; this belief is (correct/incorrect) ____________________.

---

incorrect

4–67. Both nails and hair are specialized forms of ____________________.

---

skin

4–68. At the base of each nail is a small extension of the upper layer of epidermis. This "little skin," you know from experience, is called the ____________________.

---

cuticle

4–69. The prefix "epo" means "over" or "on." You can deduce, therefore, that the term EPONYCHIUM (ep—po—NIK—i—um) applies to (the skin under the nail/the cuticle over the nail) ____________________.

---

the cuticle over the nail

4–70. The prefix "hypo" means "under." The term HYPONYCHIUM (hy—po—NIK—i—um) refers, therefore to the skin (over/under) ____________________ the free edge of the nail.

---

under

4–71. Label the parts of this drawing of the fingernail as either eponychium or hyponychium.

1. ____________________

2. ____________________

---

1. eponychium
2. hyponychium

4–72. The skin beneath the free edge of the fingernail is called (1) ____________________. The cuticle at the base of the nail is called the (2) ____________________.

---

1. hyponychium
2. eponychium

4–73. The cells of the fingernails contain keratin, a substance also present in the cells of the stratum ____________________.

---

corneum

4–74. The presence of sulfur makes keratin hard and also hardens the tissue in which the keratin is found. Thus, the keratin in the relatively soft epidermis is low in (1) ____________________, whereas the keratin in the relatively hard nails is (2) ____________________ in sulfur.

1. sulfur 2. high	**4–75.** The keratin in the nails contains a relatively (large/small) __________ amount of sulfur. The keratin in the epidermis contains a relatively (large/small) (2) __________ amount of sulfur.
1. large 2. small	**4–76.** The nails are part of the __________ system.
integumentary	**4–77.** You have learned the four components of the integumentary system. These components are: 1. __________ 3. __________ 2. __________ 4. __________
1. skin 2. glands 3. hair 4. nails (any order)	**4–78.** At some time, you must accidentally have scraped off the outer layer of your skin. The exposed tissues then proved to be (more/less) __________ sensitive to stimuli than the skin itself.
more	**4–79.** If the tissues beneath your skin were not protected by the skin, you would be excessively sensitive to stimuli. One function of the skin is to __________ underlying tissues from injurious stimuli.
protect	**4–80.** The tissues beneath the skin are generally more susceptible to infection than is the skin. A covering of skin protects these underlying tissues not only from injurious stimuli but also from __________.
infection	**4–81.** If water were not held in the underlying tissues by the skin, the underlying tissues would become too dry. The skin prevents this drying of underlying tissues by covering the tissues and preventing excessive evaporation, or loss of __________.
water	**4–82.** Two functions of the skin, then, are to (1) __________ underlying tissues from injurious stimuli and to prevent the drying of underlying tissues by preventing excessive evaporation or loss of (2) __________.
1. protect 2. water	**4–83.** In a <u>hot</u> environment the blood vessels of the skin dilate, or expand, so that more blood will reach the surface area of the skin and give off heat. As the heat is lost through the skin, the body temperature (increases/decreases) __________.
decreases	**4–84.** In a cold environment, the blood vessels constrict to prevent blood from reaching the surface and losing heat. Thus in both heat and cold, the blood vessels of the skin help to regulate the __________ of the body.

temperature	**4–85.** The nerve endings of the skin are sensitive to stimuli. Thus the skin, like the eye or ear, serves as a kind of ________________ organ.
sense	**4–86.** Through the action of the sweat glands and through the contraction and dilation of blood vessels, the skin helps to regulate the ________________ of the body.
temperature	**4–87.** Because it is better able to resist injurious stimuli than are the underlying tissues, the skin ________________ the body.
protects	**4–88.** A short review will help you to remember what you have learned in this unit. The upper layer of the skin is called the (1) ________________, and the lower layer of the skin is called the (2) ________________.
1. epidermis **2. dermis**	**4–89.** Beneath the dermis of the skin is a layer of tissue called ________________ ________________.
subcutaneous tissue	**4–90.** Epidermis may consist of four layers, but it always consists of at least two layers. The surface layer of epidermis is called the (1) ________________ ________________, and the lowest layer is called the (2) ________________ ________________.
1. stratum corneum **2. stratum germinativum**	**4–91.** The upper layer of the dermis contains small projections that rise into the epidermis. These projections are called ________________.
papillae	**4–92.** The upper layer of the dermis is called the (1) ________________ ________________, and the lower layer is called the (2) ________________ ________________.
1. papillary layer **2. reticular layer**	**4–93.** The epidermis consists of (1) ________________ tissue. The dermis consists of (2) ________________ tissue.
1. epithelial **2. connective**	**4–94.** The skin is a(n) ________________ (tissue/organ/system).
organ	**4–95.** The cells of the stratum corneum contain a horny substance called ________________.
keratin	**4–96.** The fat cells of the subcutaneous tissue are called ________________ ________________.

adipose cells	**4–97.** The epidermis is (1) ____________ (vascular/avascular). The dermis is (2) ____________ (vascular/avascular).
1. avascular **2. vascular**	**4–98.** The epidermis receives its nutrients from the blood vessels of the ____________ layer.
papillary	**4–99.** Tissue fluid is removed from the tissue spaces of the dermis by blood vessels and by ____________ ____________.
lymphatic vessels	**4–100.** The brown color of skin is caused by (1) ____________ in the deep layer of the (2) ____________.
1. melanin **2. epidermis**	**4–101.** The glands that help to regulate body temperature by secreting moisture are called the ____________ ____________.
sweat glands	**4–102.** The ceruminous glands produce (1) ____________. The sebaceous glands produce (2) ____________. The mammary glands produce (3) ____________.
1. ear wax **2. oil** **3. milk**	**4–103.** The skin opening in which a shaft of hair is enclosed is called the ____________ of the hair.
follicle	**4–104.** Hair and nails are both specialized forms of ____________.
skin	**4–105.** The skin under the nail is called the (1) ____________. The cuticle over the nail is called the (2) ____________.
1. hyponychium **2. eponychium**	**4–106.** The hard keratin of the nails contains (more/less) ____________ sulfur than the soft keratin of the epidermis.
more	**4–107.** The components of the integumentary system are: (name them) 1. ____________ 2. ____________ 3. ____________ 4. ____________

1. **skin**
2. **glands**
3. **nails**
4. **hair**

(any order)

4–108. Which of the following are functions of the integumentary system?
1. to protect underlying tissues
2. to store food and water
3. to serve as a sense organ
4. to act as a temperature regulator

all

4–109. You have completed Chapter 4.

5 The Musculoskeletal System

	5–1. The subject of this section of the program will be the MUSCULOSKELETAL (MUS—kew—lo—SKEL—e—tal) SYSTEM, which, as its name implies, is made up of the (1) ________________ and the bones of the (2) ________________.
1. muscles **2. skeleton**	**5–2.** The various movements of which the body is capable result from the action of flexible muscles pulling against the hard, inflexible bones of the ________________.
skeleton	**5–3.** Muscles are attached to the bones of the skeleton by means of the fibrous cords or bands called tendons. The large muscle in the calf of the leg is attached to the "heel bone" by the Achilles ________________.
tendon	**5–4.** Cartilage, you remember, is a somewhat flexible tissue that forms the connecting structures of the bony skeleton and often acts as a "shock absorber" between bones. Muscles are bound to bone by means of (tendons/cartilage) (1) ________________. Between some bones are shock absorbers made of (tendons/cartilage) (2) ________________.

1. tendons **2. cartilage**	**5–5.** Ligaments are the bands of tissue that bind the bones together. The bones are separated and connected by shock absorbers of (1) ____________ and are bound together by bands of (2) ____________.
1. cartilage **2. ligament (s)**	**5–6.** The human skeleton contains 206 bones. Six of these bones are located in the ears, three in each ear. The remaining ____________ (how many?) bones are divided between the AXIAL (AX—e—al) SKELETON and APPENDICULAR (ap—pen—DIK—u—lar) SKELETON.
200	**5–7.** The appendicular skeleton consists of all the bones of the limbs or appendages. The axial skeleton consists of all the remaining bones, except the six bones in the ears. The bones of the arm are part of the (1) ____________ skeleton. The bones of the head are part of the (2) ____________ skeleton.
1. appendicular **2. axial**	**5–8.** The spinal column is a part of the (1) ____________ skeleton. The bones of the foot are part of the (2) ____________ skeleton.
1. axial **2. appendicular**	**5–9.** The human body contains (2006/206) ____________ bones.
206	**5–10.** In addition to the 206 bones of the body, there are in most people a varying number of bones called SESAMOID (SES—a—moid) bones. Since the name "sesamoid" means "like a grain of sesame," you can infer that the sesamoid bones are quite (large/small) ____________.
small	**5–11.** Sesamoid bones may function as ball bearings to reduce friction in places where a muscle is joined to a bone. The sesamoids develop within the ____________ that join muscle to bone.
tendons	**5–12.** In addition to their ball-bearing function, the ____________ bones may also act as pulleys, affecting the direction of muscle pull.
sesamoid	**5–13.** Some sesamoids, such as those in the knee caps, are always present. In the joints of the hand and foot, sesamoids are usually present. In other areas of the body, sesamoids develop according to occupational demand. Thus, the number of sesamoids in the body (is/is not) ____________ the same for all humans.
is not	**5–14.** Bones may be classified in three ways, on the basis of: (1) shape, (2) structure, or (3) embryonic development. When we speak of bones as flat, long, irregular, or short, we are obviously using a classification based on ____________.

shape

5–15. A section of a bone examined under a microscope shows its structure. Structurally, the illustrated bone has an outer layer of (1) ______________ ______________ and an inner layer of (2) ______________ ______________.

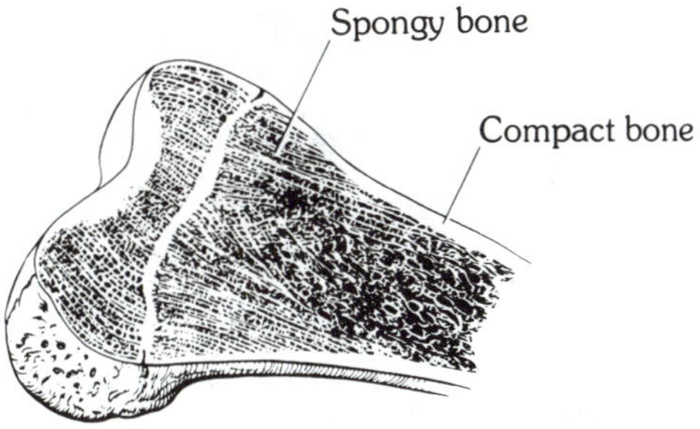

1. compact bone
2. spongy bone

5–16. Compact bone is dense and hard. Spongy bone has many small cavities filled with marrow. The classification of bones as compact or spongy is based upon their (shape/structure) ______________.

structure

5–17. Spongy bone is also called CANCELLOUS (CAN—sel—us) bone. The (inner/outer) ______________ layer of the illustrated bone is cancellous.

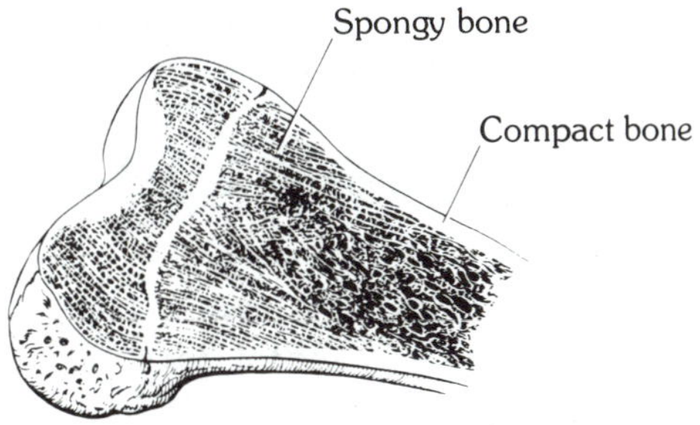

inner

5–18. All bones contain both compact and spongy tissue, but one predominates, and the bone is classified on the basis of the predominating type of tissue. Thus the classification of a bone as "spongy" indicates that the bone:
(a) contains no compact tissue.
(b) contains more spongy tissue than compact tissue.

b.

5–19. Following conception, bone, like all tissues, originates as soft tissue. Actual bone formation in the embryo occurs during the last two months of pregnancy and continues until maturity (18–21 yr.). The bones of the embryo develop in one of two ways, which we shall now discuss.
(No answer required.)

5–20. The drawings illustrate three early stages of the growth of a long bone. The long bone originates as a cartilaginous (1) ________________. Gradually the cartilage becomes (2) ________________.

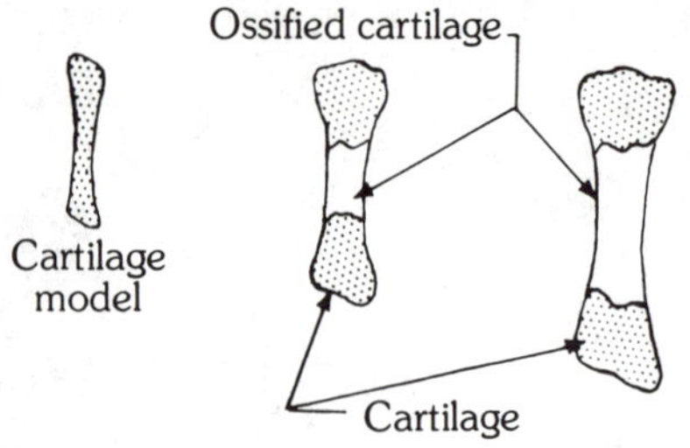

The process of bone formation is called OSSIFICATION (os—i—fi—CA—shun).

---

1. model
2. bone

5–21. The bone illustrated on the last item is, on the basis of its development, classified as CARTILAGINOUS (car—ti—LAH—ji—nus). A cartilaginous bone, as its name implies, is a bone that develops from a ________________ ________________.

---

cartilage model

5–22. The bones of the face and cranium develop by another method, as described here. First, bony spicules arise within a connective tissue (1) ________________. Then the (2) ________________ ________________ increase in number. Ultimately, they grow together or (3) ________________.

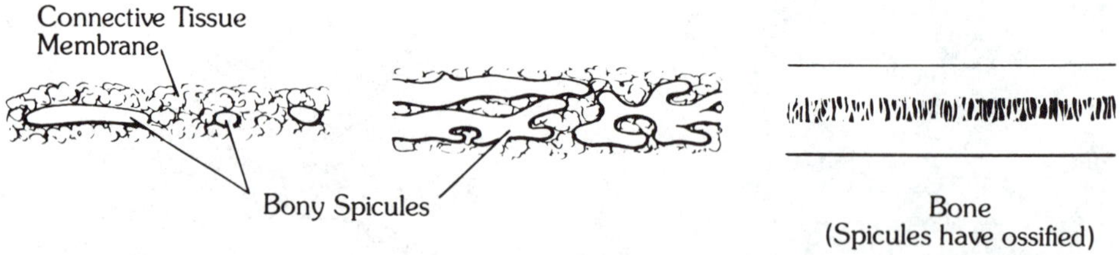

---

1. membrane
2. bony spicules
3. coalesce

5–23. The last item illustrated the development of <u>MEMBRANOUS</u> (MEM—bra—nus) bone. Membranous bone develops as (1) ________________ spicules form and ultimately coalesce within a connective tissue (2) ________________.

---

1. bony
2. membrane

5–24. Bone that develops from a cartilage model is called (1) ________________ bone. Bone that develops from bony spicules within a connective tissue membrane is called (2) ________________ bone.

---

1. cartilaginous
2. membranous

5–25. The process by which either cartilaginous bone or membranous bone develops is called ________________.

---

ossification

5–26. The classification of bones as flat, irregular, long, or short is based upon the (1) ________________ of the bones. The classification of bones as spongy (cancellous) or as compact is based upon the (2) ________________ of the bones.

1. shape
2. structure

5–27. The classification of bones as cartilaginous or as membranous is based upon the way in which the bones ____________________.

develop
(form, etc.)

5–28. Under a microscope, a longitudinal section of a typical bone looks like this:

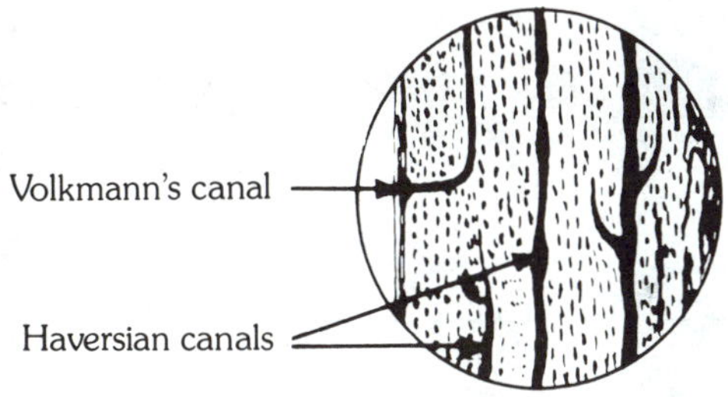

Bone tissue is not perfectly solid but is traversed by a system of ____________________.

canals

5–29. Blood vessels, lymph vessels, and nerves enter and leave a bone by means of its numerous ____________________.

canals

5–30. A cross section of bone reveals an inner cavity called a (1) ____________________ ____________________ and an outer sheath of tissue called the (2) ____________________.

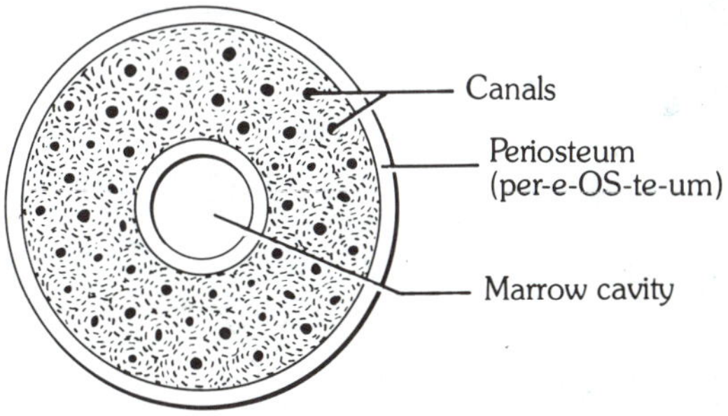

1. marrow cavity
2. periosteum

5–31. The connective tissue sheath that covers a bone is the periosteum. The connective tissue that lines the marrow cavity is the ____________________.

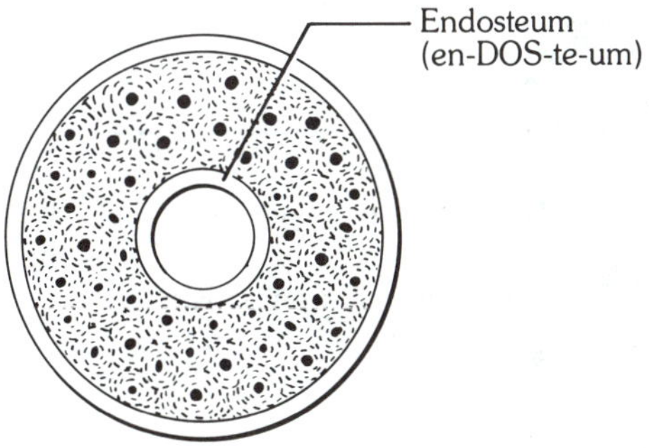

endosteum

5–32. Label the parts of this cross section of bone.

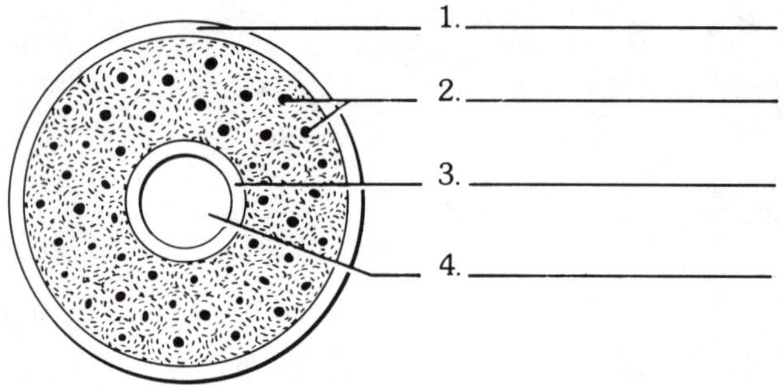

1. **periosteum**
2. **canals**
3. **endosteum**
4. **marrow cavity**

5–33. The articular end of a bone is the end that makes a joint with another bone. The periosteum covers all of a bone except its ____________ ____________.

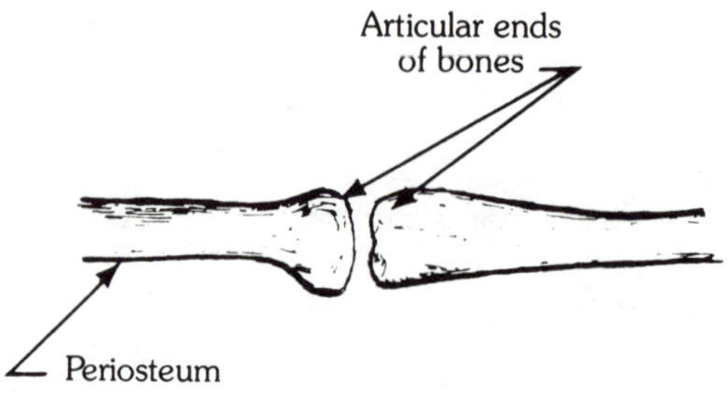

articular end (s)

5–34. The parts of a long bone (such as the bones of the arm or leg) are classified as the (1) DIAPHYSIS (di—AF—i—sis), the shaft or main body of the bone, and (2) the EPIPHYSIS (ep—IF—i—sis), the end of the bone. Label the parts of this long bone.

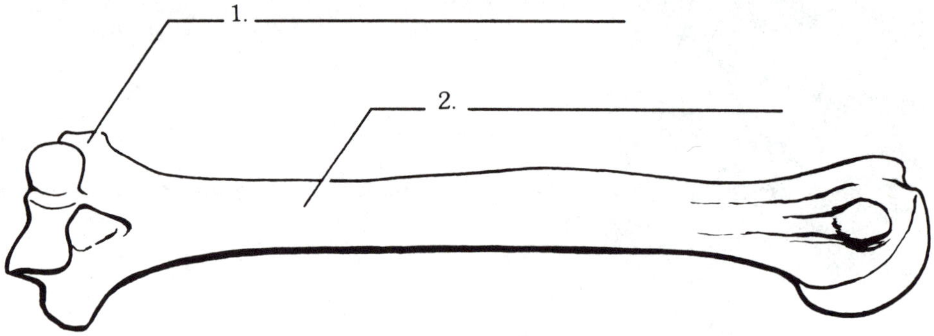

1. **epiphysis**
2. **diaphysis**

5–35. The prefix "dia" means "dividing in two." In a long bone, the ____________ is the long shaft that divides or separates the two epiphyses (plural of epiphysis).

diaphysis	**5–36.** The periosteum covers all of a bone except its articular end. In a long bone, the periosteum covers all of the (1) ________________ (epiphysis/diaphysis) but may not cover all of the (2) ________________ (epiphysis/diaphysis).
1. diaphysis 2. epiphysis	**5–37.** Bones receive their nutrients from the blood vessels, which traverse the bone canals. These blood vessels enter the bone through the outer sheath of the bone, which is called the ________________.
periosteum	**5–38.** If the periosteum is damaged, the blood vessels are disrupted, and the bone loses its blood supply. Obviously, the integrity of the periosteum is (essential to/not essential to) ________________ the health of the underlying bone.
essential to	**5–39.** The largest bones of the limbs are classified, on the basis of shape, as long bones. In the center of a long bone is a marrow cavity, which is so named because it contains bone ________________.
marrow	**5–40.** Yellow bone marrow, composed chiefly of fat cells, is found in the marrow cavities of ________________ (what shape?) bones, like the large bones of the limbs.
long	**5–41.** In the marrow cavities of long bones, there is (1) ________________ (what color?) marrow that consists chiefly of (2) ________________ cells.
1. yellow 2. fat	**5–42.** The epiphyses of long bones are composed of spongy bone filled with red marrow. The marrow in the central cavity of a long bone is (1) ________________, whereas the marrow in the spongy ends of the bone is (2) ________________.
1. yellow 2. red	**5–43.** Red marrow, found in the spongy bone of long bones and also in some flat bones, produces blood cells. One function of the bony skeleton, then, is to produce ________________.
blood (cells)	**5–44.** In addition to its function of housing marrow, which produces blood cells, the skeleton forms the framework of the body and provides attachments for skeletal ________________.
muscles	**5–45.** The bones of the cranium enclose and protect the brain. The bones of the thorax enclose and protect the heart, lungs, and other vital organs. Therefore, another function of the skeleton is ________________. (Use your own words.)

to enclose and protect the vital organs (or any similar statement).	**5–46.** A machine whose moving parts come in contact with each other must have lubrication at the points of contact. Bones come together at joints called <u>ARTICULATIONS</u> and, where movement is involved, bones also require ______________.
lubrication	**5–47.** Between the <u>mobile</u> bones of the limbs, there are articulations. There are also articulations between the <u>immobile</u> bones of the skull. At which articulations do you think there is a greater need for lubrication? a. those of the skull b. those of the limbs
b. those of the limbs	**5–48.** Articulations such as those of the limbs and spine are lubricated by <u>SYNOVIAL</u> (si—NO—vi—al) FLUID, which is contained in a cavity between bones. Taking their name from the fluid that lubricates them, these articulations are called ______________ joints.
synovial	**5–49.** The articulations of the limbs, fingers, and bones of the spine are lubricated by ______________ ______________.
synovial fluid	**5–50.** Before you go on, review what you have learned so far. The human skeleton contains (2006/206) bones ______________.
206	**5–51.** The bones of the limbs are part of the (1) ______________ skeleton. All the remaining bones in the body except the six bones in the ears are part of the (2) ______________ skeleton.
1. appendicular 2. axial If you made an error, review frames 5–6 to 5–8	**5–52.** In most people there are a varying number of small bones that may function as ball bearings in places where muscle is joined to bone or as pulleys affecting the direction of muscle pull. These small bones are called ______________ ______________.
sesamoid bones If you made an error, review frames 5–10 to 5–13.	**5–53.** The first column lists some terms used to classify bones. The second column lists the bases of bone classification. Match the items in the two columns. 1. spongy and compact bones 2. flat, round, long, short, and irregular bones 3. membranous and cartilaginous bones a. embryonic development b. shape c. structure

1. c
2. b
3. a
If you made an error, review frames 5–14 to 5–27.

5–54. Label the parts in the drawing of the cross section of a bone.

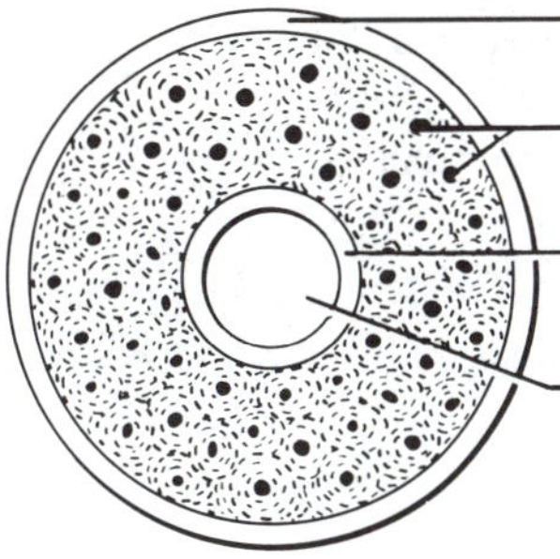

1. periosteum
2. canals
3. endosteum
4. marrow cavity
If you made an error, review frames 5–28 to 5–32.

5–55. Label the two parts of this drawing of a long bone as either diaphysis or epiphysis.

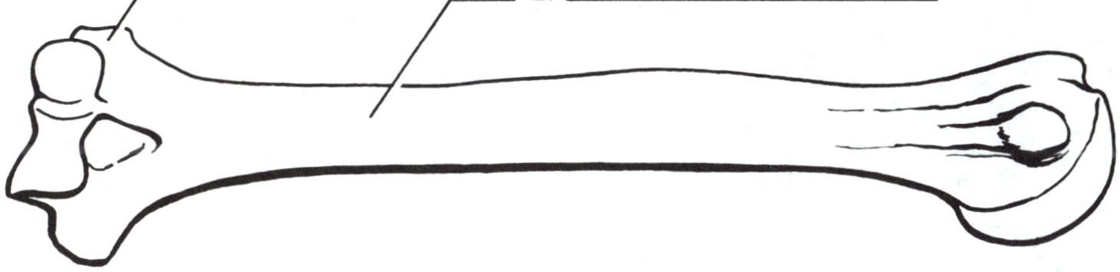

1. epiphysis
2. diaphysis
If you made an error, review frames 5–33 to 5–35.

5–56. Articulations, such as those of the limbs and spine, are lubricated by a fluid called ____________ ____________.

synovial fluid
If you made an error, review frames 5–46 to 5–49.

5–57. Now that you have examined the structure, functions, and classification of bones, consider how these bones are arranged to form the skeleton, which supports the body.

(No answer required.)

5–58. The skull consists of 28 bones; six of these bones are located in the ears; eight of these bones form the CRANIUM (KRA—ni—um), which houses the brain, and the remaining fourteen form the face. The terms skull and cranium are not synonyms. Both the (1) ____________ and the face are parts of the (2) ____________.

1. cranium
2. skull

5–59. The bones of the cranium form the CRANIAL VAULT. The cranial vault houses the ____________.

brain

5–60. Except for the mandible, which has a ball and socket articulation, bones of the skull are joined together by serrated joints called ________________.

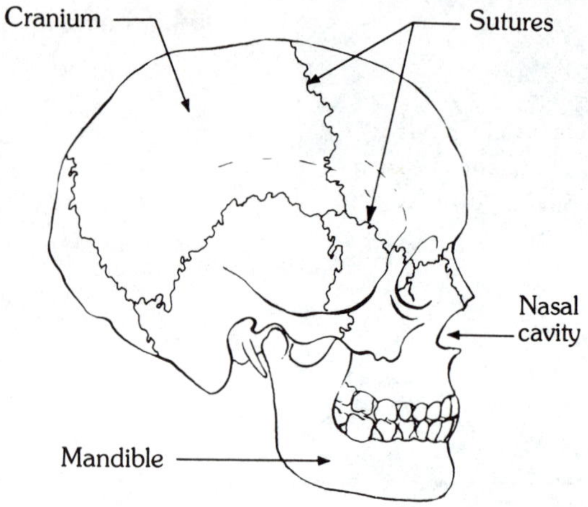

sutures

5–61. The bones of the skull are immovable <u>except</u> for the "jawbone," called the ________________, which is movable.

mandible

5–62. Below the skull is the spine or ________________ ________________.

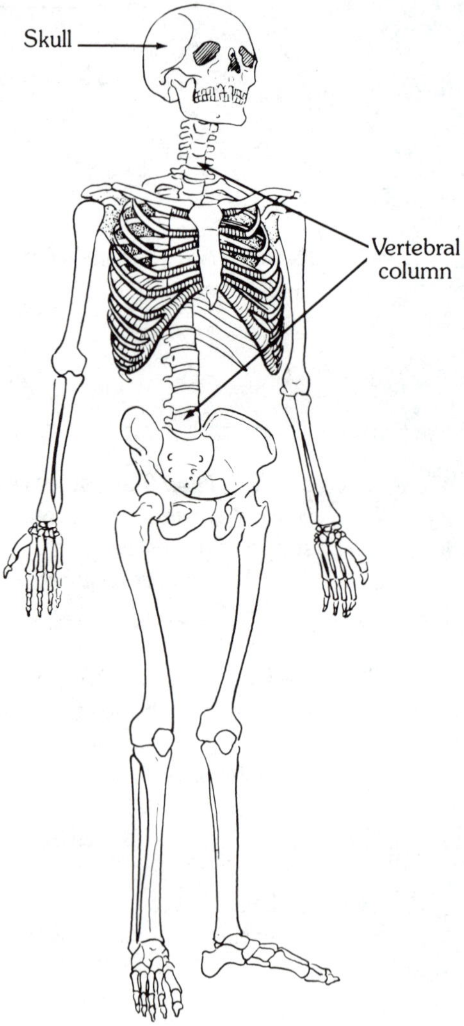

vertebral column

5–63. The vertebral column of an adult consists, usually, of 33 separate bones called VERTEBRAE (VER—te—bray), although the number may vary by one in some individuals. The articulations of these vertebrae make the spine (flexible/inflexible) ________________.

Regions of the Vertebral Column

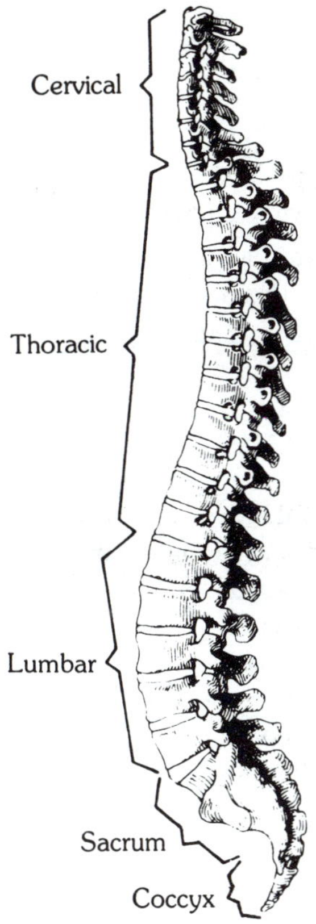

---

flexible

5–64. Look at the drawing of the vertebral column, and use it to answer this and the following items.

The vertebral column is divided into five regions. The region just below the skull derives its name from the Latin word "cervix" meaning "neck" and is called the ________________ region.

---

cervical

5–65. The THORAX (THOR—ax) is the chest cavity. That section of the vertebral column in the region of the chest is called the ________________ region.

---

thoracic

5–66. Just below the skull, at the neck, is the (1) ________________ region of the vertebral column. Below that, in the chest area, is the (2)________________ region of the vertebral column.

---

1. cervical
2. thoracic

5–67. Below the thoracic region of the vertebral column is the ________________ region, its name meaning "loin."

lumbar

5–68. Below the lumbar region, the vertebrae have fused to form a single spade-shaped bone about which there arose a curious superstition. The bone is called the SACRUM, which means "sacred." It was so named because people once believed that the sacrum alone did not disintegrate after death and served as the basis for the resurrected body.

(No answer required.)

5–69. Below the lumbar region, the vertebrae have fused to form a single bone, called the ________________.

---

sacrum

5–70. At the bottom of the vertebral column, more fused vertebrae form the COCCYX (COCK—six). Coccyx means "cuckoo." The coccyx is so named because it was thought to resemble the beak of a cuckoo.

(No answer required.)

5–71. Below the sacrum is the ________________.

---

coccyx

5–72. It's quite easy to remember the names of the regions of the spine if you know what their names mean (and remember that the "sacred bone" is above the "cuckoo's beak"). Match the names on the left with their meanings on the right.

1. sacrum	a. chest
2. coccyx	b. neck
3. cervical	c. sacred
4. lumbar	d. loin
5. thoracic	e. cuckoo

---

1. sacrum–c. sacred
2. coccyx–e. cuckoo
3. cervical–b. neck
4. lumbar–d. loin
5. thoracic–a. chest

5–73. Label this drawing of the regions of the spine:

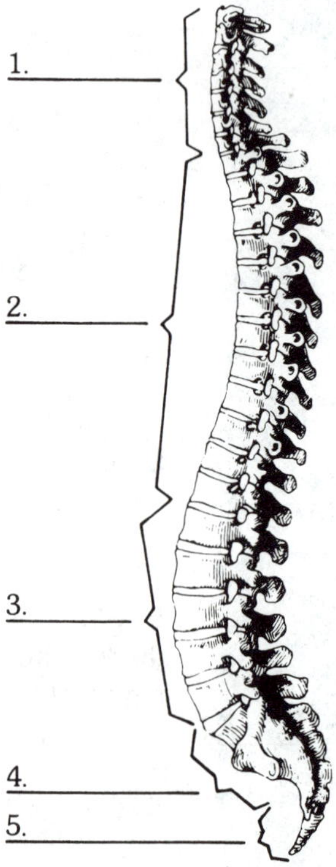

1. cervical region 2. thoracic region 3. lumbar region 4. sacrum 5. coccyx	**5–74.** The cervical, thoracic, and lumbar vertebrae are separate, movable bones, and so are called TRUE VERTEBRAE. The vertebrae of the sacrum and coccyx have fused together and so are called FALSE VERTEBRAE. The upper three regions of the spine are composed of (true/false) ________________ vertebrae.
true	**5–75.** The true vertebrae make up the (1) ________________, (2) ________________, and (3) ________________ regions of the vertebral column.
1. cervical 2. thoracic 3. lumbar (any order)	**5–76.** The term "false vertebrae" refers to the fused vertebrae of the (1) ________________ and (2) ________________.
1. sacrum 2. coccyx	**5–77.** A typical vertebra consists of a rounded (1) ________________ at the front and a spiny (2) ________________ at the back.
1. body 2. arch	**5–78.** The spinal cord is housed in a canal called the ________________ ________________, which runs through the center of each vertebra.

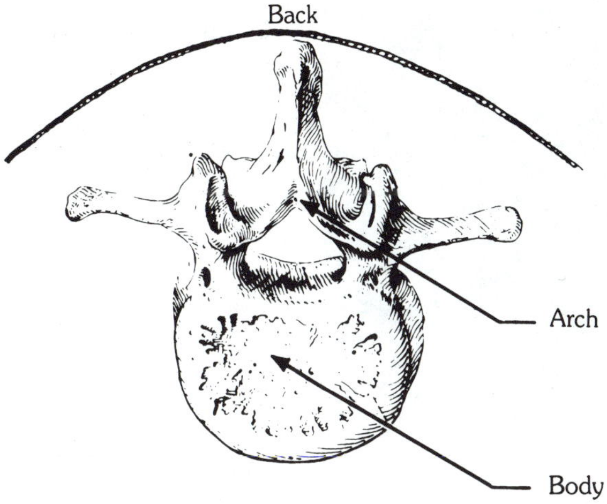

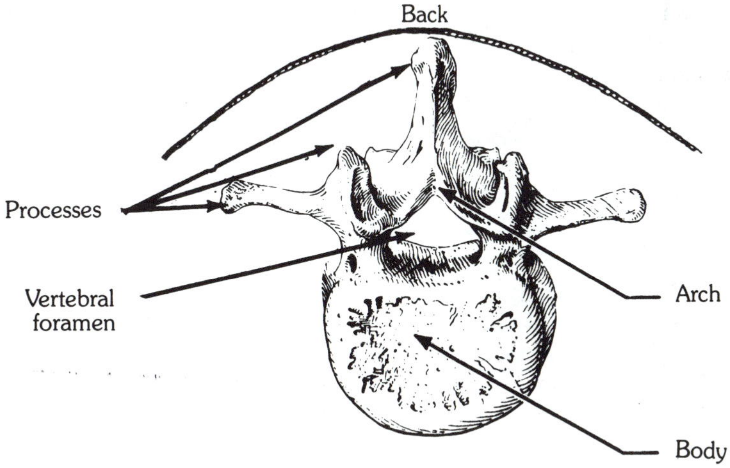

vertebral foramen

5–79. Muscles and ligaments are attached to the spiny projections called ________________ on the arch of the vertebra.

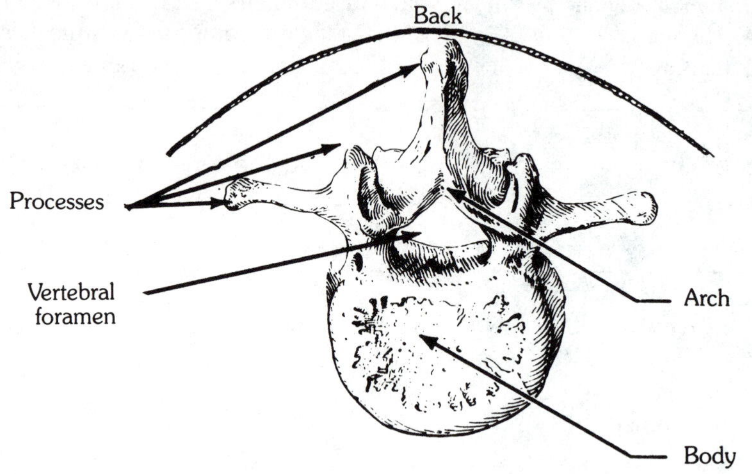

processes

5–80. Label this drawing of a typical vertebra:

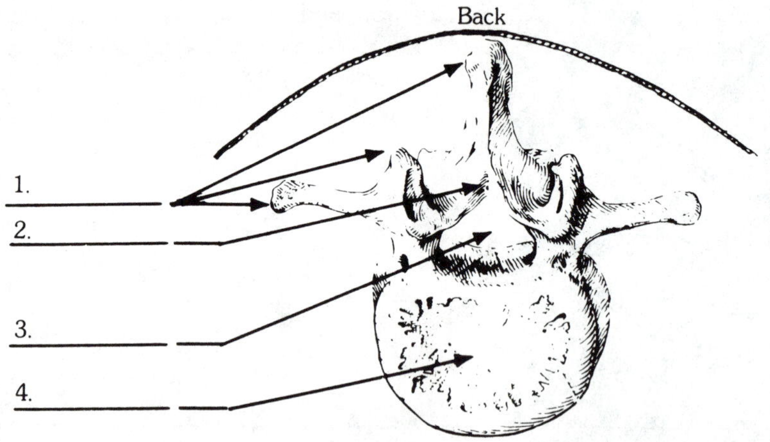

1. processes
2. arch
3. vertebral foramen
4. body

5–81. Between the vertebrae are DISCS of cartilage, which act as shock absorbers. A person who has a split or ruptured disc may require an operation on his ________________.

spine (or back or vertebral column)	**5–82.** The "shock absorbers" between the vertebrae are called ____________.

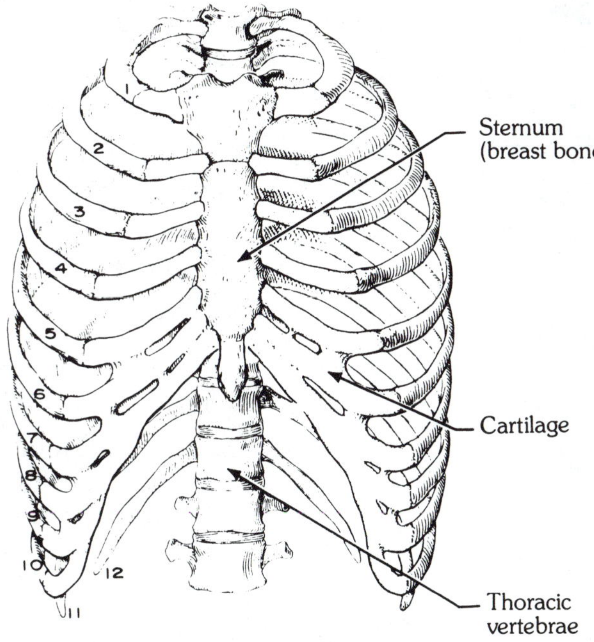

The Thorax (Rib Cage)

discs	**5–83.** Refer to the drawing of the thorax (in the preceding frame), and use it to answer this and the following items. The ribs articulate with the ____________ vertebrae to form the thorax.
thoracic	**5–84.** Those ribs that articulate with the STERNUM (breast bone) as well as with the thoracic vertebrae are called TRUE RIBS or VERTEBROSTERNAL RIBS. There are ____________ (how many?) true ribs.
seven	**5–85.** The lowest five ribs are called FALSE RIBS because they do not articulate with the ____________.
sternum	**5–86.** Because they articulate with the cartilage of the seventh rib the, ______, ______, and ______ (what numbers?) ribs are called VERTEBROCHONDRAL (VER—te—bro—KON—dral) RIBS.
8th, 9th, 10th	**5–87.** The front ends of the FLOATING RIBS are not articulated with the sternum or the cartilage of other ribs. The lowest ____________ (how many?) ribs are floating ribs.

two

5–88. In the following ten frames you will see how the remaining bones of the body combine to form the skeleton. The next ten frames are intended to give you a general concept of skeletal stucture; you will not be expected to remember the names of all the bones mentioned in these frames.

(No answer required.)

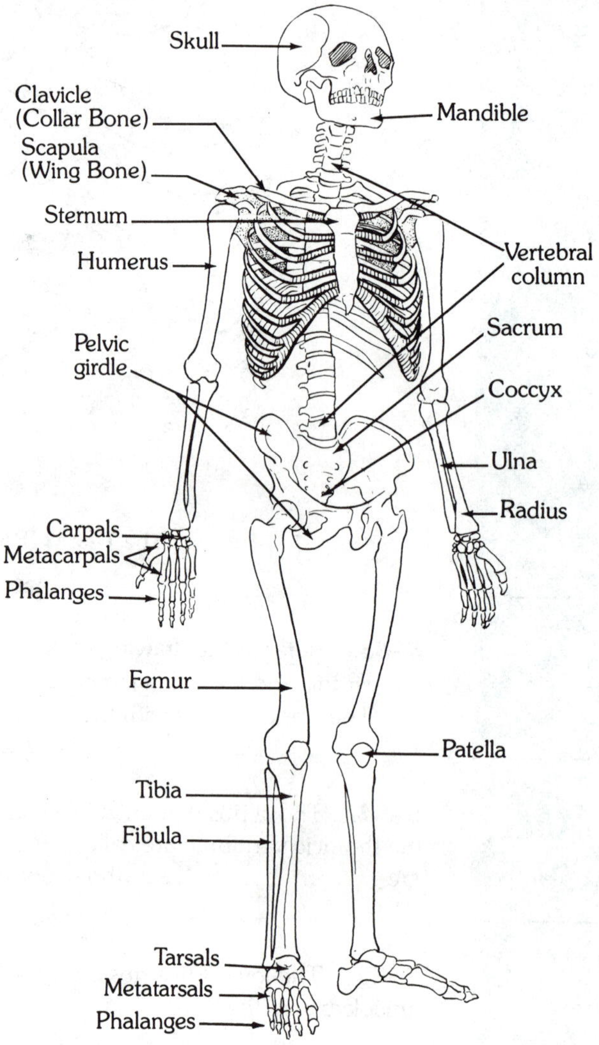

Skeleton

5–89. Refer to the drawing of the skeleton and use it to answer this and the following items.

The coalition of several bones, including the sacrum and coccyx, forms the ______________ ______________.

pelvic girdle

5–90. The bones of the pelvic girdle articulate with the (1) ______________ ______________ above and with the thigh bones or (2) ______________ below.

1. vertebral column
2. femurs

5–91. The hip joint is formed by the articulation of the bones of (1) ________________ ________________ with the (2) ________________ below.

1. pelvic girdle
2. femur

5–92. Below the knee, the main bone of the leg is the (1) ________________, which articulates with the (2) ________________ in the thigh above.

1. tibia
2. femur

5–93. In addition to the tibia, the lower leg contains a smaller bone, the ________________, which articulates with the tibia.

fibula

5–94. The lower extremity of the skeleton is completed by the ankle bones or (1) ________________, the foot bones or (2) ________________, and the toe bones or (3) ________________.

1. tarsals
2. metatarsals
3. phalanges

5–95. In the upper part of the body, the "collar bone" or (1) ________________ articulates with the "wing bone" or (2) ________________.

1. clavicle
2. scapula

5–96. The bony structure of the arm is somewhat similar to that of the leg. The upper arm contains a single large bone called the ________________.

humerus

5–97. The humerus articulates with two bones, the (1) ________________ and the (2) ________________ in the lower arm.

1. radius
2. ulna
(either order)

5–98. The radius and ulna articulate with the bones of the wrist, which are called (1) ________________. Below the carpals are the bones of the hand, called (2) ________________, and the bones of the fingers called (3) ________________.

1. carpals
2. metacarpals
3. phalanges

5–99. Consider the muscles that combine with the skeleton to form the musculoskeletal system. In previous units, you learned that the body contains three kinds of muscle: smooth muscle, which lines the internal organs; cardiac muscle, which is found in the heart; and skeletal muscle, which allows the body to ________________.

move

5–100. The musculoskeletal system functions to support the body and to allow it to move. The principal components of the musculoskeletal system are bones and ________________ (what kind?) muscles.

skeletal

5–101. Skeletal muscles contain bundles of fibers held together by connective tissue and abundantly supplied with blood vessels. Skeletal muscles, in other words, are (vascular/avascular) ________________.

vascular

5–102. The numerous nerves present in skeletal muscle permit voluntary movement of the muscles. A muscle may be unable to move if the ________________ in that muscle are damaged.

nerves

5–103. Muscles may be attached directly to a bone, or else they may be attached indirectly, by means of a ________________, like the Achilles tendon in the lower leg.

tendon

5–104. Typically, a muscle is attached at two points: an ORIGIN, which acts as a fixed point or anchor, and an INSERTION, which is the movable end of the muscle. When a muscle contracts, the movable end or (1) ________________ is brought closer to the anchor or (2) ________________.

1. insertion
2. origin

5–105. The BICEPS (BI—seps), the large muscle in the upper arm, is anchored to a point in the shoulder; its movable end joins the radius in the forearm. Thus the biceps has its (1) ________________ at the shoulder and its (2) ________________ at the radius.

1. origin
2. insertion

5–106. Most skeletal muscles move on the principle of leverage. A lever is a barlike structure that moves on a fixed point or fulcrum. In this drawing of a see-saw, A is the (1) ________________ and B is the (2) ________________.

The muscle is the force that acts to pull the lever, the bone.

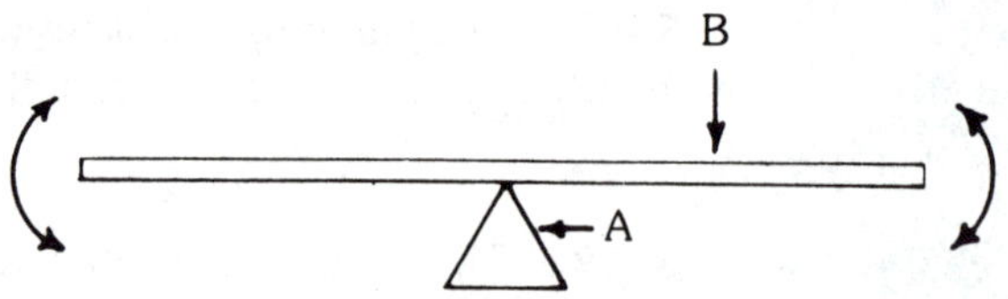

1. fulcrum
2. lever

5–107.

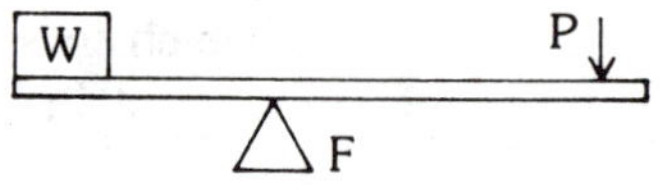

First-Class Lever

Abbreviations

F = Fulcrum
P = Force
W = Weight or Resistance

This diagram of a first-class lever shows that the ________________ lies between the point where force (P) is applied and the point where resistance (W) is met.

fulcrum

5–108. A shows the triceps muscle extending the arm. B shows the gastrocnemius muscle raising the weight of the body on the toes. (A/B) ________ illustrates a muscle that acts like a first-class lever.

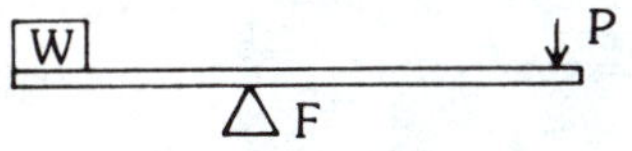

First-Class Lever

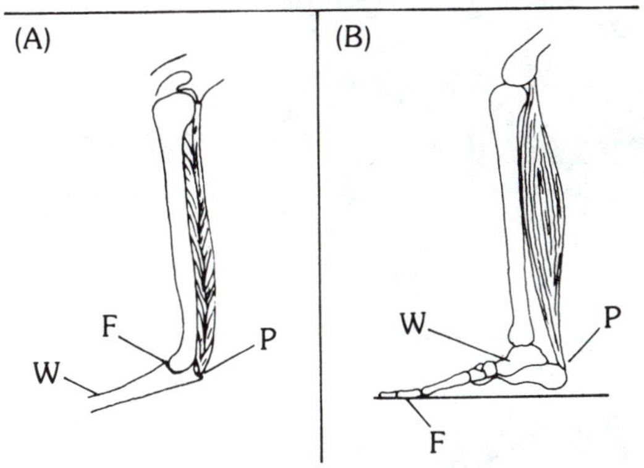

A

5–109.

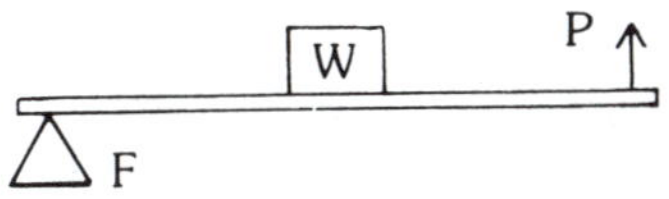

Second-Class Lever

Abbreviations

D = Fulcrum
P = Force
W = Weight or Resistance

A second-class lever is one whose point of ________ lies between the fulcrum and the point where force is applied.

resistance

5–110. A shows the biceps muscle flexing the arm at the elbow. B shows the gastrocnemius muscle raising the body on the toes. (A/B) ________ illustrates a muscle that acts like a second-class lever.

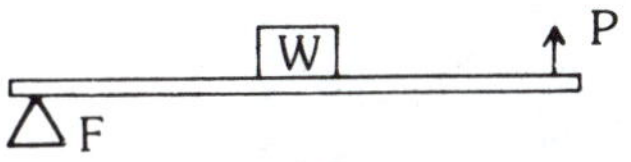

Second-Class Lever

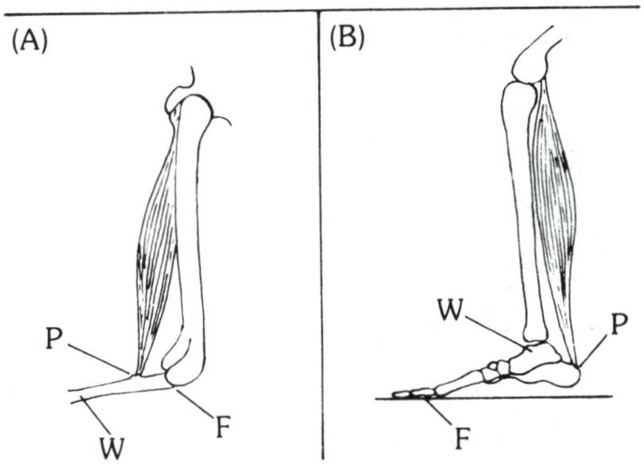

B

5–111.

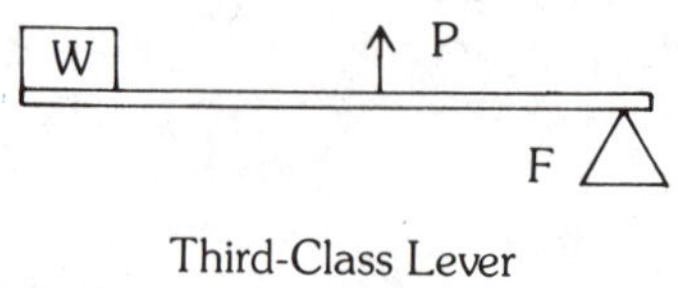

Abbreviations

F = Fulcrum
P = Force
W = Weight or Resistance

In a third-class lever, the point where ____________ is applied lies between the fulcrum and the point of resistance.

force

5–112. A shows the triceps muscle extending the arm at the elbow. B shows the biceps muscle flexing the arm at the elbow.
(A/B) ____________ illustrates a muscle that acts like a third-class lever.

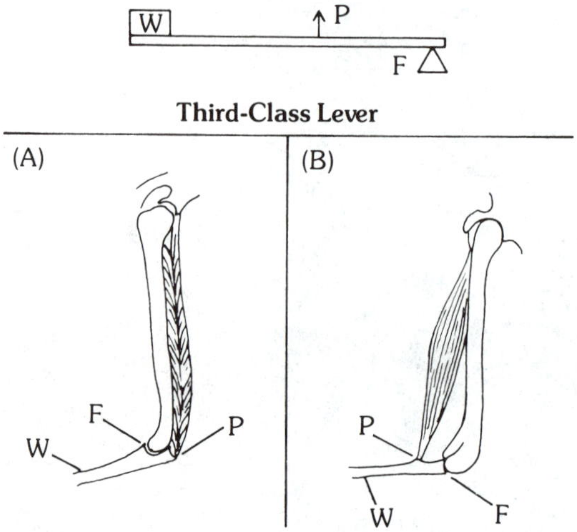

B

5–113. Skeletal muscles act like levers; that is, they act on the principle of ____________.

leverage

5–114. By observing a specific example of muscle movement, we can best understand some general principles. When the arm is flexed at the elbow, the ____________ muscle contracts or shortens.

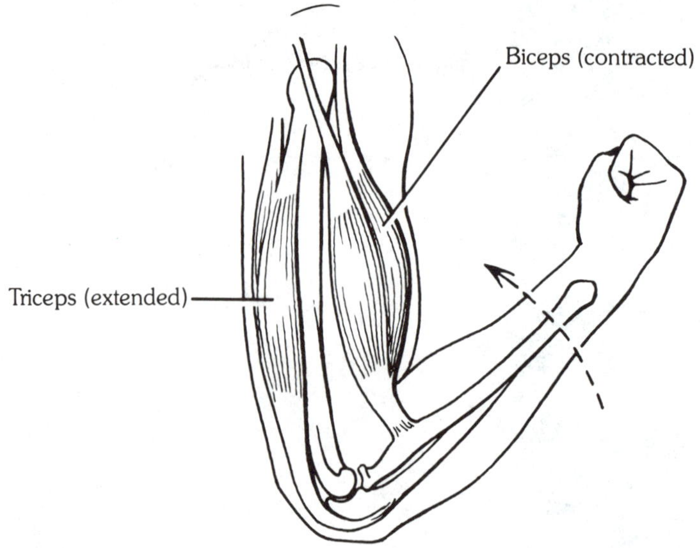

biceps

5–115. During flexion, the biceps muscles contracts and the ____________ muscle extends, influencing the speed and degree of contraction and flexion.

triceps

5–116. Flexion of the arm results <u>chiefly</u> from the contraction of the biceps muscle, but the speed and degree of flexion are influenced by the triceps muscle and by other muscles in the arm. Thus flexion results from the action of (one muscle/a group of muscles) ____________.

a group of muscles

5–117. The example of muscle movement in flexion of the arm illustrates a general principle that is true of all muscles. Body movement results <u>not</u> from the action of an individual muscle, but from the action of a ____________ of muscles.

group

5–118. Although a group of muscles acts together to move a part of the body, one muscle in the group usually bears the principal responsibility for the action. Flexion of the arm results principally from the contraction of the ____________.

biceps

5–119. The muscle that bears the principal responsibility for the movement of a group of muscles is called the <u>PRIME MOVER</u> of the group. When the arm is flexed, the ____________ is the prime mover.

biceps

5–120. <u>Extension</u> of the arm results principally from the action of the triceps. When the arm is extended, the triceps is the ____________ ____________.

prime mover

5–121. The ANTAGONIST of a muscle is another muscle that acts with an opposite effect. In flexion, the antagonist of the biceps muscle is the ____________.

triceps

5–122. When the movement desired is extension of the arm, the triceps is the prime mover and the biceps is the ____________.

antagonist

5–123. When the arm is flexed, certain groups of muscles called FIXATION MUSCLES hold the shoulder and other body parts steady to facilitate the action. Whenever any part of the body is moved, other parts are held steady by the action of ____________ muscles.

fixation

5–124. When someone lifts a heavy object, his legs, back, neck, and so on are held steady by ____________ ____________.

fixation muscles	**5–125.** When the fingers are being flexed by the flexor muscles in the forearm, certain muscles called SYNERGISTS (SIN—er—jistz) prevent the unnecessary action of bending at the wrist. Synergists function by preventing ______ muscle action.
unnecessary	**5–126.** The muscles that hold or fix body parts steady are called (1) ______ muscles. The muscles that prevent unnecessary movements are called (2) ______.
1. fixation 2. synergists	**5–127.** EMERGENCY MUSCLES are muscles that act to produce extra force or speed in a motion. When a hot object touches the sole, flexion of the foot is aided by leg and even thigh muscles acting as ______ muscles.
emergency	**5–128.** The DELTOID (DEL—toid) muscle caps the shoulder like an epaulet. The fibers at the front of the deltoid flex the arm; the fibers at the rear of the deltoid extend the arm. Different parts of the same muscle (can/cannot) ______ have opposite effects.
can	**5–129.** The RECTUS MUSCLES, which extend from the sternum to the pelvic girdle, can flex the body. The ABDOMINAL MUSCLES can produce the same effect, namely, flexion of the body. This example illustrates that two different muscles can have the ______ effect.
same	**5–130.** Unless a muscle has both (1) an intact nerve supply and (2) exercise, the muscle will waste or ATROPHY (A—tro—fe). Prolonged fixation of a broken bone in a plaster cast results in DISUSE ATROPHY, which is due to lack of ______.
exercise	**5–131.** In addition to exercise, a muscle must have an intact nerve supply, or ______ will result.
atrophy	**5–132.** Muscles are sheathed in a dense fibrous tissue called DEEP FASCIA (FA—sha). Deep fascia encloses blood and lymphatic vessels and serves to separate the individual ______ that it sheathes.
muscles	**5–133.** Another name for the subcutaneous tissue beneath the skin is SUPERFICIAL FASCIA. The muscles are invested in sheaths of (1) ______ ______. The layer of subcutaneous tissue beneath the dermis is called the (2) ______ ______.
1. deep fascia 2. superficial fascia	**5–134.** SUBSEROUS (sub—SE—rus) FASCIA lines the thoracic cavity. The lining of the abdominal cavity, like that of the thoracic cavity, is also composed of ______ fascia.

subserous

5–135. Match the items in these two columns:

1. deep fascia
2. subserous fascia
3. superficial fascia

a. subcutaneous tissue
b. forms the investing sheaths of muscles
c. lines the thoracic and abdominal cavities

1. b
2. c
3. a

5–136. You have reached the end of Chapter 5.

6 The Respiratory System

1. oxygen
2. carbon dioxide

respiration

1. inspiration
2. expiration

6–1. In order to carry on its life functions, the human body must constantly receive oxygen from the air and must, in turn, eliminate carbon dioxide. RESPIRATION is the process through which the body accomplishes this intake of (1) ________________ and elimination of (2) ________________ ________________.

6–2. In this unit you will learn about the RESPIRATORY SYSTEM, which is composed of all the organs involved in the vital process of ________________.

6–3. The terms INSPIRATION and EXPIRATION refer to breathing in and to breathing out. Ordinary unstrained respiration is called QUIET RESPIRATION and consists of two phases, QUIET (1) ________________ in which air is inhaled and QUIET (2) ________________ in which air is exhaled.

6–4. It is, of course, possible to breathe through one's mouth. Usually, however, human beings breathe through their ________________.

noses (nostrils)	**6–5.** The nostrils, which are lined with short, stout hairs, are the outermost limits of the respiratory system. As air is inhaled through the (1) ____________, it is cleaned and filtered by the lining of (2) ____________.
1. nostrils 2. hair(s)	**6–6.** The dilated part of the nostrils, just inside its opening, is called the VESTIBULE (VES—ti—bewl). The lining of hair is found only in the portion of the nostril just inside its opening, that is, in the ____________.
vestibule	**6–7.** Above the vestibule, the nostrils are lined with a layer of mucous membrane from which grow hairlike projections of microscopic size called CILIA (SIL—ee—a). The cilia, like the hairs in the vestibule, serve to ____________ the air that passes through them.
clean (or filter, etc.)	**6–8.** Large particles of matter are removed from the air as it passes the hairs in the vestibule. Smaller particles pass into the upper nostril where they become entrapped in ____________ secreted by the mucous membrane lining.
mucus	**6–9.** Some of the particles entrapped in mucus move down and out of the vestibule; the remainder pass along a channel leading to the mouth. The mucus is moved toward the mouth by the beating of the hairlike ____________.
cilia	**6–10.** As much as a quart of mucus may be secreted by the nasal membranes in one day. The fluid passes into the mouth where it can be expectorated or ____________ out.
spit	**6–11.** Before air enters the lungs, it must be warmed and moistened, as well as cleaned. The mucous membrane in the nostrils helps not only to clean air but to (1) ____________ it and (2) ____________ it as well.
1. warm 2. moisten (either order)	**6–12.** To help you form your own mental picture of the Respiratory System, look at the diagram of the Respiratory System. (Page 108) (No response required.)
	6–13. The two nostrils are separated by a wall of cartilage and bone called the SEPTUM (SEP—tum). In the diagram, the arrow pointing to the septum is numbered ____________.
1	**6–14.** A person who has broken his nose has usually broken the bony plate separating the nostrils; he has broken his ____________.

The Respiratory System

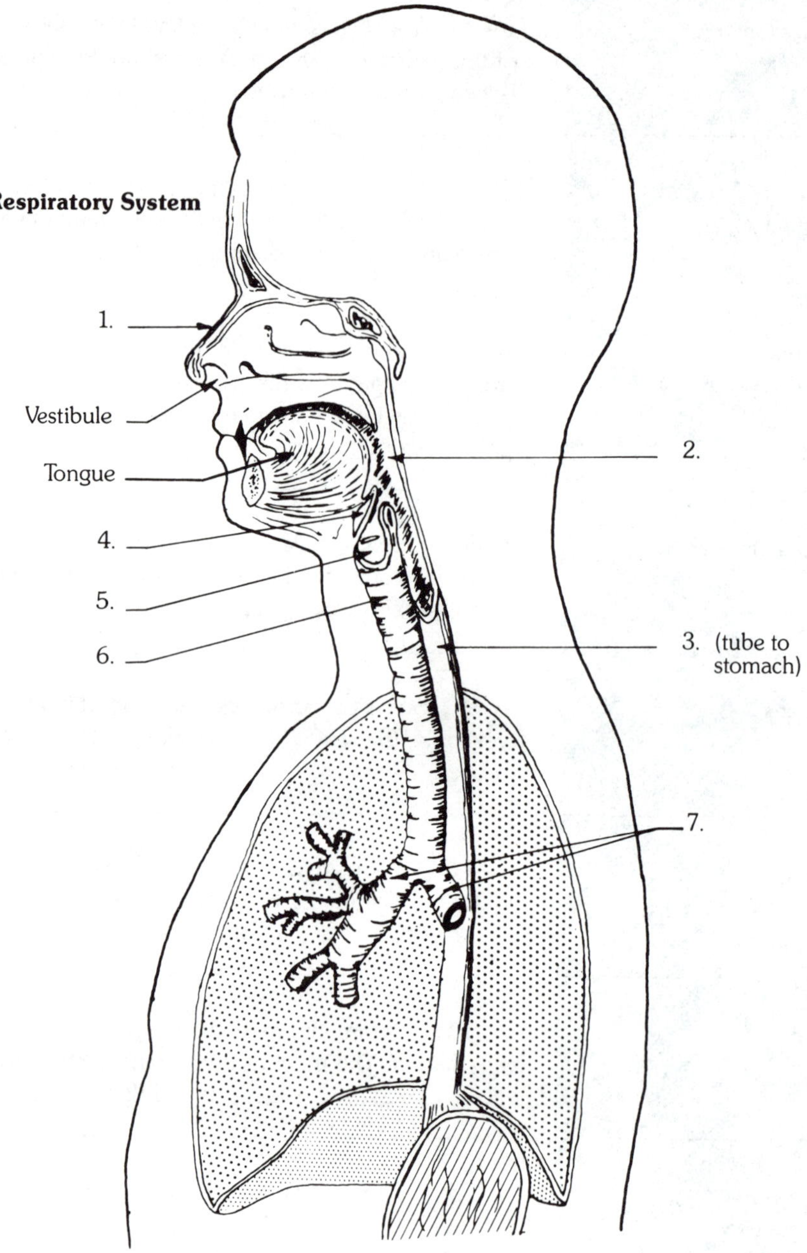

septum

6–15. In the bone surrounding the nasal cavity are eight small air pockets called SINUSES (SY—nuh—sez). Because the sinuses have openings leading to the nasal cavity, their secretions drain into the ________________ ________________.

nasal cavity

6–16. Air pressure within the sinuses is equalized by the presence of passageways to the ________________ ________________.

nasal cavity	**6–17.** Both air from the nostrils and food from the mouth pass into a single tube called the PHARYNX (FAR—inks). In the diaphragm, the arrow that points to the pharynx is numbered ________________. (Refer to the diagram located on page 108.)
2	**6–18.** The pharynx serves as a passageway for both (1) ________________ and (2) ________________.
1. air **2. food** **(either order)**	**6–19.** Food reaches the ESOPHAGUS (e—SOF—a—gus) after passing through the tube called the ________________.
pharynx	**6–20.** The esophagus is the immediate extension of the pharynx; through it, food passes into the stomach. In the diagram, the arrow which points to the esophagus is numbered ________________. (Refer to the diagram located between frames 12 and 13.)
3	**6–21.** Whereas food passes directly through the pharynx into the esophagus, air takes a different route when it leaves the pharynx. In the diagram, the arrow numbered 4 points to the opening of the tube into which air passes. Air passages into a tube (in back/in front) ________________ of the esophagus.
in front	**6–22.** Air passes through the LARYNX (LAR—inks) or "voice box," where vocal sounds originate. The front part of the larynx is the "Adam's apple." In the diagram, the arrow pointing to the larynx is numbered ________________.
5	**6–23.** At the opening of the larynx is a disc of cartilage called the EPIGLOTTIS (ep—i—GLOT—is). In the diagram the arrow pointing to the epiglottis is numbered ________________.
4	**6–24.** When food is swallowed, an automatic nerve muscle action closes the epiglottis and prevents food from entering the ________________.
larynx	**6–25.** Occasionally, the automatic closure does not occur quickly enough, and food becomes lodged in the larynx, producing a choking sensation. Normally, however, food does not enter the larynx because the ________________ closes to block its passage.

epiglottis

6–26. Label the numbered items in this drawing.

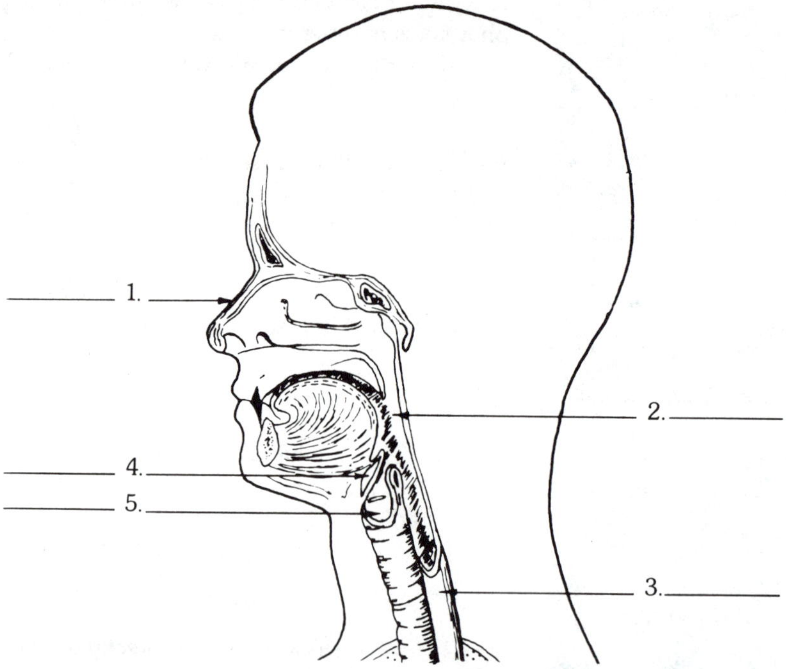

(1) septum
(2) pharynx
(3) esophagus
(4) epiglottis
(5) larynx

6–27. The tube through which both air and food pass is called the ________________.

pharynx

6–28. The tube that both acts as an air passage and produces vocal sounds is called the ________________.

larynx

6–29. The larynx sits on top of a tube called the TRACHEA (TRA—ke—a) which is commonly referred to as the "windpipe." In the diagram (page 108), the arrow pointing to the trachea is numbered ________________.

6

6–30. The pharynx, larynx, and trachea, like the nostrils, are lined with hairlike ________________, which clean and filter incoming air.

cilia

6–31. In the chest, the trachea forks into a right and left BRONCHUS (BRON—kus), leading to the right and left lungs, respectively. In the diagram (page 108), the arrow pointing to the bronchi (plural of bronchus) is numbered ________________.

7

6–32. Air passes from the windpipe or (1) ________________ into right and left (2) ________________.

1. trachea
2. bronchi

6–33. The ridge where the bronchi begin is known as the CARINA (ka—RE—na). Irritation of the carina triggers the cough reflex. Thus the "cough reflex center" in the brain receives impluses from the (1) ________________, which is located at the juncture of the two (2) ________________.

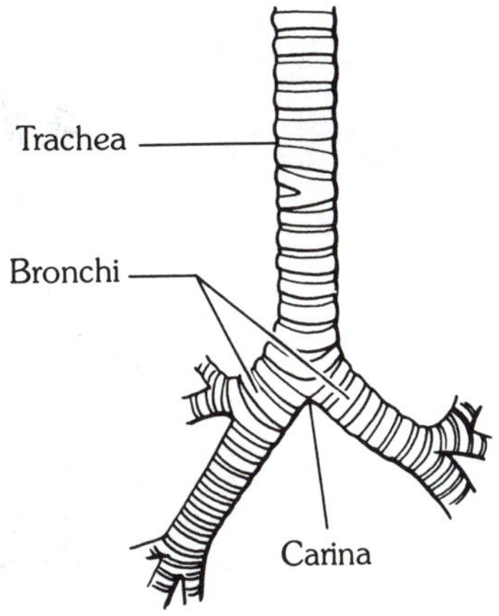

1. carina
2. bronchi

6–34. The trachea and bronchi are composed in part of cartilage rings. Within the lungs, the bronchi branch into the increasingly smaller and smaller bronchi. The smallest bronchi, barely visible without a microscope, still contain ________________ rings.

cartilage

6–35. The bronchi continue to branch into microscopic tubes that no longer contain cartilage but are composed of muscle tissue. These tiniest branches of the bronchi are called BRONCHIOLES (BRON—ke—olz); they differ from the bronchi in that they contain no (1) ________________ rings but are composed only of (2) ________________ tissue.

1. cartilage
2. muscle

6–36. Label the parts of this drawing.

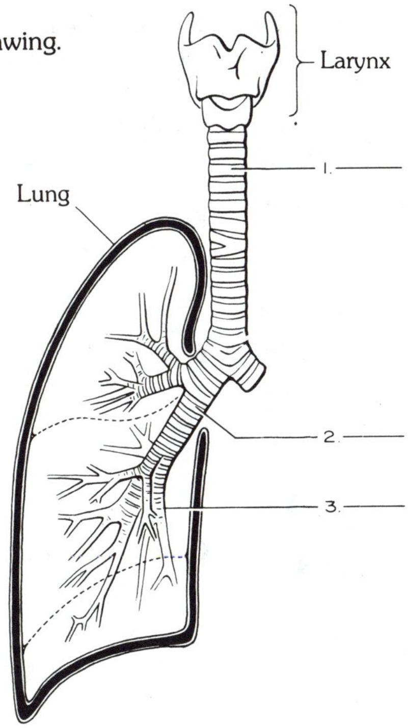

1. trachea
2. bronchus
3. bronchiole

6–37. Within the lungs, the bronchioles terminate in small air sacs called ____________________.

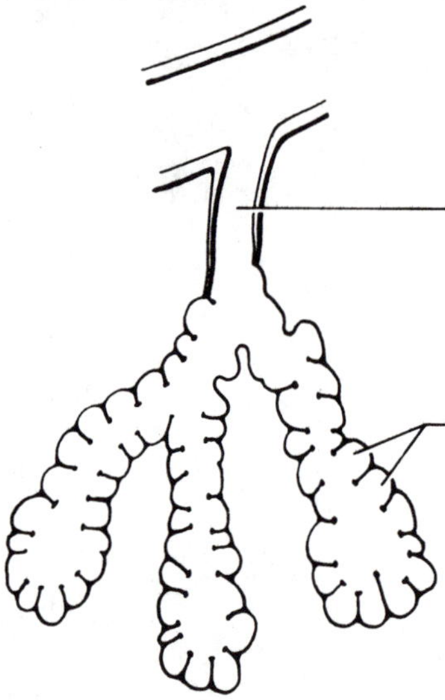

alveoli

6–38. Oxygen is carried through the blood stream to all parts of the body. Oxygen passes into the (1) ____________________ stream from the air sacs called (2) ____________________ in the lungs.

(We shall discuss this process in more detail later in the program.)

1. blood
2. alveoli

6–39. The alveoli are located in the primary organs of respiration, the two ____________________.

lungs

6–40. The lungs are housed in the thoracic cavity. They are covered by a serous membrane (a membrane that secretes a watery fluid) called the PLEURA (PLOO—ra). The thoracic cavity is also lined by a ____________________ like that which covers the lungs.

pleura

6–41. Because it is lined by pleura, the ____________________ cavity is also called the PLEURAL CAVITY.

thoracic

6–42. The thoracic cavity is divided into a right and left cavity by a partition called the MEDIASTINUM (me—de—as—TI—num). The mediastinum contains all the major organs of the thoracic cavity except the lungs. One lung is located on either side of the ____________________.

mediastinum

6–43. The heart and great vessels, the trachea, part of the esophagus, and many smaller structures are all part of the ____________________.

mediastinum

6–44. The lungs are asymmetrical. The heart projects into the left thoracic cavity and creates a notch called the ________________ ________________ in the left lung.

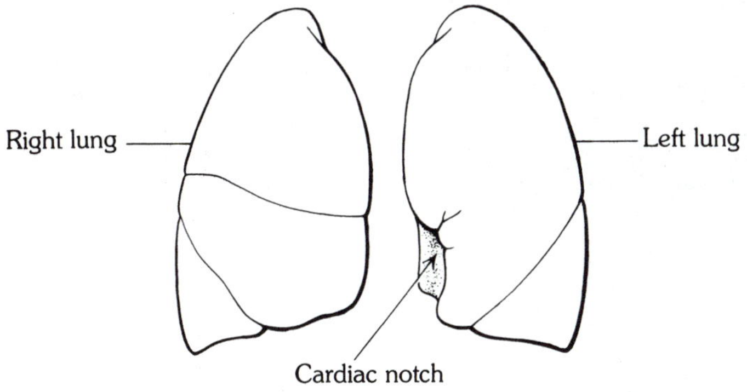

cardiac notch

6–45. Deep fissures in the lungs divide them into LOBES. In the right lung there are (1) ________________ (how many?) lobes. In the left lung there are (2) ________________ lobes.

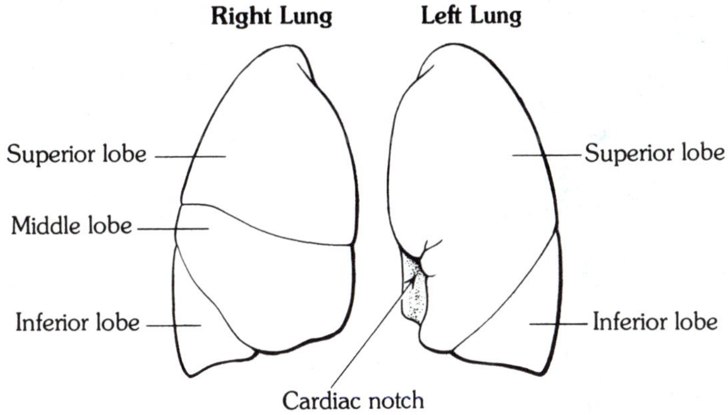

1. three
2. two

6–46. The upper and lower lobes of each lung are called the (1) ________________ and (2) ________________ lobes, respectively. In the right lung, there is also a (3) ________________ lobe.

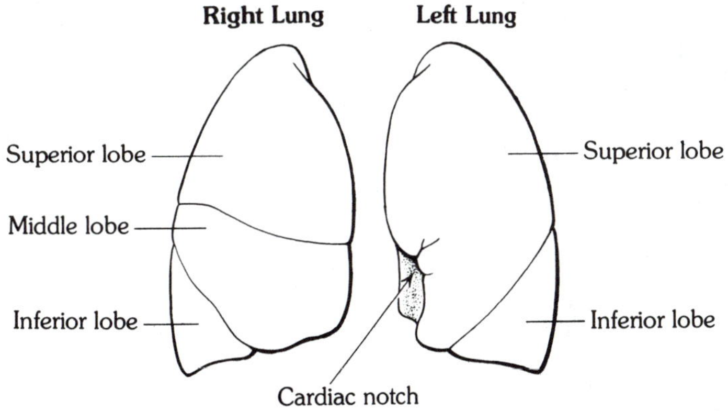

1. superior
2. inferior
3. middle

6–47. The superior lobe of a lung is the (1) ________________ lobe. The inferior lobe is the (2) ________________ lobe.

1. upper
2. lower

6–48. The lungs rest upon a large muscle called the ________________.

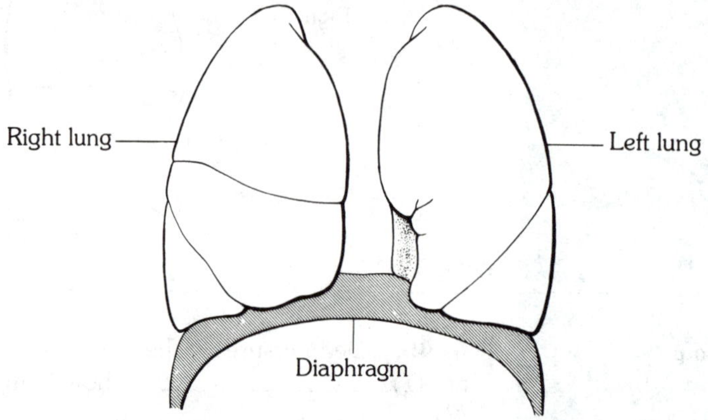

diaphragm

6–49. During inspiration, the diaphragm contracts and moves down, thus (increasing/decreasing) ________________ the volume of the thoracic cavity.

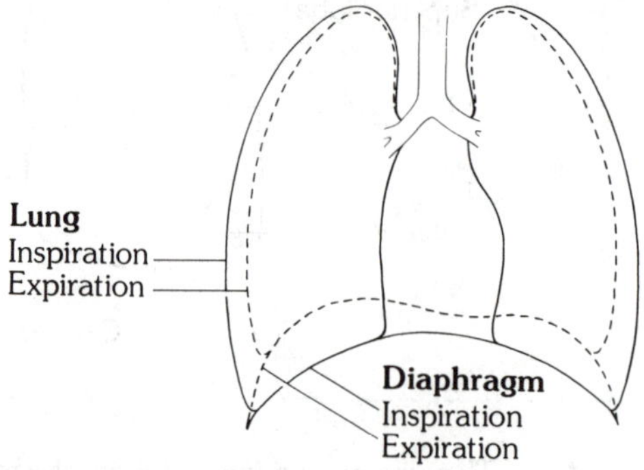

increasing

6–50. During expiration, the diaphragm relaxes and moves up, thus (increasing/decreasing) ________________ the volume of the thoracic cavity.

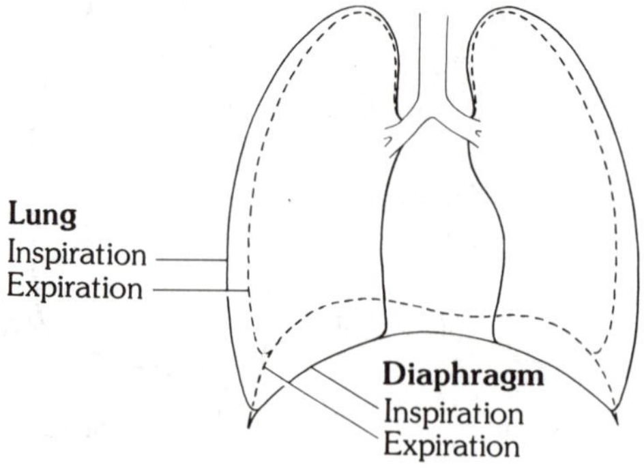

decreasing	**6–51.** The diaphragm moves down and increases the size of the thoracic cavity during (1) ________. It relaxes and moves up to decrease the size of the thoracic cavity during (2) ________.
1. inspiration 2. expiration	**6–52.** You are probably aware that you can, to some extent, control respiration (by "holding your breath"). However, you do not normally have to think about breathing; it is an unconscious process. Thus, although some voluntary control of respiration is possible, respiration is normally (a voluntary/an involuntary) ________ process.
an involuntary	**6–53.** The volume of the thoracic cavity is greater during (inspiration/expiration) (1) ________ than it is during (inspiration/expiration) (2) ________.
1. inspiration 2. expiration	**6–54.** Respiration results from pressure changes as well as from muscle action. Pressure is the force exerted upon a unit of a given surface. For example, atmospheric pressure is the ________ that the air or atmosphere exerts upon surfaces.
force	**6–55.** At sea level, atmospheric pressure is about 14.7 lb per square inch ($10^5 N/m^2$). Thus, on a window measuring 3 feet by 9 feet, atmospheric pressure exerts a force of more than 28 tons. If, as is usually the case, the pressures on both sides of the window are equal, the window will not break. (No answer required.)
	6–56. Air is a gas. Whenever possible, gases tend to move in a way that equalizes their pressures. If we have an open container, we can expect that the pressure inside the container will be (equal to/different from) ________ the pressure outside the container.
equal to	**6–57.** Exhausting air from a container lowers its internal pressure and creates a partial vacuum. If we open the container again, air will rush in to fill the vacuum, and the internal pressure will again ________ the external pressure.
equal (be equal to)	**6–58.** A law of physics (Boyle's Law) states that in a closed space, at a constant temperature, the pressure of a gas varies inversely with its volume. Thus, when the volume of such a gas decreases, its pressure increases. And when the volume of such a gas increases, its pressure ________.

decreases

6–59. Suppose you blow some air into a balloon and close the neck of the balloon; then, with your hand, you force the air into a smaller and smaller area of the balloon. When you do this, you are (increasing/decreasing) ________________ the volume of air in the balloon.

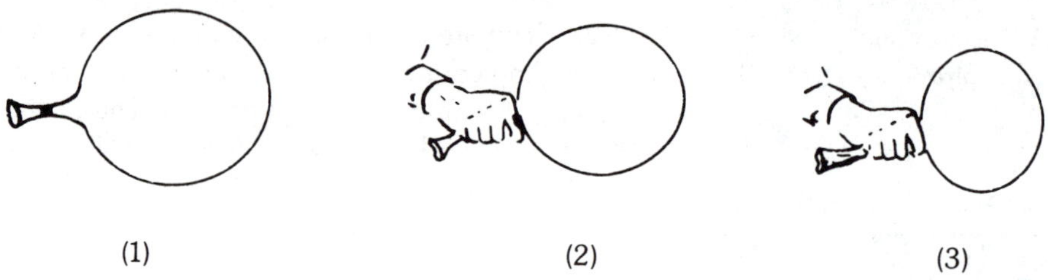

decreasing

6–60. If you continue to force the air into a smaller and smaller space, the increased internal pressure will cause the balloon to break, because, as you (1) ________________ the volume of air, you (2) ________________ the pressure.

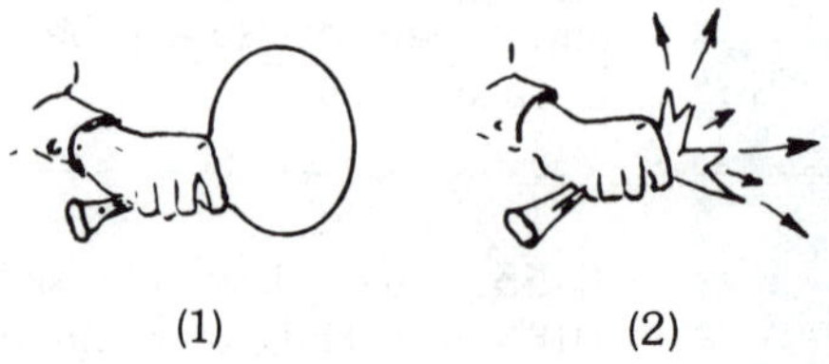

1. decrease(d)
2. increase(d)

6–61. The thorax is a closed space whose size, hence volume of air, is altered by movements of the large muscle called the ________________.

diaphragm

6–62. When the diaphragm contracts and moves down during inspiration, the size of the thoracic cavity (decreases/increases) ________________.

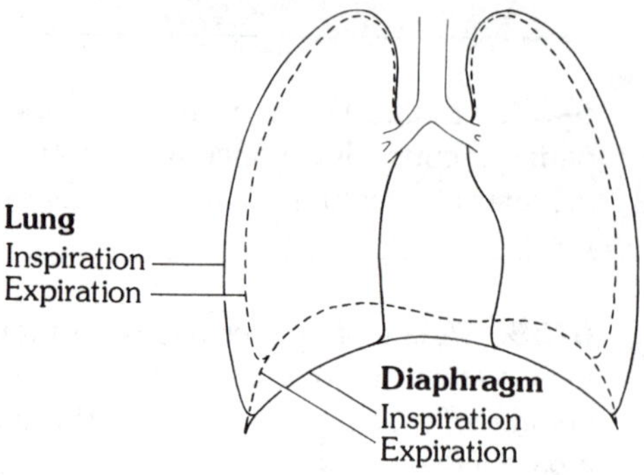

increases

6–63. When the volume of the thorax increases during inspiration, the pressure within the thorax naturally:

a. increases (go on to frame 64)
b. decreases (go on to frame 65)

	6–64. You said that when the volume of the thorax increases, the pressure within the thorax increases. Wrong. The thorax is a closed space; so its pressure changes in inverse ratio with its volume. When the volume of the thorax increases, the pressure decreases. Conversely, when the volume decreases, the pressure ________________. (Go on to frame 65.)
increases	**6–65** You said that when the volume of air in the thorax increases, the pressure decreases. You're right. Conversely, when the volume of air decreases, the pressure ________________.
increases	**6–66.** The pressure within the thorax is called intrathoracic pressure, and it is usually less than atmospheric pressure. During expiration, intrathoracic pressure increases, but it never equals ________________ ________________. The only exceptions to this are during coughing or when one bears down, at which time the pressure is temporarily greater.
atmospheric pressure	**6–67.** Because it is usually less than atmospheric pressure, ________________ pressure creates a slight vacuum within the thorax.
intrathoracic	**6–68.** Because intrathoracic pressure creates a slight vacuum, it exerts a suction on the elastic lungs, so that they never fully collapse but always remain somewhat expanded. When intrathoracic pressure decreases, the suction increases, and the lungs ________________ still further.
expand	**6–69.** During inspiration, as the diaphragm contracts, intrathoracic volume (1) ________________; intrathoracic pressure (2) ________________, and the lungs (3) ________________.

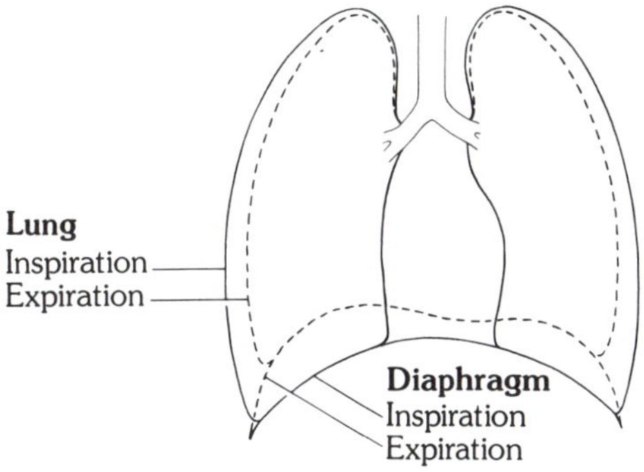

1. increases
2. decreases
3. expand

6–70. During expiration, as the diaphragm relaxes, intrathoracic volume (1) ________________, intrathoracic pressure (2) ________________, and the lungs partly (3) ________________.

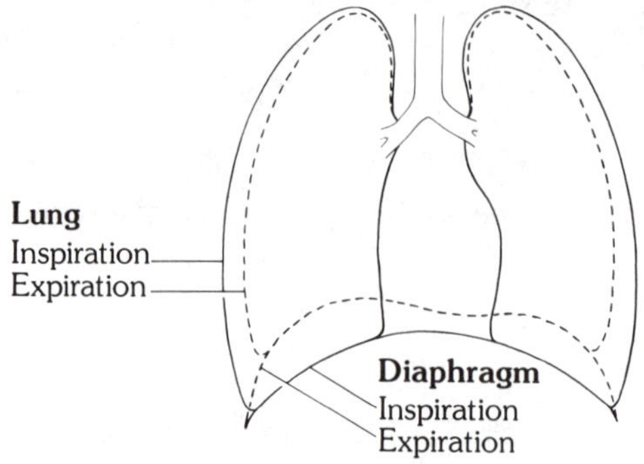

1. decreases
2. increases
3. collapse (or any synonym)

6–71. The lungs are not closed like the thorax; they open into air passages. When the lungs expand during inspiration, ________________ rushes in to fill them.

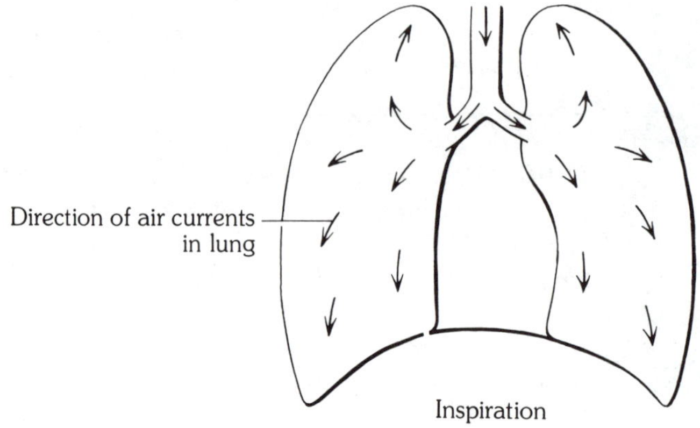

air

6–72. When air fills the lungs during inspiration, the pressure within the lungs increases until it equals atmospheric pressure. The pressure within the (1) ________________ <u>never</u> equals atmospheric pressure, but the pressure within the (2) ________________ equals atmospheric pressure during inspiration.

1. thorax
2. lungs

6–73. During inspiration, intrathoracic volume (1) ________________, intrathoracic pressure (2) ________________, and the lungs (3) ________________. At the same time, air rushes into the lungs, and the pressure in the lungs (4) ________________.

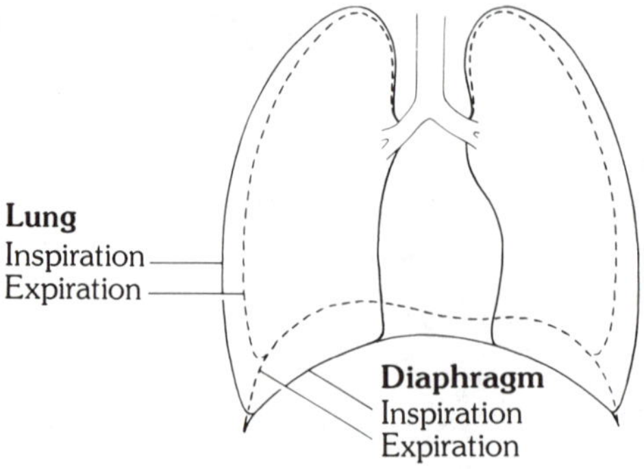

1. increases
2. decreases
3. expand
4. increases

6–74. During expiration, intrathoracic volume decreases, and intrathoracic pressure increases, forcing the air (into/out of) ________________ the lungs.

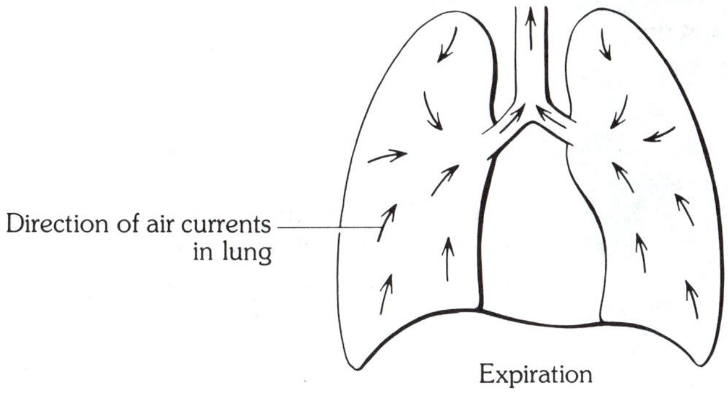

out of

6–75. When air is drawn into the lungs during inspiration, the pressure within the lungs increases. Conversely, when air is forced out during expiration, the pressure within the lungs ________________.

decreases

6–76. During expiration, intrathoracic volume (1) ________________, intrathoracic pressure (2) ________________, and the lungs partly (3) ________________. At the same time, air is forced out of the lungs, so the pressure in the lungs (4) ________________.

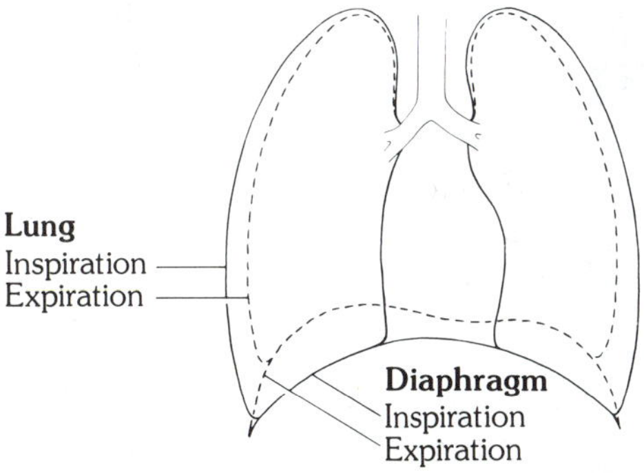

1. decreases
2. increases
3. collapse
4. decreases

6–77. Describe the changes in volume and pressure within the thorax and lungs during expiration. (Use your own words.)

Intrathoracic volume decreases; intrathoracic pressure increases; the lungs collapse, and the pressure within the lungs decreases.

6–78. Describe the changes in volume and pressure within the thorax and lungs during inspiration. (Use your own words.)

Intrathoracic volume increases; intrathoracic pressure decreases; the lungs expand, and the pressure within the lungs increases.	**6–79.** Normal unstrained respiration is called QUIET RESPIRATION and consists of two phases: QUIET INSPIRATION and ________________ ________________.
quiet expiration	**6–80.** The amount of air inhaled and exhaled during quiet respiration averages about 500 ml for the adult male. During one respiratory cycle only about 350 ml of this reaches the alveoli, whereas the remaining air, called DEAD SPACE AIR, remains in the respiratory passages. Normally, the respiratory passages contain 150 ml of ________________ ________________ ________________.
dead space air	**6–81.** Dead space air is air which does not reach the (1) ________________, but remains in the (2) ________________ ________________.
1. alveoli (lungs) **2. respiratory passages**	**6–82.** The amount of air inhaled and exhaled during each cycle of quiet respiration is called TIDAL VOLUME. For a normal adult male, ________________ ________________ averages 500 ml of air.
tidal volume	**6–83.** Tidal volume is the amount of air exchanged per breath during ________________ respiration.
quiet	**6–84.** With maximal effort, the quantity of air exchanged is greatly increased. The total amount of air that can be forcibly expelled from the lungs after a maximal inspiration is called the VITAL CAPACITY. Because the vital capacity of an adult male averages 4000 ml, vital capacity is (greater than/less than) ________________ tidal volume, which is 500 ml.
greater than	**6–85.** The air that can be forcibly expelled from the lungs after a maximal inspiration is called the ________________ ________________.
vital capacity	**6–86.** During inspiration, some air remains in the respiratory passages as dead space air. The remainder of the inspired air passes into the air sacs called ________________ in the lungs.
alveoli	**6–87.** Oxygen passes from the alveoli into the blood stream. The alveoli are enclosed by tissue, as are the blood vessels. Thus, to get from the alveoli into the blood stream, oxygen must pass through a wall of ________________.
tissue (s)	**6–88.** When carbon dioxide passes out of the blood stream, it too must pass through ________________.

tissue (s)

6–89. Oxygen and carbon dioxide pass through tissues by a process called diffusion. Thus we say that oxygen diffuses out of the alveoli into the blood stream and that carbon dioxide ______________ out of the blood stream into the alveoli.

---

diffuses

6–90. Consider the mechanics of diffusion. Suppose we have a container like that in the illustration. The container is closed except for a gas hose at point A. If we pump gas into the container, the gas will momentarily be concentrated where it enters at point ______________.

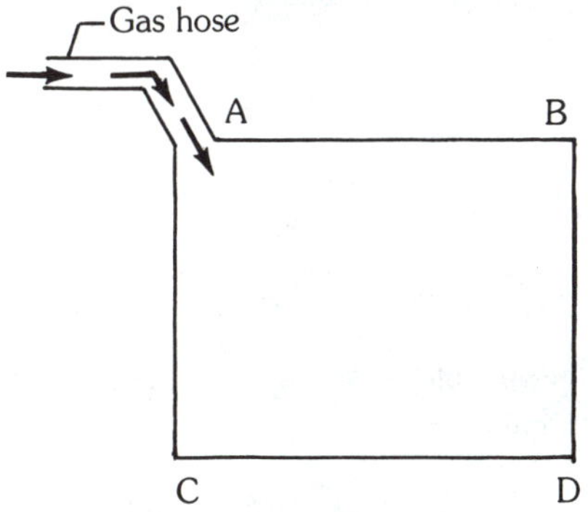

A

6–91. The "concentration" of a gas refers to the number of gas molecules present at a given point. When we say that a gas is more concentrated at point A than at point B, we mean that there are more ______________ ______________ present at point A than at point B.

---

gas molecules

6–92. Diffusion refers to the tendency of gas molecules to move from an area of high concentration to an area of low concentration. Thus if gas enters at point A and if gas molecules are briefly concentrated there, the molecules:

a. will remain concentrated at A.
b. will move throughout the container to B, C, D, and other points.

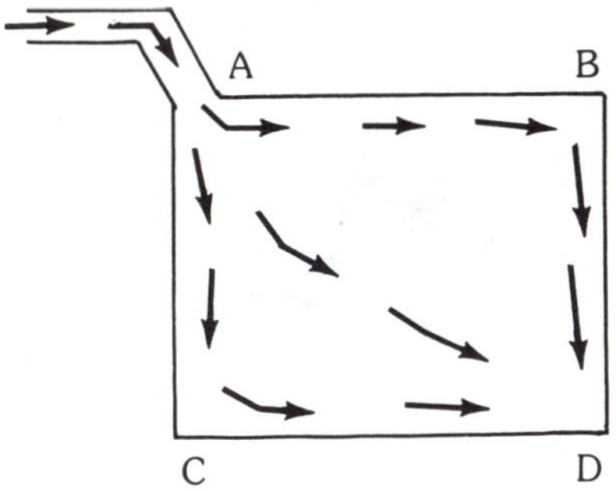

---

b. will move throughout the container to B, C, D, and other points.

6–93. The gas molecules in a given space will move until the concentration of molecules is equal at all points within the space. Thus, in a closed space, we (find/do not find) ______________ a higher concentration of gas molecules at any one point than at another point.

---

do not find

6–94. A PERMEABLE MEMBRANE is a membrane through which gas molecules can pass. Because oxygen and carbon dioxide can pass through the tissues of the alveoli and blood vessels, these tissues are ______________ membranes.

permeable

6–95. The principle of diffusion applies to gases separated by a permeable membrane. The illustration shows a container with two compartments separated by a permeable membrane. If gas molecules are concentrated on Side A, the molecules will tend to move through the (1) ________________ ________________ to side (2) ________________.

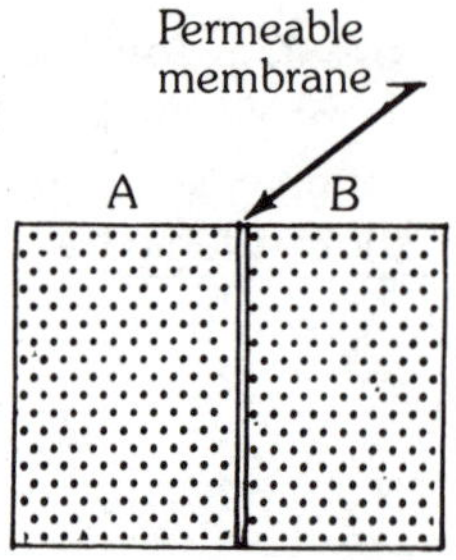

1. (permeable) membrane
2. B

6–96. Suppose that in the illustrated container we have a gas that is a mixture of both oxygen (O_2) and carbon dioxide (CO_2). On side A there are more oxygen molecules than on side B. Although the gas is a mixture, oxygen molecules will still diffuse independently over to side ________________.

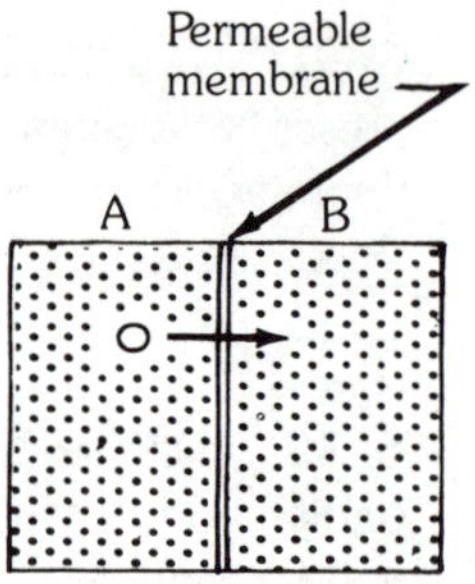

B

6–97. If the mixture of carbon dioxide and oxygen has a greater concentration of carbon dioxide molecules on side B than on side A, the (1) ________________ ________________ molecules will diffuse independently over to side (2) ________________.

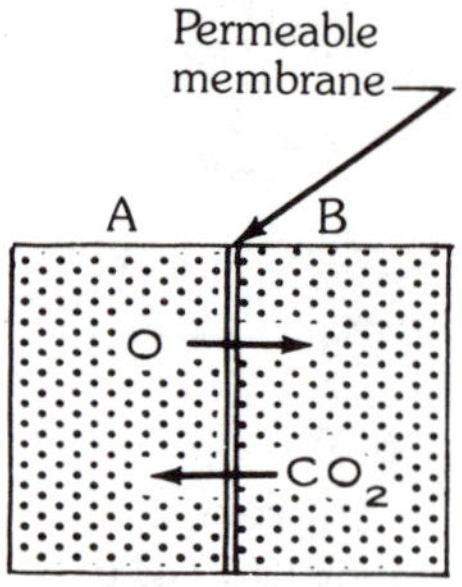

1. carbon dioxide
2. A

6–98. Thus we see that in a mixture of gases, the molecules of each gas diffuse independently. If one gas in a mixture has a higher concentration on side A of a membrane than on side B, the molecules of that gas will ________________ ________________________________. (Use your own words.)

Diffuse independently to side B, or move to the other side or any similar statement.	**6–99.** The human body constantly uses oxygen and gives off carbon dioxide. The body receives oxygen through the inspiration of air. Air is a mixture of gases that has a higher concentration of ________________ than of carbon dioxide.
oxygen	**6–100.** When blood enters the capillaries of the lungs, it has a low concentration of oxygen whereas the air in the alveoli of the lungs contains a <u>high</u> concentration of oxygen. Therefore the oxygen diffuses out of the (1) ________________ into the (2) ________________.
1. alveoli (lungs) **2. blood**	**6–101.** The blood entering the capillaries of the lungs has a high concentration of <u>carbon dioxide</u>, whereas the air in the alveoli has a low concentration of carbon dioxide. Therefore, the carbon dioxide diffuses out of the (1) ________________ into the (2) ________________.
1. blood **2. alveoli**	**6–102.** The transfer of oxygen into the blood and of carbon dioxide into the alveoli takes place by the process of ________________.
diffusion	**6–103.** Once the carbon dioxide from the blood reaches the alveoli, it traverses the same route followed by incoming air during inspiration. The carbon dioxide is forced up and out through the respiratory passages during (inspiration/expiration) ________________.
expiration	**6–104.** The oxygen that has passed into the bloodstream is carried to the cells throughout the body. The membranes surrounding these cells, like the membranes of the capillaries, are ________________ membranes.
permeable	**6–105.** When the oxygen blood reaches the cells, there is a higher concentration of oxygen in the blood than in the cells, and a higher concentration of carbon dioxide in the cells than in the blood. Therefore, oxygen diffuses into the (1) ________________ and carbon dioxide diffuses into the (2) ________________.
1. cells **2. blood**	**6–106.** After giving up its oxygen to the cells, blood carries a high concentration of carbon dioxide and a low concentration of oxygen. The blood then returns to the alveoli of the lungs where, once more, (1) ________________ diffuses out of the capillaries and (2) ________________ diffuses into the blood.
1. carbon dioxide **2. oxygen**	**6–107.** Blood that leaves the lungs has a high concentration of ________________. The blood is carried away from the lungs to the heart by the PULMONARY VEIN, the only vein in the body that carries oxygen-rich blood.

oxygen

6–108. After it enters the heart from the lungs, the blood is pumped to the body, carrying (1) ________________ to all the tissues. The vessels that transport blood from the (2) ________________ to the tissues of the body are called ARTERIES (AR—ter—eez).

1. oxygen
2. heart

6–109. Arterial blood that is traveling from the heart to the body cells has a high concentration of (oxygen/carbon dioxide) ________________.

oxygen

6–110. VEINS carry blood from the cells of the body back to the heart. VENOUS (VEE—nus) BLOOD, therefore, is blood that is being carried away from the (1) ________________ back to the (2) ________________.

1. cells
2. heart

6–111. VENOUS blood has given up its oxygen to the cells and has, therefore, a (high/low) ________________ concentration of oxygen. Venous blood also has a high concentration of carbon dioxide.

low

6–112. Blood contains a protein molecule called HEMOGLOBIN (he—mo—GLO—bin). It is this protein molecule that carries the oxygen in the blood. The term OXYHEMOGLOBIN (OX—e—he—mo—GLO—bin) refers to the molecule of (1) ________________ when it is combined with (2) ________________.

1. hemoglobin
2. oxygen

6–113. Oxyhemoglobin is, of course, present in (arterial/venous) ________________ blood.

arterial

6–114. When the hemoglobin molecule has given up its oxygen to the body tissue, it is called REDUCED HEMOGLOBIN. Reduced hemoglobin is present in (arterial/venous) ________________ blood.

venous

6–115. Because tissue metabolism is perpetual, the body's production of carbon dioxide is also perpetual. The ACIDITY of the blood is in part dependent upon the total amount of carbon dioxide present in the blood. The *p*H of the blood must remain within a narrow range (7.35–7.45). The condition known as ACIDOSIS (as—i—DO—sis) results if the lungs become incapable of eliminating ________________ ________________.

carbon dioxide

6–116. The inability of the lungs to eliminate carbon dioxide results in (1) ________________. In this condition, the blood *p*H would be (greater than/less than) (2) ________________ the normal range.

1. acidosis
2. less than

6–117. A condition known as ALKALOSIS (al—ka—LO—sis) results when the lungs eliminate too much ________________ ________________. In this condition, the blood *p*H would be (greater than/less than) the normal range.

1. **carbon dioxide**
2. **greater than**

6–118. When the lungs eliminate too much carbon dioxide, the resultant condition is (1) ________________. When they eliminate too little carbon dioxide, the resultant condition is (2) ________________.

Because lung function affects the amount of carbon dioxide in the blood, proper lung action aids in the regulation of the body acid-base (3) ________________.

---

1. **alkalosis**
2. **acidosis**
3. **balance**

6–119. Diseases that impair the ability of the lungs to transfer oxygen into the blood may produce HYPOXIA (hi—POX—e—a), in which too little oxygen enters the blood, or ANOXIA (a—NOX—e—a), in which no ________________ enters the blood.

---

oxygen

6–120. Some body tissues are very sensitive to a decrease in oxygen. Brain tissue can be permanently damaged by total oxygen lack in three to five minutes. The brain is the body organ most seriously injured by oxygen decrease or

________________.

---

hypoxia

6–121. The regulation of the respiratory cycle is controlled by RESPIRATORY CENTERS in the brain. Respiration is thus controlled by the central nervous system.

(No answer needed.)

6–122. The respiratory centers respond chiefly to the carbon dioxide content of the blood. An increase in the carbon dioxide content of the blood increases the rate of respiration, whereas a decrease in the carbon dioxide content ________________ the rate of respiration. This is one of the regulatory mechanisms of the respiratory system in maintaining acid-base balance.

---

decreases

6–123. Respiration depends chiefly upon the response of the respiratory centers to the (oxygen/carbon dioxide) ________________ ________________ content of the blood.

---

carbon dioxide

6–124. A short review will help you remember what you have learned in this section.

The outermost limits of the respiratory system are the nostrils. The vestibule or dilated portion of each nostril is lined with (1) ________________, which serve to (2) ________________ incoming air.

---

1. **hairs**
2. **filter**

6–125. Above the vestibule, the passages of the respiratory system are lined with hairlike ________________, which also help to clean incoming air.

cilia

6–126. Label the numbered parts of this drawing.

The Respiratory System

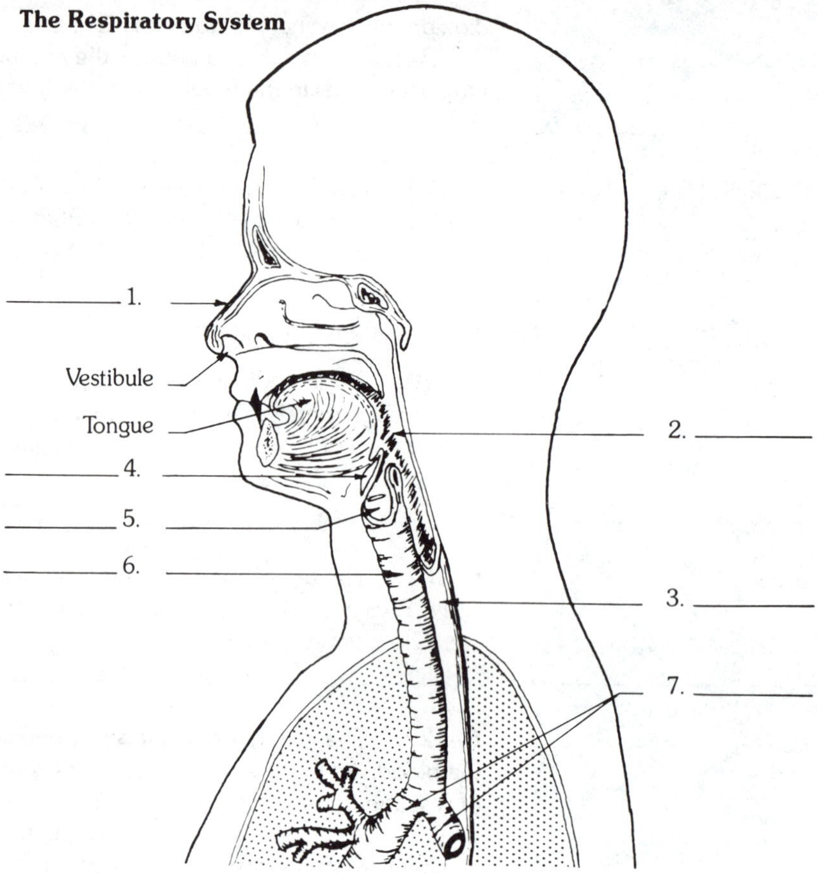

1. septum
2. pharynx
3. esophagus
4. epiglottis
5. larynx
6. trachea
7. bronchi

6–127. In the lungs, the bronchi branch into minute tubes called (1) ________________. These tubes terminate in tiny air sacs called (2) ________________.

1. bronchioles
2. alveoli

6–128. The lungs are enclosed in a covering called the (1) ________________. The lungs rest upon a large muscle called the (2) ________________.

1. pleura
2. diaphragm

6–129. The top lobe of each lung is called the (1) ________________ lobe. The bottom lobe is called (2) ________________ lobe. The right lung also has a (3) ________________ lobe.

1. superior
2. inferior
3. middle

6–130. During inspiration, the diaphragm (1) ________________ (contracts/relaxes). During expiration, the diaphragm (2) ________________ (contracts/relaxes).

1. contracts 2. relaxes	**6–131.** During inspiration, volume of the thoracic cavity (1) ____________ and intrathoracic pressure (2) ____________.
1. increases 2. decreases	**6–132.** During expiration, the volume of the thoracic cavity (1) ____________ and intrathoracic pressure (2) ____________.
1. decreases 2. increases	**6–133.** During inspiration, the lungs (1) ____________ as air is (drawn in/pushed out) (2) ____________. During expiration, the lungs (3) ____________ as air is (4) ____________ (drawn in/pushed out).
1. expand 2. drawn in 3. collapse 4. pushed out	**6–134.** Intrathoracic pressure is always (equal to/less than) ____________ atmospheric pressure.
less than	**6–135.** Oxygen passes into the bloodstream from the ____________ in the lungs.
alveoli	**6–136.** Oxygen and carbon dioxide pass through the tissue membranes of the alveoli and capillaries in a process called ____________.
diffusion	**6–137.** Diffusion occurs because gas molecules tend to pass from an area of (1) ____________ concentration to an area of (2) ____________ concentration.
1. high 2. low	**6–138.** Oxygen in the blood is carried by a protein molecule called ____________.
hemoglobin	**6–139.** The rate of respiration depends chiefly on the amount of ____________ in the blood.
carbon dioxide	**6–140.** Now that you have completed Chapter 6, you will want to take a break before beginning Chapter 7, the Circulatory System.

7

The Circulatory System

Part 1

Cardiovascular System

7–1. The CIRCULATORY SYSTEM, the subject of this unit, is the transportation system of the body. It is composed of the CARDIOVASCULAR (CAR—de—o—VAS—kew—lar) SYSTEM, which circulates blood, and the LYMPHATIC (lim—FAT—ik) SYSTEM, which circulates LYMPH (LIMF). Both the cardiovascular and the lymphatic systems are members of the ____________________ ____________________.

circulatory system

7–2. The larger part of this unit will be devoted to the cardiovascular system. The word "cardiovascular" can be broken into two parts: "cardio," meaning "heart," and "vascular," meaning "vessel." The cardiovascular system, as its name implies, includes the (1) ____________________ and the blood (2) ____________________.

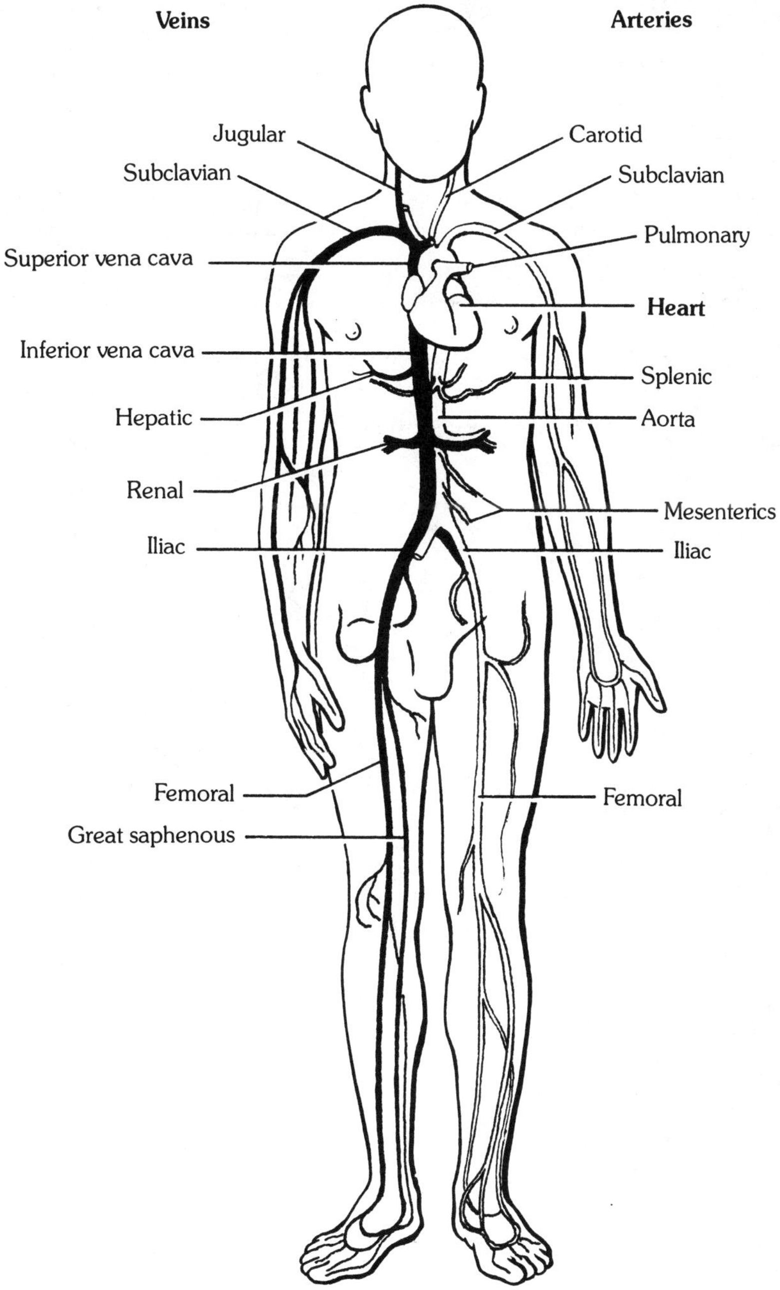
Veins
Arteries
Jugular
Subclavian
Superior vena cava
Inferior vena cava
Hepatic
Renal
Iliac
Femoral
Great saphenous
Carotid
Subclavian
Pulmonary
Heart
Splenic
Aorta
Mesenterics
Iliac
Femoral

1. heart
2. vessels

7–3. The cardiovascular system transports ____________________ throughout the body.

blood

7–4. The heart is the "pump" of the ____________________ system.

cardiovascular

7–5. The system of blood vessels in the body is sometimes called the VASCULAR TREE. The analogy stems from the fact that from a single large vessel, the AORTA (a—OR—ta), grow branches that become progressively smaller until they terminate in tiny "twigs." The main trunk of the vascular tree is the large ____________________.

aorta

7–6. The smallest blood vessels in the body are called CAPILLARIES (CAP—il—a—reez). The capillaries are the "twigs" of the ____________________ ____________________.

vascular tree

7–7. The vascular tree has a double system of branches. The (1) ____________________ carry blood to the heart. The (2) ____________________ carry blood away from the heart.

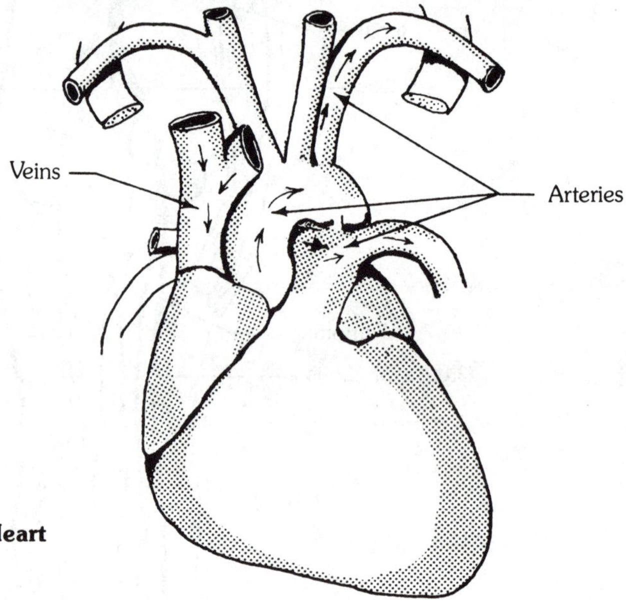

Heart

1. veins
2. arteries

7–8. The trunk of the vascular tree, the aorta, carries blood away from the heart. Technically speaking, the aorta is a(n) ____________________.

artery

7–9. The connection between the arteries and the veins is accomplished by the "twigs" of the vascular tree, which are the ______________.

capillaries

7–10. The walls of arteries are composed of three layers called TUNICS. The walls of veins are thinner than those of arteries but are also composed of three ______________. (Use technical term.)

tunics

7–11. The innermost tunics of the veins and arteries are composed of a specialized epithelium called ENDOTHELIUM (en—do—THEE—le—um). The two outer tunics are missing from capillaries, which are composed only of ______________ like that which lines the veins and arteries.

endothelium

7–12. Arteries are classified as SMALL, MEDIUM, and LARGE. Large arteries are composed chiefly of elastic tissue, whereas small and medium arteries are composed chiefly of muscular tissue. On the basis of the predominating tissue, small and medium arteries are called (1) ______________ (muscular/elastic) arteries, and large arteries are called (2) ______________ (muscular/elastic) arteries.

1. muscular
2. elastic

7–13. The term "muscular arteries" refers to (1) ______________ and (2) ______________ arteries. The term "elastic arteries" refers to (3) ______________ arteries.

1. small
2. medium
3. large

7–14. The smallest arteries are called ARTERIOLES (ar—TER—e—olz). Arterioles, like larger arteries, have an inner tunic of endothelium and ______________ (how many?) outer tunics.

two

7–15. The smallest arteries are called (1) ______________, and they, like other arteries, carry blood (2) ______________ (to/away from) the heart.

1. arterioles
2. away from

7–16. Although the walls of veins are proportionately thinner than those of arteries, the veins, like the arteries, are composed of three ______________.

tunics

7–17. Veins, like arteries, are classified as small, medium, and large and as VENULES (VEN—yewls), which are very (small/large) ______________ veins.

Answers	Frames
small	**7–18.** Very small veins are called (1) ________________, and they, like other veins, conduct blood (2) ________________ (to/away from) the heart.
1. venules 2. to	**7–19.** The twigs of the vascular tree, the capillaries, do <u>not</u> have the outer two ________________ found in the arteries and veins.
tunics	**7–20.** Lacking the outer tunics, the (1) ________________ are composed only of (2) ________________ like that which lines the rest of the cardiovascular system.
1. capillaries 2. endothelium	**7–21.** As you know, the transfer of oxygen and carbon dioxide between the blood and tissues takes place through the capillary walls. Nutrients and waste products are also transferred between the (1) ________________ and the (2) ________________ through the (3) ________________ walls.
1. blood 2. tissues (either order) 3. capillary	**7–22.** The total system of capillaries in a particular area is called the CAPILLARY BED. In order to meet the demands of all the body cells for food and oxygen, the capillary bed is extensive. It has been estimated that approximately 60,000 miles of capillaries form the ________________ ________________ of skeletal muscle alone.
capillary bed	**7–23.** A brief review. (1) ________________ carry blood to the heart. (2) ________________ carry blood away from the heart to all parts of the body. The tiniest blood vessels, through which food and oxygen pass into the tissues, are called (3) ________________.
1. veins 2. arteries 3. capillaries	**7–24.** Both veins and arteries are composed of an inner lining of (1) ________________ and two outer (2) ________________. Capillaries are composed only of (3) ________________.
1. endothelium 2. tunics 3. endothelium	**7–25.** The "pump" of the cardiovascular system is the ________________.
heart	**7–26.** The heart is composed chiefly of muscle. The muscle of the heart is called the MYOCARDIUM (my—o—KAR—de—um). A connective tissue sac encloses the heart muscle or ________________.
myocardium	**7–27.** Contractions of the heart muscle, called the ________________, pump blood throughout the body.

myocardium

7–28. The heart has four chambers. The top chambers are the right atrium and the left (1) ________________. The bottom chambers are the right and left (2) ________________.
(Note: The drawings of the heart in this program are schematics designed to show the directions in which blood flows. They are not exact representations of the way the heart looks.)

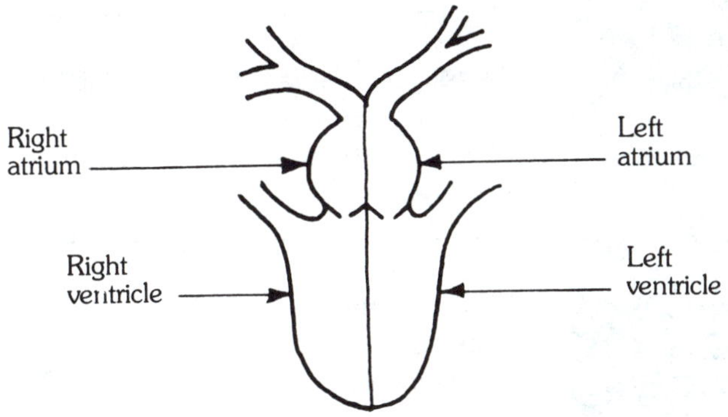

1. atrium
2. ventricles

7–29. The plural of "atrium" is "atria." In the heart, the two atria are located (above/below) ________________ the two ventricles.

above

7–30. The major pumping force of the cardiovascular system comes from the lower two chambers of the heart, called the ________________.

ventricles

7–31. The arrows in this and the following drawings represent the direction in which blood circulates through the heart. Blood from the body enters the right atrium through two veins called the (1) ________________ ________________ ________________ and the (2) ________________ ________________ ________________.

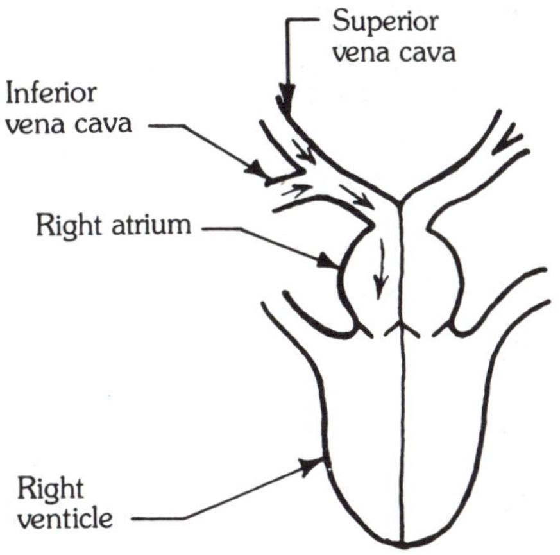

1. superior vena cava
2. inferior vena cava

7–32. Blood from the body passes through the superior vena cava into the ________________ ________________ of the heart.

right atrium	**7–33.** "Superior" means "above," and "inferior" means "below." These terms, applied to the venae cavae (plural of vena cava), describe the areas drained by the two veins. Blood from the upper part of the body drains into the (1) ______ ______ ______, whereas blood from the lower part of the body drains into the (2) ______ ______ ______.
1. superior vena cava 2. inferior vena cava	**7–34.** The superior and inferior venae cavae drain all of the body except the lungs. Thus, the blood that passes through the venae cavae into the right atrium has given up its oxygen to the tissues and is (rich/poor) ______ in oxygen.
poor	**7–35.** Between the right atrium and right ventricle is a valve called the (1) ______ ______. When this valve opens, blood passes from the right atrium into the (2) ______ ______.
1. tricuspid valve 2. right ventricle	**7–36.** Blood from the body enters the (1) ______ ______ of the heart. The blood then passes through the (2) ______ valve into the (3) ______ ______ of the heart.
1. right atrium 2. tricuspid 3. right ventricle	**7–37.** A powerful contraction of the right ventricle forces blood through the (1) ______ valve into the (2) ______ ______, which carries blood to the lungs.
1. pulmonic 2. pulmonary artery	**7–38.** Blood from the right ventricle enters the lungs by passing through the (1) ______ valve into the (2) ______ ______, which leads to the lungs.
1. pulmonic 2. pulmonary artery	**7–39.** Blood that has passed through the pulmonary artery into the capillaries of the lungs gives off (1) ______ ______ and receives (2) ______ in the process of respiration.

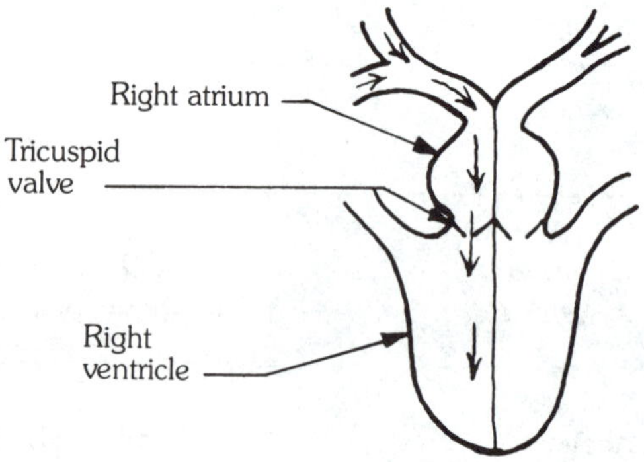

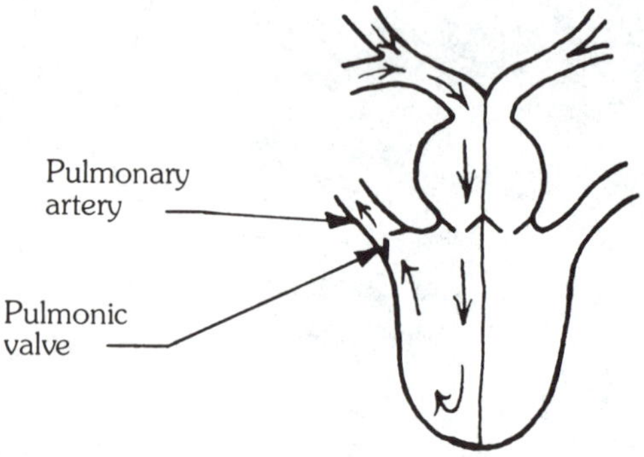

1. carbon dioxide
2. oxygen

7–40. Oxygen-rich blood from the lungs enters the <u>left atrium</u> of the heart through two veins, one for each lung, called the

____
________________.

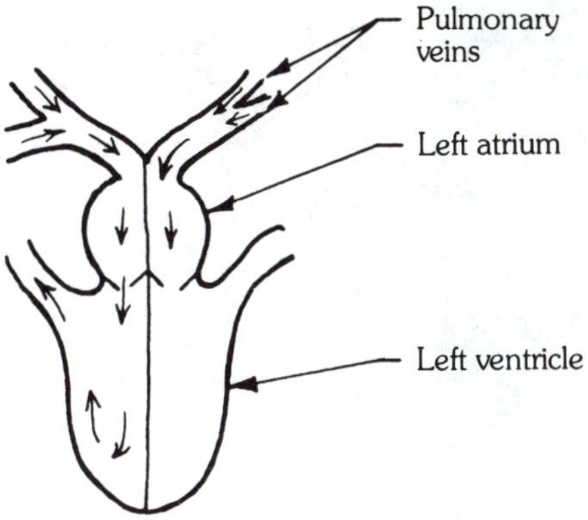

pulmonary veins

7–41. Blood from the lungs passes through the pulmonary veins into the (1) ________________ ________________ of the heart. The blood entering this chamber of the heart is (2) ________________ (rich/poor) in oxygen.

1. left atrium
2. rich

7–42. Between the left atrium and the left ventricle is a valve called the (1) ________________
________________.

When this valve opens, blood passes through it into the (2) ________________
________________ of the heart.

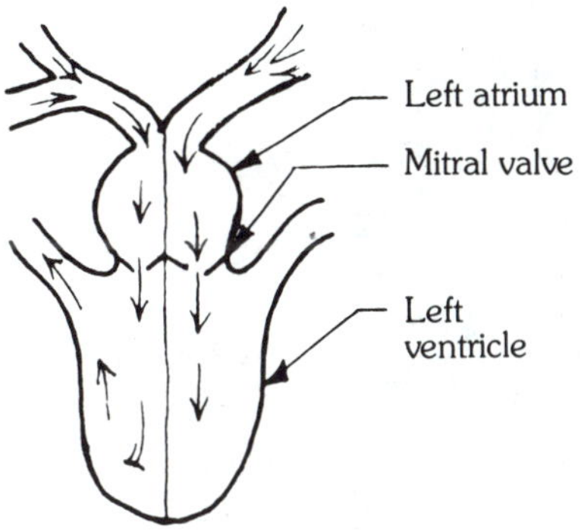

1. mitral valve
2. left ventricle

7–43. Blood from the lungs enters the (1) ________________ ________________ of the heart by passing through the two (2) ________________ ________________.

1. left atrium
2. pulmonary veins

7–44. Blood from the left atrium then passes through the (1) ________________ valve into the (2) ________________ ________________ of the heart.

1. mitral
2. left ventricle

7–45. A powerful contraction of the left ventricle then pushes the blood from the heart through the aortic valve and into the main artery of the body, the

________________.

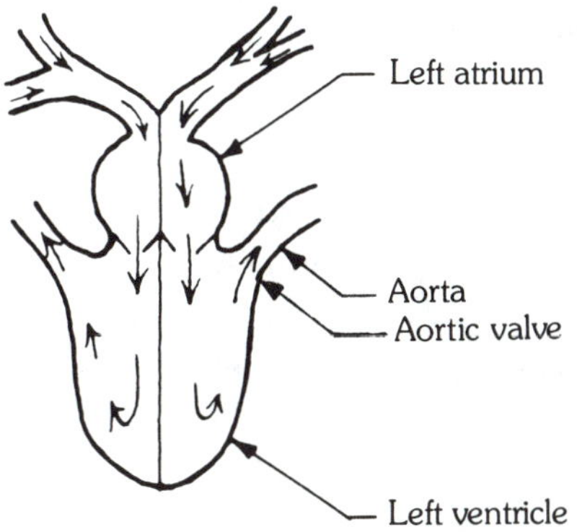

aorta	**7–46.** Oxygen-rich blood from the heart reaches the entire body through branches of the main artery called the ____________.
aorta	**7–47.** Let's review the location of the valves in the heart. The <u>tricuspid</u> valve is between the (1) ____________ ____________ and the (2) ____________ ____________. The <u>mitral</u> valve is between the (3) ____________ ____________ and the (4) ____________ ____________. The <u>pulmonic</u> valve is between the (5) ____________ artery and the (6) ____________ ____________. The <u>aortic</u> valve is between the (7) ____________ ____________ and the (8) ____________.

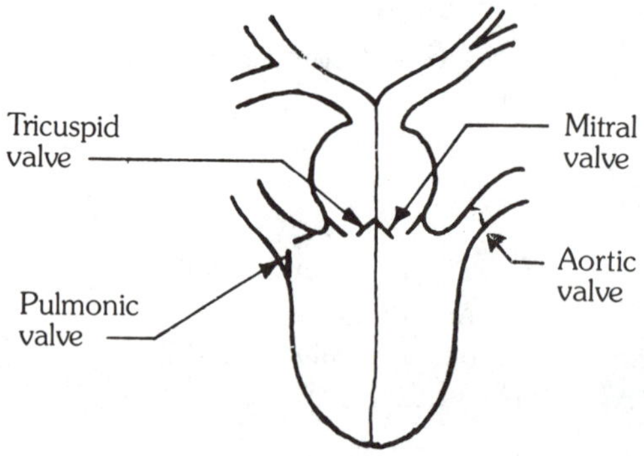

1. right atrium **2. right ventricle** **3. left atrium** **4. left ventricle** **5. pulmonary** **6. right ventricle** **7. left ventricle** **8. aorta**	**7–48.** Now, a brief review of the circulation of blood through the heart. Oxygen-poor blood from all of the body except the lungs enters the (1) ____________ ____________ of the heart through the (2) ____________ ____________ ____________ and the (3) ____________ ____________ ____________.
1. right atrium **2. superior vena cava** **3. inferior vena cava**	**7–49.** Blood from the right atrium passes through the (1) ____________ valve into the (2) ____________ ____________ of the heart.
1. tricuspid **2. right ventricle**	**7–50.** A contraction of the right ventricle pushes the blood through the (1) ____________ valve into the (2) ____________ ____________, which carries the blood to the (3) ____________.
1. pulmonic **2. pulmonary artery** **3. lungs**	**7–51.** After oxygenation in the lungs, blood reenters the heart by passing through the two (1) ____________ ____________. The blood from the lungs enters the (2) ____________ ____________ of the heart.

1. pulmonary veins **2. left atrium**	**7–52.** Oxygen-rich blood from the left atrium passes through the (1) ________________ valve into the (2) ________________ ________________ of the heart. A contraction of the heart then forces the blood out into the (3) ________________.
1. mitral **2. left ventricle** **3. aorta**	**7–53.** Oxygen-rich blood is carried throughout the body by arteries, all of which are branches of a main artery called the ________________.
aorta	**7–54.** The brain is very richly supplied with blood. The richness of the blood supply to the brain reflects the importance of a constant supply of oxygen to the brain. Since brain tissue may be permanently damaged by oxygen deprivation in as short a time as three to five minutes, it is clear that (1) ________________ is of immense importance to the (2) ________________.
1. oxygen **2. brain**	**7–55.** Because it can be permanently damaged by oxygen lack in as little as three to five minutes, the ________________ is the organ that is most sensitive to oxygen deprivation.
brain	**7–56.** The tissues of the heart are nourished by blood from two major arteries, the RIGHT and LEFT CORONARY ARTERIES. In its need for a constant supply of oxygen, the ________________ is second only to the brain.
heart	**7–57.** Consider what happens when we change the pressure inside a container. To begin with, we have an open container, so the pressure inside it is ________________ ________________ the pressure outside it.

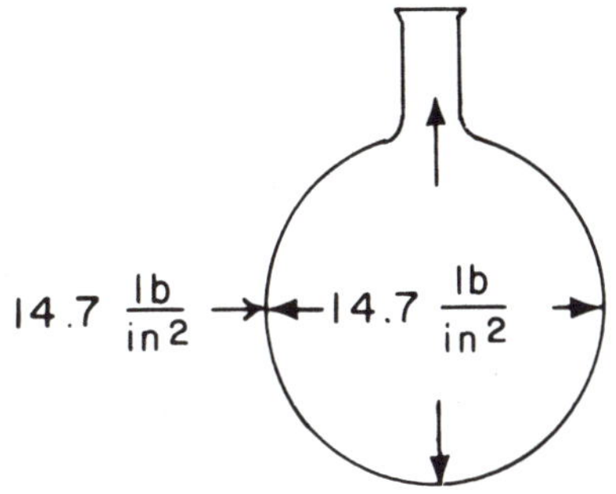

equal to

7–58. Our container is a collapsible rubber bottle. We stop the neck of the container and insert a glass tube connected to a vacuum pump. As we pump out some of the air from the container, the inside pressure (1) ________________ (increases/decreases), and the container begins to (2) ________________ (expand/collapse).

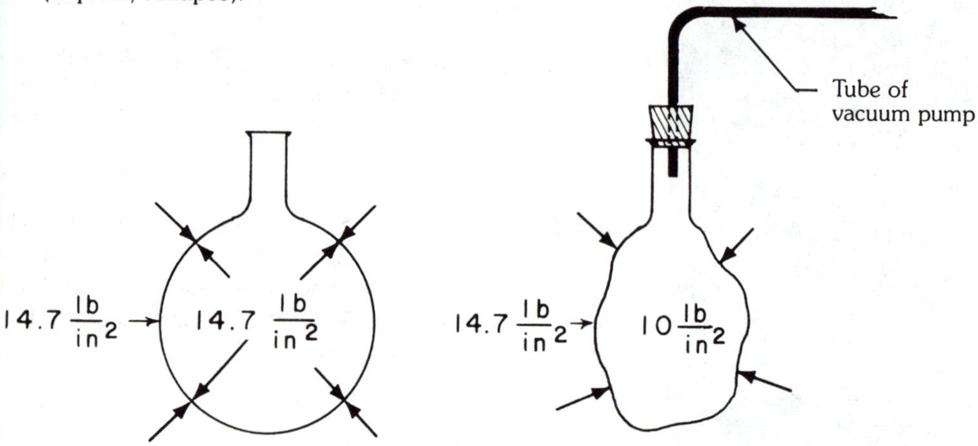

Open Container **Air Partly Exhausted**

1. decreases
2. collapse

7–59. Diseases like ARTERIOSCLEROSIS (ar—TER—e—o—skler—O—sis), "hardening of the arteries," reduce the oxygen supply to the heart and predispose the heart to damage and death from ________________ or oxygen lack.

hypoxia

7–60. Like the arteries that supply the brain, the coronary arteries are interconnected so that an obstruction in one artery may not result in death from anoxia. In the event of such an obstruction, the entire myocardium may (but does not always) still receive adequate blood through the unimpaired ________________ artery.

coronary

7–61. The myocardium is drained by a venous system (a system of veins) that more or less parallels its arterial system. Similarly, the other parts of the body receive blood through (1) ________________ systems and are drained by nearly parallel (2) ________________ systems.

1. arterial
2. venous

7–62. You have already learned about the circulation of blood through the heart. The following frames will introduce you to the mechanisms which control this circulation.

(No answer needed.)

7–63. The circulation of blood is controlled by the alternate contraction and relaxation of the heart muscle, called the ________________.

myocardium

7–64. The alternate contraction and relaxation of the myocardium constitutes the CARDIAC CYCLE. Blood circulation is maintained by the ________________ ________________.

cardiac cycle	**7–65.** The cardiac cycle consists of two phases. The period during which the myocardium contracts is called (1) ________________. The period during which the myocardium relaxes is called (2) ________________.

Cardiac Cycle

Systole (contraction)	Diastole (relaxation)

1. systole **2. diastole**	**7–66.** SYSTOLE (SIS—to—le), during which the myocardium (1) ________________, and DIASTOLE (di—AS—to—le), during which the myocardium (2) ________________, are the two phases of the (3) ________________ ________________.
1. contracts **2. relaxes** **3. cardiac cycle**	**7–67.** The chambers of the heart do <u>not</u> go into systole simultaneously. ATRIAL SYSTOLE occurs when the (1) ________________ contract. VENTRICULAR SYSTOLE occurs when the (2) ________________ ________________.

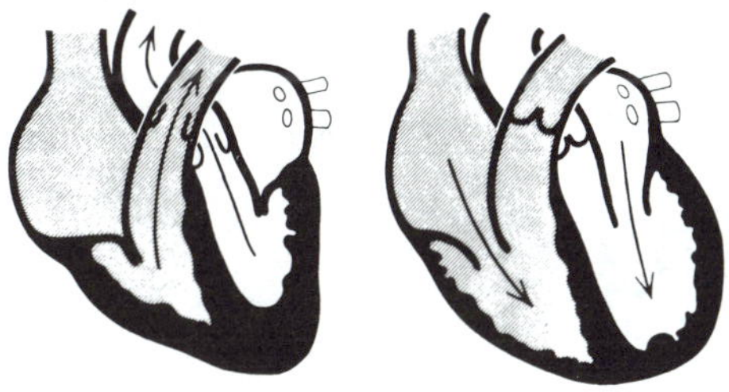

1. atria **2. ventricles contract**	**7–68.** When the ventricles contract, the atria relax; when the atria contract, the ventricles relax. During atrial systole, the ventricles (contract/relax) ________________ or go into diastole.
relax	**7–69.** During diastole, the chambers of the heart are resting and preparing for the pumping effort of systole. While they relax, they are filling with blood. Blood enters the chambers of the heart when they (1) ________________ (contract/relax) and is pumped out when they (2) ________________ (contract/relax).
1. relax **2. contract**	**7–70.** The chambers of the heart fill with blood during (1) ________________ and pump out the blood during (2) ________________.

1. diastole
2. systole

7–71. Systole is initiated by the spread of an electrical impulse from a specialized node of heart muscle tissue called the CARDIAC PACEMAKER. The cardiac pacemaker is located in the ________________ ________________ of the heart.

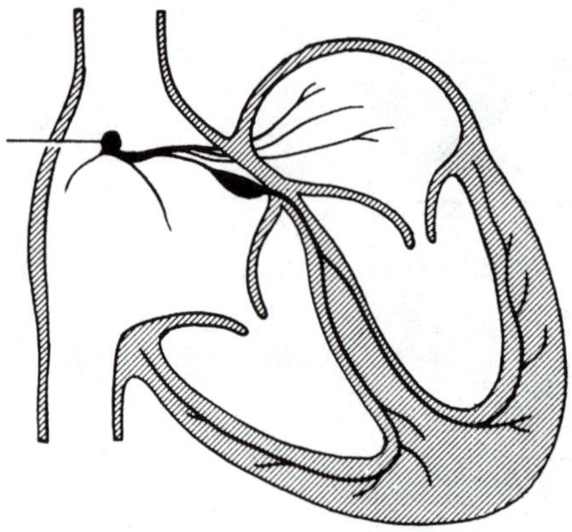

right atrium

7–72. The node of tissue that initiates atrial systole is called, in technical terminology, the SINOATRIAL (SI—no—A—tre—al) NODE, or simply the SA NODE. The (1) ________________ ________________ node is also referred to as the (2) ________________ pacemaker.

1. SA
2. cardiac

7–73. Systole begins when an electrical impulse is transmitted by the (1) ________________ ________________ node, also called the (2) ________________ ________________, which is located in the (3) ________________ ________________ of the heart.

1. SA
2. cardiac pacemaker
3. right atrium

7–74. When the electrical impulse initiating systole leaves the SA node or carciac pacemaker, the first chamber of the heart to go into systole is the chamber in which the cardiac pacemaker is located, namely the ________________ ________________.

right atrium

7–75. Within .02 of a second, the impulse spreads from the right atrium to the left atrium, which then goes into systole.

Although the two atria do not contract simultaneously, their contractions occur so close together that they can be discussed as if they contracted at the same time. Atrial systole, involving both atria, takes only about .08 of a second. The term "atrial systole" refers to the contraction of both the (1) ________________ and the (2) ________________ atria.

1. right
2. left
(either order)

7–76. During the period of diastole that precedes atrial systole, the atria relax and fill with blood. Thus, at the beginning of atrial systole, both atria are filled with ________________.

blood

7–77. At the beginning of atrial systole, the right atrium is filled with oxygen-poor blood from the (1) ________________, whereas the left atrium is filled with oxygen-rich blood from the (2) ________________.

1. body 2. lungs	**7–78.** As the two atria contract in atrial systole, the oxygen-poor blood in the right atrium is pumped into the (1) ________ ________, and the oxygen-rich blood in the left atrium is pumped into the (2) ________ ________.
1. right ventricle 2. left ventricle	**7–79.** When the atria go into systole and pump their blood into the two ventricles, the ventricles are relaxing; that is, they are in ________.
diastole	**7–80.** When the electrical impulse that originated at the SA node or cardiac pacemaker has traversed the two atria, it spreads to another node of special heart muscle tissue, called the ATRIOVENTRICULAR (A—tre—o—ven—TRIK—u—lar) NODE, or simply the AV NODE. The AV node, like the SA node, is located in the ________ ________.

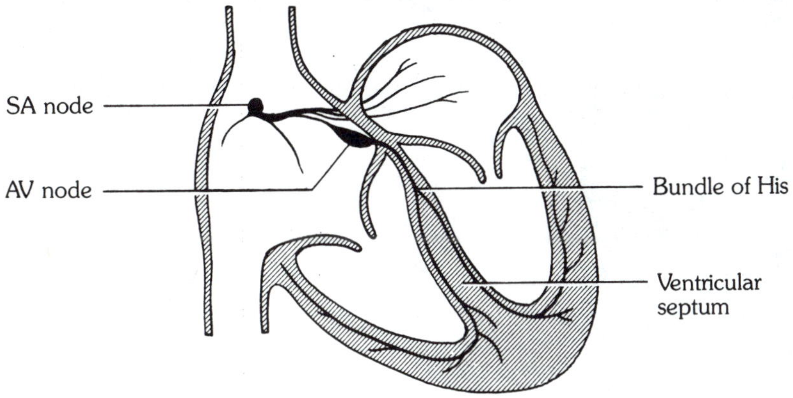

right atrium	**7–81.** The atrioventricular, or AV node, is located in the right (1) ________, but it transmits impulses to the two lower chambers of the heart, the (2) ________.
1. atrium 2. ventricles	**7–82.** Systole begins when an electrical impulse is transmitted from the cardiac pacemaker or (1) ________ node. When this impulse has traversed the atria, it reaches the (2) ________ node, which transmits the impulse to the ventricles.
1. SA 2. AV	**7–83.** At the AV node, the impulse is delayed while the ventricles receive whatever blood continues to enter from the atria. The impulse is then transmitted to the specialized conducting system of the ventricles, and the ventricles contract or go into ________.

systole

7-84. Separating the right from the left ventricle is a wall of muscle called the ______________ ______________.

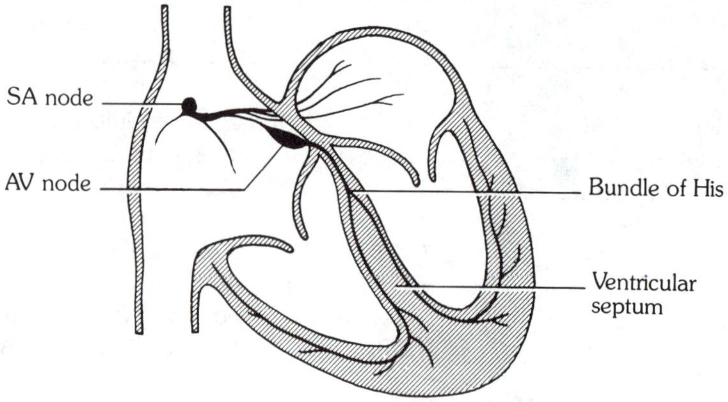

ventricular septum

7-85. The impulse from the AV node is carried through the ventricular septum to the walls of the ventricles by two branches of the ______________ ______________.

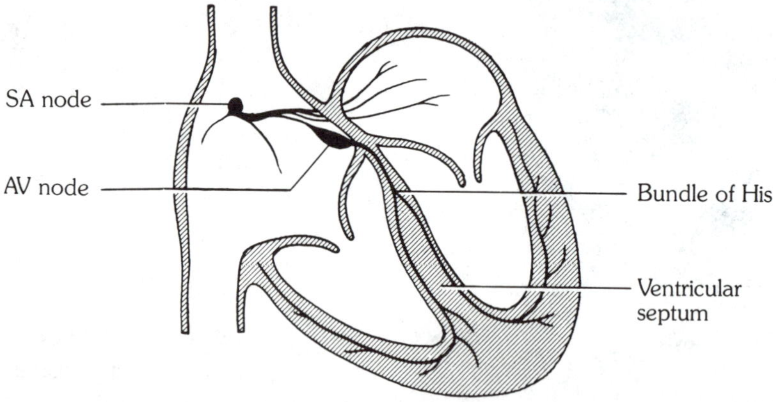

bundle (of) His

7-86. When the impulse transmitted by the bundle of His reaches the ventricles, the ventricles contract in systole. Because the ventricular systole is preceded by ventricular diastole, the ventricles are filled with ______________ when they go into systole.

blood

7-87. As the ventricles go into systole, the atria go into ______________ and begin to fill with blood.

diastole

7-88. When the ventricles go into systole, the right ventricle contains (1) ______________ (oxygen-rich/oxygen-poor) blood from the body, and the left ventricle contains (2) ______________ (oxygen-rich/oxygen-poor) blood from the lungs.

Answers	Frames
1. oxygen-poor 2. oxygen-rich	**7–89.** During ventricular systole, the oxygen-poor blood in the right ventricle is pumped up through the pulmonary artery and goes to the ________ to pick up oxygen.
lungs	**7–90.** During ventricular systole, the oxygen-rich blood in the left ventricle is pumped out through the ________ to be carried to the cells of the body.
aorta	**7–91.** After ventricular systole, the ventricles again relax, or go into (1) ________. At the same time, the two atria begin to go into (2) ________.
1. diastole 2. systole	**7–92.** Atrial systole begins when an electrical impulse is transmitted by the (1) ________ node or (2) ________ ________.
1. SA 2. cardiac pacemaker	**7–93.** The impulse originated by the SA node is transmitted next to the ________ node where it is temporarily delayed.
AV	**7–94.** During atrial systole, blood is pumped from the (1) ________ into the (2) ________.
1. atria 2. ventricles	**7–95.** During atrial systole, when blood from the atria is being pumped into the ventricles, the ventricles are in ________.
diastole	**7–96.** At the end of the atrial systole, the atria go into (1) ________, and the impulse that has been delayed at the AV node is transmitted to the (2) ________ ________ ________, which is the conducting system of the ventricles.
1. diastole 2. bundle of His	**7–97.** The impulse transmitted to the ventricles by branches of the bundle of His causes the (1) ________ to go into (2) ________.
1. ventricles 2. systole	**7–98.** During ventricular systole, blood from the right ventricle is pumped into the (1) ________, and blood from the left ventricle is pumped into the (2) ________.
1. pulmonary artery (lungs) 2. aorta (body)	**7–99.** The ability of the heart to supply the body's need for blood depends partly upon its ability to fill with an adequate blood supply during diastole. Without an adequate period of rest in diastole, the heart cannot receive an adequate blood supply and cannot, therefore, supply adequate ________ to the body.

blood	**7–100.** Hypoxia and other diseases that produce abnormally high rates of heart action naturally shorten the period of diastole, and, after a critical level, tend to (increase/decrease) ______________ the heart's intake and output of blood.
decrease	**7–101.** Some diseases, such as rheumatic fever, may damage the impulse-conducting system of the heart. Damage to the AV node may cause the atria and ventricles to act independently and uncooperatively. Here, too, the ventricles may begin to contract before they are fully filled with blood, and the body (will/will not) ______________ receive an adequate supply of blood.
will not	**7–102.** Efficient functioning of the cardiovascular system is obviously of vital importance to the health of the body. The phrase "the cardiovascular system" implies only one system, but scientists recognize two separate systems for the circulation of blood. (No answer required.)
	7–103. As you know, arteries carry blood away from the heart. Most of the arteries in the body are filled with blood that has just been oxygenated in the lungs and has, therefore, a (high/low) ______________ oxygen content.
high	**7–104.** The pulmonary artery carries oxygen-poor blood from the right ventricle to the lungs. Therefore, the pulmonary artery differs from the other arteries in that it carries oxygen-(1) ______________ blood instead of oxygen-(2) ______________ blood.
1. poor 2. rich	**7–105.** Most of the veins in the body carry oxygen-poor blood back to the heart. The pulmonary vein, however, transports blood from the lungs into the left atrium. Unlike other veins, the pulmonary vein carries oxygen-______________ blood.
rich	**7–106.** The term ARTERIAL BLOOD refers to oxygen-rich blood. The term VENOUS BLOOD refers to oxygen-poor blood. Most arteries in the body transport (1) ______________ (arterial/venous) blood, but the pulmonary artery leading to the lungs transports (2) ______________ (arterial/venous) blood.
1. arterial 2. venous	**7–107.** Most veins in the body carry (1) ______________ blood, but the pulmonary vein leading from the lungs to the left atrium carries (2) ______________ blood.
1. venous 2. arterial	**7–108.** On the basis of the difference in the oxygen content of the blood in veins and arteries, the cardiovascular system is subdivided into two systems. Thus, the veins and arteries of all of the body except the lungs are considered to be members of one system, whereas the veins and arteries of the ______________ are considered to be members of another system.

lungs

7-109. SYSTEMIC (sis—TEM—ik) CIRCULATION is the circulation of blood through all of the (1) ____________ except the lungs. PULMONIC CIRCULATION is the circulation of blood through the (2) ____________.

1. body
2. lungs

7-110. The head and feet, for instance, receive blood through the (1) ____________ circulation. The lungs receive blood through the (2) ____________ circulation.

1. systemic
2. pulmonic

7-111. The terms "blood pressure" and "pulse" are frequently used in connection with the cardiovascular system. The folowing frames will be devoted to a consideration of these terms.
(No answer required.)

7-112. When the left ventricle contracts, blood pulses through the arteries. Certain large arteries are located close to the surface of the skin, and the pulsation of blood through these ____________ can be felt through the skin.

arteries

7-113. "Pulse" refers to the pulsation of (1) ____________ through large arteries. The pulse can be felt when the (2) ____________ ____________ contracts and forces blood through the arteries.

1. blood
2. left ventricle

7-114. Although it is commonly felt through the radial artery in the wrist, the (1) ____________ can be felt in any part of the body where an (2) ____________ is close to the skin.

1. pulse
2. artery

7-115. If the left ventricle is not filled with blood when it contracts, little blood will be forced through the arteries, and the pulse will be (weak/strong) ____________.

weak

7-116. Pulse depends partly upon the presence of an adequate supply of (1) ____________ in the (2) ____________ ____________ when it contracts.

1. blood
2. left ventricle

7-117. If there is an obstruction between the left ventricle and the artery in which the pulse is sought, the pulse will be weak or absent. Therefore, the pulse also depends upon the absence of (1) ____________ between the left ventricle and the (2) ____________ in which the pulse is to be felt.

1. obstruction
2. artery

7-118. If the left ventricle does not contract with sufficient force to pump all of its blood into the arteries, the pulse will be weak. Pulse also depends upon the amount of (1) ____________ with which the (2) ____________ ____________ contracts.

1. force
2. left ventricle

7–119. Pulse depends upon:

1. The presence of an adequate (1) ________________ supply in the (2) ________________ ________________.
2. The absence of (3) ________________ between the (4) ________________ ________________ and the (5) ________________ in which the pulse is sought.
3. The (6) ________________ with which the (7) ________________ ________________ contracts.

1. blood
2. left ventricle
3. obstruction
4. left ventricle
5. artery
6. force
7. left ventricle

7–120. The cardiovascular system contains a series of blood vessels through which blood is pumped from the heart. "Blood pressure" is the pressure of the (1) ________________ within these blood (2) ________________.

1. blood
2. vessels

7–121. The total diameter of all the capillaries in the capillary bed fed by an artery is much larger than the diameter of the artery itself. For this reason, the blood pressure in the arteries is higher than the pressure in the ________________.

capillaries (capillary bed)

7–122. Blood pressure is higher in the arteries than it is in the capillaries. Blood pressure (is/is not) ________________ the same in all the blood vessels of the body.

is not

7–123. Usually, when physicians measure blood pressure, they measure pressure in the arteries. Therefore, the term "blood pressure" when used alone refers to the pressure in the ________________.

arteries

7–124. When the left ventricle contracts during systole, ARTERIAL PRESSURE (blood pressure in the arteries) rises. When the left ventricle relaxes in diastole, the arterial pressure diminishes. Arterial pressure is higher during ventricular (1) ________________ (systole/diastole) than during ventricular (2) ________________ (systole/diastole).

1. systole
2. diastole

7–125. SYSTOLIC PRESSURE is blood pressure during ventricular systole. DIASTOLIC PRESSURE is blood pressure during ventricular diastole. In any individual, (1) ________________ pressure is always lower than (2) ________________ pressure.

1. diastolic
2. systolic

7–126. The textbook figure for the normal blood pressure of a resting young adult is 120/80. Naturally, the higher figure, 120, is a measure of (1) ________________ pressure, and the lower figure, 80, is a measure of (2) ________________ pressure.

Answers	Frames
1. systolic 2. diastolic	**7–127.** If the left ventricle does not contract forcefully enough to pump all of its blood into the arteries, the blood pressure will be low. Blood pressure, like the pulse, depends partly upon the (1) ________ with which the (2) ________ ________ contracts.
1. force 2. left ventricle	**7–128.** The walls of the arteries are somewhat elastic. This elasticity permits them to expand and recoil as blood pulses through them. If the arteries "harden," they lose their elasticity, and the blood pressure rises above normal. Blood pressure, then, depends partly upon the elasticity of the walls of the ________.
arteries	**7–129.** Two forces that influence blood pressure are: 1. The (1) ________ with which the (2) ________ ________ contracts, and 2. The (3) ________ of the walls of the (4) ________.
1. force 2. left ventricle 3. elasticity 4. arteries	**7–130.** Small arteries, arterioles, are capable of expanding and constricting so as to change the size of the lumen (central cavity). When the arterioles constrict and resist the blood flow, blood pressure in the arteries rises. When the arterioles relax, the blood flows through more freely, and blood pressure in the arteries (increases/decreases) ________.
decreases	**7–131.** The resistance provided by the arterioles influences the pressure of blood in the arteries. Thus, another factor influencing blood pressure is the (1) ________ of the (2) ________ to the flow of blood from the arteries.
1. resistance 2. arterioles	**7–132.** The "viscosity" of blood refers to its fluidity (relative to water). Thick or viscous fluids flow more slowly than thin, aqueous fluids and require more pressure to move them through tubes. Blood pressure, therefore, is partly dependent upon the (1) ________ of the (2) ________.
1. viscosity (thickness) 2. blood	**7–133.** Two factors that influence blood pressure are: 1. The resistance of (1) ________ to the flow of blood from the arteries, and 2. The thickness of (2) ________ of the (3) ________.
1. arterioles 2. viscosity 3. blood	**7–134.** No significant pressure can be achieved within a system of tubes if the system is not filled with fluid. If a great deal of blood is lost from the vascular system (through hemorrhage, for instance), the blood pressure will drop. Blood pressure depends also upon the volume of ________ in the vascular system.

blood	**7–135.** Blood pressure is influenced by: 1. The (1) ________________ with which the (2) ________________ ________________ contracts. 2. The (3) ________________ of the walls of the (4) ________________. 3. The resistance of the (5) ________________ to the flow of blood from the arteries.
1. force **2. left ventricle** **3. elasticity** **4. arteries** **5. arterioles**	**1–136.** Blood pressure also depends upon: 1. The ________________ of blood. 2. The ________________ of blood.
1. volume **2. viscosity** **(either order)**	**7–137.** Infection, drugs, and nervous influences may decrease the tone of the arterioles so that they offer too <u>little</u> resistance to blood from the arteries. In this case, blood will flow too freely through the arteries, and blood pressure will be (high/low) ________________.
low	**7–183.** During an accident in which a large amount of blood is lost, the blood volume decreases, and the blood pressure (increases/decreases) ________________.
decreases	**7–139.** Constriction of the arterioles may very gravely affect the blood pressure, producing <u>ARTERIAL HYPERTENSION</u>. In arterial hypertension, the constricted arterioles prevent blood from flowing through the arteries. Thus, the blood volume in the arteries increases, and blood pressure (rises/drops) ________________.
rises	**7–140.** The following frames will be devoted to a discussion of blood, its composition and its functions. An average-size adult male weighing 150 lb (68 kg) has approximately 5 quarts (4½ liters) of blood in his body. (No answer needed.)
	7–141. Recall from Chapter 3 that blood is composed of a group of cells (plus fluid), each fulfilling specific functions, and all functioning together. Blood, therefore, like skin and bone, is a ________________.
tissue	**7–142.** Blood functions primarily in carrying food and the vital gas (1) ________________ to all cells of the body and in removing from the cells the waste products of metabolism, such as the waste gas (2) ________________ ________________.

1. oxygen 2. carbon dioxide	**7–143.** When the body is exposed to cold, the arterioles near the skin help to maintain body temperature by constricting, thus reducing the surface area of blood from which heat may be lost. When the body is exposed to heat, these same arterioles expand to produce a larger surface flow of blood from which ________________ may be lost.
heat	**7–144.** Thus, one function of the blood, assisted by the vessels which carry it, is to maintain body ________________.
temperature	**7–145.** Blood has other indispensable functions; it carries water, electrolytes, and hormones to the tissues. Blood serves as the transportation system of the body, transporting not only oxygen and carbon dioxide, but also (1) ________________, (2) ________________, and (3) ________________.
1. water 2. electrolytes 3. hormones	**7–146.** More than half of the blood is a watery fluid called PLASMA (PLAZ—ma), and the remainder consists of <u>formed</u> elements that can be viewed under a microscope. The formed elements are classified as (1) ________________ ________________ ________________, (2) ________________ ________________ ________________, and (3) ________________.

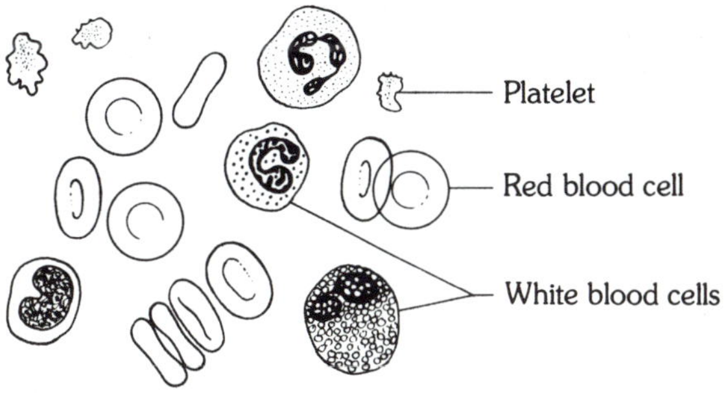

1. red blood cells 2. white blood cells 3. platelets	**7–147.** The capillaries through which red blood cells travel are often smaller than the cross-sectional diameters of the red blood cells themselves. An internal elastic framework maintains the shape of the cells but permits them to bend and twist as they pass through the small ________________.
capillaries	**7–148.** The red blood cells contain a protein molecule called HEMOGLOBIN (HE—mo—glo—bin) that has an affinity for oxygen. Because they contain (1) ________________, the red blood cells are able to carry (2) ________________ to the body cells.
1. hemoglobin 2. oxygen	**7–149.** The red color of the blood is produced by the protein molecule. (1) ________________, carried in the (2) ________________ ________________ ________________.

1. hemoglobin 2. red blood cells	**7–150.** Approximately 2–1/2 million new red blood cells enter the blood every second, and an equal number are lost every second. The continual production of red blood cells is the function of the red ________________ contained in the bones.
marrow	**7–151.** Red blood cells are atypical cells because they have no nucleus. White blood cells, on the other hand, contain organelles, inclusions, and a more or less central nucleus. Thus, white blood cells are (typical/atypical) ________________ cells.
typical	**7–152.** (1) ________________ blood cells are typical nucleated cells. (2) ________________ blood cells are atypical cells that have no nuclei.
1. white 2. red	**7–153.** White blood cells appear to be largely inactive in the blood stream. Most of their activities are carried on outside the blood stream in the tissues of the skin and other organs. When body tissues are invaded by bacteria, the white blood cells enter the diseased (1) ________________ and ingest the (2) ________________ .
1. tissue 2. bacteria	**7–154.** Some white blood cells produce ANTIBODIES, that can combine with invading foreign proteins. The antibodies help to fight infection. Thus, the white blood cells counteract disease in two ways: they ingest (1) ________________ , and they produce (2) ________________ .
1. bacteria 2. antibodies	**7–155.** Red blood cells are called ERYTHROCYTES (er—ITH—ro—sites), and white blood cells are called LEUKOCYTES (LU—ko—sites). (1) ________________ ingest bacteria and produce antibodies. (2) ________________ carry oxygen in the bloodstream.
1. leukocytes 2. erythrocytes	**7–156.** Use technical terms to answer this: (1) ________________ carry hemoglobin and are atypical cells without nuclei. (2) ________________ are typical nucleated cells that are inactive in the blood stream but active in the body tissues.
1. erythrocytes 2. leukocytes	**7–157.** Platelets are also present in the blood. When a blood vessel is injured, platelets act at the site of the injury to plug the tear. When an injury to a vessel is very small, the plug formed by ________________ is sufficient to prevent bleeding from the wounded vessel.
platelets	**7–158.** If an injury to a vessel is too large for a platelet "plug" to prevent bleeding, a more complicated process of blood clotting takes place. When a blood vessel tear is larger, special proteins work with the ________________ to help the blood to clot.

platelets	**7–159.** You recall that more than half of the blood is a watery fluid called ________________.
plasma	**7–160.** Plasma contains a large concentration of proteins. Among the plasma proteins are special ones called ACCESSORY or CLOTTING FACTORS. These are important in the process of blood clotting. Factors are special substances, mainly proteins, that act early in the process of blood ________________.
clotting	**7–161.** Two major plasma proteins that act in blood clotting are PROTHROMBIN and FIBRINOGEN. Plasma contains special substances called (1) ________________. Two important plasma proteins that act in blood clotting are the accessory or clotting factors called (2) ________________ and (3) ________________.
1. factors **2. prothrombin** **3. fibrinogen**	**7–162.** Prothrombin is a plasma ________________ that is converted to an enzyme called THROMBIN, which functions in blood clotting.
protein	**7–163.** You should recall that enzymes are special (1) ________________ that regulate metabolic activities. The enzyme formed from prothrombin is (2) ________________.
1. proteins **2. thrombin**	**7–164.** Thrombin converts the proten fibrinogen to FIBRIN. In order to form a clot, fine threads of fibrin join together to build a meshlike network of ________________ threads.
fibrin	**7–165.** Blood cells, platelets, and plasma become trapped in the ________________ ________________ to form the blood clot.
fibrin mesh	**7–166.** A blood clot is composed of blood cells, (1) ________________ and (2) ________________ trapped in a fibrin mesh. The clot prevents blood loss from the damaged vessel.
1. platelets **2. plasma**	**7–167.** The last step in preventing blood loss occurs when the torn edges of the damaged vessel are pulled together by the ________________ threads of the clot.
fibrin	**7–168.** Prothrombin and fibrinogen are just two of the ________________ that are involved in blood clotting. At this time, twelve specialized clotting factors have been identified!

factors	**7–169.** A person who has a disturbance in one or more of the clotting factors has a disease called HEMOPHILIA (he—mo—FIL—i—a). The majority of hemophiliacs have a deficiency in Factor 8. (No response required.)
	7–170. To review: 1. blood clotting involves the actions of substances called (1) ____________; 2. two major plasma protein factors are (2) ____________ and (3) ____________; 3. blood cells, platelets, and plasma become trapped in a (4) ____________ ____________ to form the blood clot.
1. factors **2. fibrinogen** **3. prothrombin** **4. fibrin mesh**	**7–171.** The remainder of the program will be devoted to a discussion of the LYMPHATIC SYSTEM that, together with the cardiovascular system, forms the ____________ system of the body.
circulatory	**7–172.** In the unit on the respiratory system, you learned that gas molecules will move through a permeable membrane, moving from an area of high concentration to an area of low concentration. The process by which the gas molecules move in this way is called ____________.
diffusion	**7–173.** By a similar process, called OSMOSIS (os—MO—sis), water molecules can move through a membrane. The membrane through which water molecules pass in the process of osmosis is called a SEMIPERMEABLE MEMBRANE because some large particles dissolved in the water will not pass through it, but ____________ molecules and some small dissolved particles will pass through it.
water	**7–174.** Let us illustrate osmosis. To begin with, we have a tank divided into two sections (A and B) by a (1) ____________ ____________. The tank contains plain water. The water level on both sides of the membrane (2) ____________ (is/is not) the same.

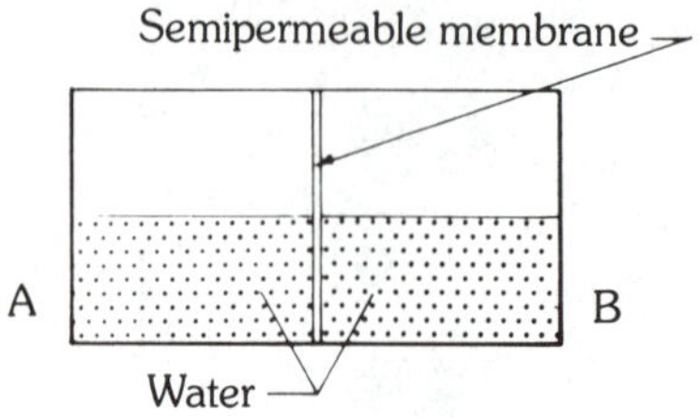

1. semipermeable membrane
2. is

7–175. Now we add some sugar to the water on side B of the membrane, but not to the water on side A. The water level on side B of the membrane (1) ________________ (rises/drops) because (2) ________________ (sugar/water) molecules pass through the membrane from side A to side B.

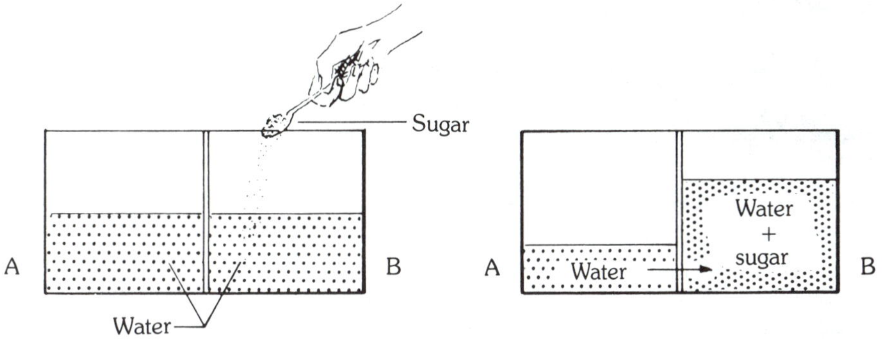

1. rises
2. water

7–176. When a semipermeable membrane separates two solutions containing different concentrations of dissolved particles, water molecules tend to pass through the membrane in an attempt to equalize the concentrations of the particles on both sides of the membrane. The process involved is called osmosis and refers to the passage of (1) ________________ molecules through a (2) ________________ membrane.

1. water
2. semipermeable

7–177. As the capillaries traverse the tissues of the body, water and electrolytes osmose through the semipermeable membrane of the capillaries into the spaces between the cells of the ________________.

tissues

7–178. The resulting fluid is called TISSUE FLUID. Tissue fluid is formed when, by the process of osmosis, (1) ________________ passes through the walls of the (2) ________________ into spaces between the cells of the tissue.

1. water
2. capillaries

7–179. Tissue fluid helps to nourish tissues. It loses its oxygen and nutrients into the tissues and acquires waste products from the tissues. The tissue fluid itself then becomes a ________________ product.

waste

7–180. Tissue fluid that has given up its nutrients and oxygen to the tissues and has acquired the waste products of metabolism then passes (by osmosis) into the LYMPHATIC (lim—FAT—ik) VESSELS. The tissue fluid combines with other elements in the lymphatic vessels to form LYMPH (limf). The lymphatic vessels carry the (1) ________________ back to its origin in the (2) ________________ stream.

1. lymph
2. blood

7–181. Because lymph carries away the waste products of tissue metabolism, lymph is not carried to the tissues; it is only carried ________________ from the tissues.

away	**7–182.** The drainage of lymph through the ________ vessels is a one-way flow, not a circulation.
lymphatic	**7–183.** The walls of the large arteries and veins are tissues. They are nourished partly by tissue fluid that, as a waste product, must be drained away. Hence, the walls of large (1) ________ and (2) ________ are drained by a system of (3) ________ vessels.
1. veins **2. arteries** **(either order)** **3. lymphatic**	**7–184.** The lymphatic system originates as small blind lymphatic capillaries in the skin. These capillaries course centrally as they converge and form progressively larger vessels that ultimately empty into the great (arteries/veins) ________, which carry blood back to the heart.
veins	**7–185.** All of the organs of the body, with the exception of the central nervous system, the eye, and inner ear possess a lymphatic system. In which of these is a lymphatic system absent? a. the walls of the lungs b. the eye c. the walls of arteries
b. the eye	**7–186.** To review, the cardiovascular system is composed of the (1) ________ and the blood (2) ________.
1. heart **2. vessels**	**7–187.** The function of the vessels is to provide pathways for the distribution of nutrients and (1) ________ to all body cells, and to collect waste products from the cells. The vessels of the (2) ________ system aid the blood vessels in collecting and transporting waste products away from the cells.
1. oxygen **2. lymphatic**	**7–188.** The blood vessels are able to carry out their functions because the ________ acts as a pump to circulate the blood.
heart	**7–189.** The functions of the blood are carried out by specialized cells. Write the name of the particular blood cell beside its function: 1. carries oxygen to tissues and cells (1) ________ 2. destroys bacteria that could be harmful to the body (2) ________ 3. acts in clotting blood (3) ________
1. erythrocytes **2. leukocytes** **3. platelets**	**7–190.** When platelets are not adequate in stopping blood loss from damaged vessels, a group of substances called ________ work to form a blood clot.

factors

7–191. You have completed Part 1 of Chapter 7. Part 2 will continue your study of the circulatory system by discussing the ways that the body protects itself from injury and disease.

Part 2

Protection from Disease

7–192. In Part 1 of this chapter you learned that the circulatory system has two major systems, the (1) ________________ system and the (2) ________________ system.

1. cardiovascular
2. lymphatic

7–193. The lymphatic system is <u>not</u> a true circulation because it carries lymph <u>only</u> in the direction ________________ ________________ the cells.

away from

7–194. Both the cardiovascular and the ________________ systems are important in helping the body protect itself from disease, by transporting special cells to the part of the body invaded by foreign substances and removing waste products formed in the destruction of those substances.

lymphatic

7–195. The special cells that the cardiovascular system carries to tissues to fight invasion by foreign substances are the white blood cells or ________________.

leukocytes

7–196. The body has three lines of defense against invasion. Just as an army defends its territory with many different strategies, the body's army is organized to combat disease.

(No response required.)

7–197. When the enemy attacks, the army puts up a wall of defenders. The body's first wall of (1) ________________ is composed of the intact skin surface and the protective, (2) ________________ (tissue-type) lining of body openings, such as the mouth and nose.

1. defense
2. epithelial

7–198, If the enemy penetrates the first line of defense, the army must mobilize its behind-the-lines resources and begin attacking the invaders.

Likewise, in the body, once the first line of defense, the ________________ or the epithelial lining is broken, the second line of defense begins its attack.

skin

7–199. The special forces of the army engage in hand-to-hand combat with the enemy. The special ________________ of the body attack the harmful particles of the invaders.

cells

7–200. The cells involved in attacking and destroying foreign substances are called ________________ or WBC.

leukocytes

7–201. Some leukocytes ingest and destroy the invading particles by a process called PHAGOCYTOSIS (FAY—go—cy—TOE—sis). Phago- means to swallow; -cytosis refers to cells. So, foreign material is destroyed by the process of ________________.

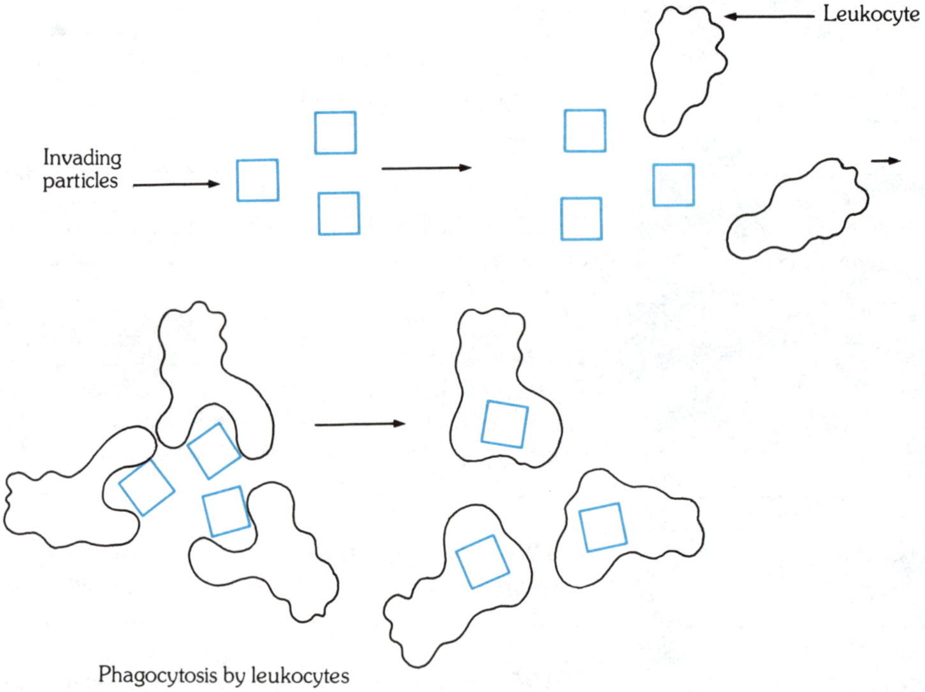

Phagocytosis by leukocytes

phagocytosis

7–202. The leukocytes that are involved in the process of destroying invaders are the NEUTROPHILS (NEW—tro—fils) and TISSUE MACROPHAGES (MAC—ro—fag—es).

The cells involved in phagocytosis are (1) ________________ and (2) ________________ ________________.

1. neutrophils
2. tissue macrophages

7–203. Neutrophils and tissue macrophages are special types of ________________ or WBC.

leukocytes

7–204. You recall that leukocytes are produced in bone marrow. Neutrophils are produced in ________________ ________________ and stored there until they are called on to destroy invading foreign substances.

bone marrow

7–205. Tissue macrophages are produced in (1) ______________________. However, once formed, they travel to special tissues of the body. The cells remain there waiting to engage invaders in the process of (2) ______________________.

1. bone marrow
2. phagocytosis

7–206. You recall from Chapter 6 that the respiratory system must "clean" the air to prevent bacteria from entering the bloodstream. In the alveolar tissue, there are tissue ______________________ that ingest any bacteria that penetrate the upper respiratory tract.

macrophages

7–207. Also, lymph collects foreign particles that could not pass into the blood stream. As the lymph travels along the lymphatic vessels, it passes through glands called LYMPH NODES. Tissue (1) ______________________ located in the lymph (2) ______________________ phagocytize foreign particles.

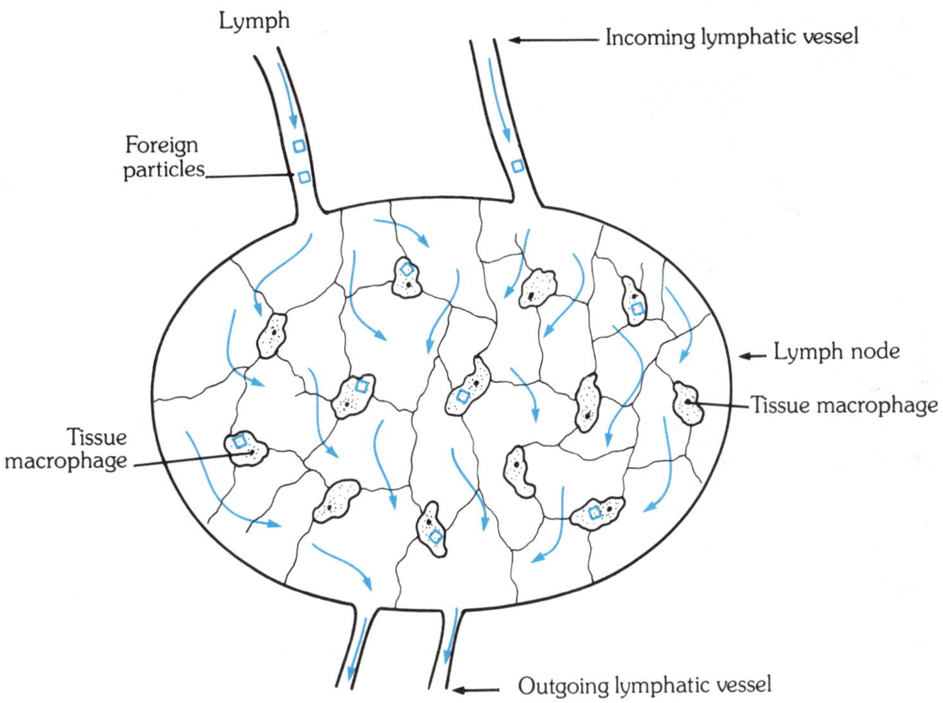

LYMPH NODES contain many tissue macrophages that can destroy foreign particles in lymph.

1. macrophages
2. nodes

7–208. Although the second line of defense, the process of phagocytosis by (1) ______________________ and tissue (2) ______________________ is very powerful, it cannot destroy all types of foreign substances.

1. neutrophils
2. macrophages

7–209. In a battle there are times when highly specialized troops are needed to defeat the enemy. The body has very particular abilities to fight invaders by specialized (1) ________________.
The third line of (2) ________________ is called IMMUNITY.

1. cells
2. defense

7–210. Special cells that act in immunity are produced by white blood cells called LYMPHOCYTES (LIM—fo—sites). The cells that the ________________ produce are called ANTIBODIES.

lymphocytes

7–211. Lymphocytes are another special type of ________________, or WBC.

leukocytes

7–212. Instead of being stored in bone marrow or at tissue sites, as are the neutrophils and tissue macrophages, the (1) ________________ are housed in lymphoid tissue, especially the (2) ________________ nodes.

1. lymphocytes
2. lymph

7–213. Foreign substances that invade the body stimulate the lymphocytes to produce ________________. The foreign substances are called ANTIGENS (AN—ti—jens).

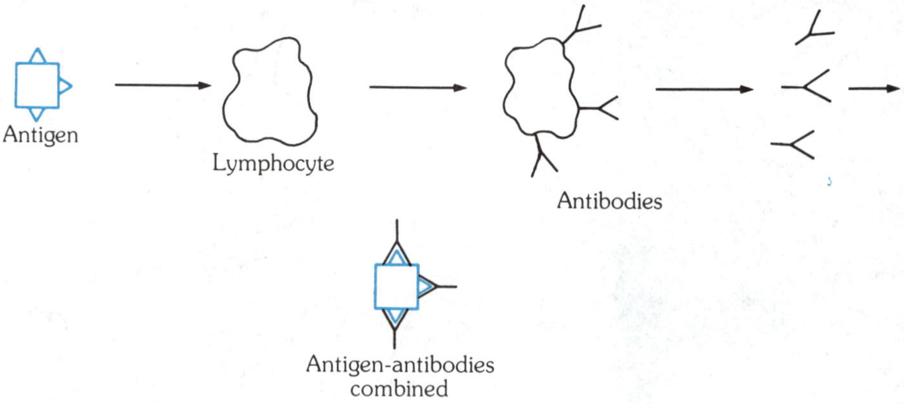

antibodies

7–214. The immune response is often referred to by the names of the substances involved. The immune response is sometimes called the ________________-antibody response.

antigen

7–215. The ________________ that is formed in response to an antigen is capable of destroying only that invader, or antigen.
Refer to the illustration in frame 7—213.

antibody

7–216. However, after particular antibodies have been formed, the cells that produced the antibody, the (1) ________________, develop a "memory" for the antigen. In future encounters with the same (2) ________________, antibodies will be quickly formed to fight against the invader.

1. lymphocytes
2. antigen

7–217. An example of an antigen-antibody, or (1) ________________ response is vaccination.

When you received the polio vaccine, you were given a harmless dose of the antigen responsible for polio. That one set of vaccinations was strong enough to stimulate your lymphocytes to produce (2) ________________ to polio organisms. You did not get polio from the injection and you were protected for many years by the ability of the (3) ________________ to "remember" the antigen.

---

1. immune
2. antibodies
3. lymphocytes

7–218. If the body is invaded by bacteria, or is injected with the antigen, antibodies are formed to ________________ it from disease.

---

protect

7–219. Viruses are also foreign substances capable of inflicting illness on the body. The body's ________________ response to viral invasion is carried out by substances called INTERFERONS (in—ter—FEAR—ons).

---

immune

7–220. Interferons are substances produced by the cells that are attacked. The virus is destroyed by the ________________.

---

interferons

7–221. Like neutrophils and antibodies that are transported by the circulatory system to various parts of the body, the (1) ________________ are carried to other areas to provide protection from (2) ________________ invasion.

---

1. interferons
2. viral

7–222. To review, the body has ________________ (how many?) lines of defense that help to protect it from illness.

---

three

7–223. The first line of defense is provided by the intact (1) ________________ surface of the (2) ________________ lining of the body openings.

---

1. skin
2. epithelial

7–224. When the first line of defense is penetrated, the body has a second response that can destroy foreign substances by the process of ________________.

---

phagocytosis

7–225. The second line of defense is provided by cells that phagocytize harmful invaders. The cells are (1) ________________ and tissue (2) ________________, both special types of leukocytes.

---

1. neutrophils
2. macrophages

7–226. The third line of defense works against specific invaders, or (1) ________________, by forming (2) ________________, in the process called the immune response.

1. antigens 2. antibodies	**7–227.** A second type of immune reaction is the production of ________________ by cells attacked by viruses.
interferons	**7–228.** The proper functioning of the immune system depends on the ability of the body to recognize invaders as ________________ substances.
foreign	**7–229.** An important use of antigen-antibody information is seen in cross-matching blood. Blood typing is the first step in cross-matching blood. Each person's blood type tells vital facts about the types of antigens and (1) ________________ his red (2) ________________ cells and plasma fluid contain, respectively.
1. antibodies 2. blood	**7–230.** There are four blood types, called type A, type B, type AB, and type O. A person with type-A blood has A–antigens in his red (1) ________________ cells. Persons with type-B blood have (2) ______-________________ in their red blood cells. In type-AB blood the red blood cells contain both (3) ________ and (4) __________ antigens. Finally, in type-O blood, the red blood cells contain neither A nor B antigens.
1. blood 2. B-antigens 3. A 4. B	**7–231.** In the following column identify the kind or kinds of antigens in the red blood cells of persons with the listed blood type. BLOOD TYPE — ANTIGEN IN RBC A — 1. ________________ B — 2. ________________ AB — 3. ________________ O — 4. ________________
1. A-antigens 2. B-antigens 3. A and B antigens 4. neither A nor B antigens	**7–232.** When a person has a particular antigen in his red blood cells, he has antibodies ________________ (different from/same as) the antigens.
different from	**7–233.** His antibodies must be different from the antigens. If these substances were of the same kind, the person's RBC would be ________________.
destroyed (phagocytized)	**7–234.** When a person has type-A blood he has antibodies to type-B blood. Since type-O blood has (1) ________________ (how many?) antigens, the plasma of a person with type-O red blood cells contains both (2) ________________ and B antibodies.

1. no 2. A	**7–235.** In the following column identify the kind or kinds of antibodies in the plasma of a person with the listed blood type. BLOOD TYPE — ANTIBODIES IN PLASMA A — 1. B — 2. AB — 3. O — 4.
1. B 2. A 3. none 4. A and B	**7–236.** In a person with type-AB blood, there can be no antibodies for red blood cells because type AB contains both A and B ________.
antigens	**7–237.** It is important to know someone's blood (1) ________ before he receives a transfusion of blood from another person. If the wrong blood type is given he could receive (2) ________ that his own antibodies would destroy.
1. type 2. antigens	**7–238.** The immune system is helpful to the body because it recognizes and destroys foreign substances, such as bacteria, ________, and antigens.
viruses	**7–239.** Sometimes the immune response is harmful. As you know, the body contains many different types of tissues. In most cases each person's immune system recognizes the tissues as part of his own body. However, if the immune system considers the tissues as foreign substances, it begins to destroy them. (No response required.)
	7–240. When the ________ system destroys the body's own tissue, the response is called an AUTOIMMUNE DISEASE. Auto- refers to "self," so the autoimmune response is literally a reaction in which the body destroys itself.
immune	**7–241.** The study of autoimmune diseases is a very important area of medical science. Some diseases, such as rheumatic fever, are known to be caused by ________ disorders. Other diseases, such as multiple sclerosis, are being studied to determine whether they are caused by an autoimmune response.
autoimmune	**7–242.** Another way that the body's immune response may be harmful occurs when a person receives tissue from another person's body. When tissue is transplanted, the receiver's immune system will recognize the new tissue as a ________ substance and try to destroy it.

foreign	**7–243.** Transplanting tissue is a rapidly developing medical practice. Persons have received kidneys, bone marrow, and hearts from donors. The success of the transplant depends on the ability of the receiver's immune system to tolerate the new tissue. (No response required.)
	7–244. There are many medications that act to slow down the body's ________________ response. Persons with autoimmune diseases and those who receive tissue transplants can be helped by these drugs.
immune	**7–245.** To review, the immune system protects the body from disease by forming ________________ to fight the invading antigens.
antibodies	**7–246.** Antibodies are formed to fight and destroy particular ________________, or invaders.
antigens	**7–247.** When the invader is a virus, the immune response of the cells is to produce ________________ to destroy the virus.
interferons	**7–248.** In most cases, the immune system (1) ________________ the body from disease. However, if the body's own tissues become antigens, the immune response is to form antibodies against them. This response is called an ________________ disease.
1. protects **2. autoimmune**	**7–249.** You have completed Chapter 7.

8 The Nervous System

Part 1 The Central Nervous System

Review of Nerve Tissue

8–1. You will recall that the structural and functional unit of nerve tissue—the neuron—was described in some detail in Chapter 3 of this course. A review of nerve tissue is important to the understanding of how impulses are received and transmitted by your body's communication center, the ________________ system.

nervous

8–2. A neuron is composed of a cell body and fibers that can vary in length from less than an inch to over a yard long. There are two types of processes, or fibers, extending from the neuron's cell body: the axon and the dendrites. Label the two extensions of the cell body's cytoplasm:

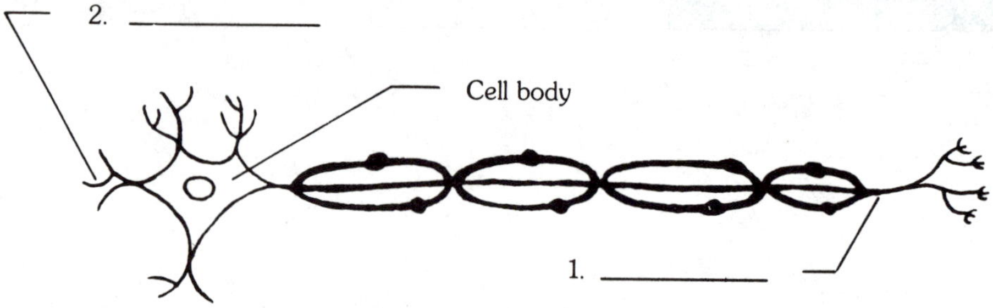

1. axon
2. dendrite

8–3. The axon is a fiber along which nerve impulses move away from the cell body. Dendrites, which are also fibers, conduct impulses ____________ the cell body.

toward

8–4. Axons may be encased in:

1. A sheath of protein-lipid material—MYELIN
2. A sheath of cellular material (the NEUROLEMMA, sometimes called SCHWANN sheath)
3. Or a combination of both sheaths.

When an axon is encased by both sheaths, the neurolemma is outermost and is separated from the axon by the protein-lipid ____________ sheath.

myelin

8–5.

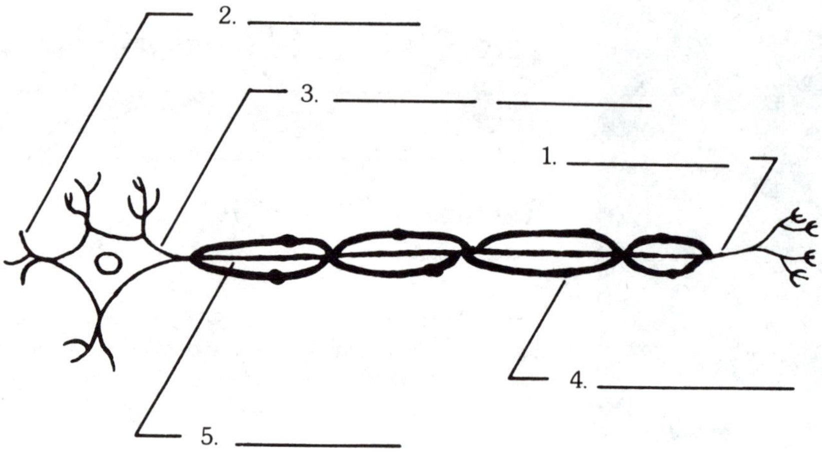

Label the blank spaces.

1. axon
2. dendrite
3. cell body
4. neurolemma (Schwann sheath)
5. myelin

8–6.

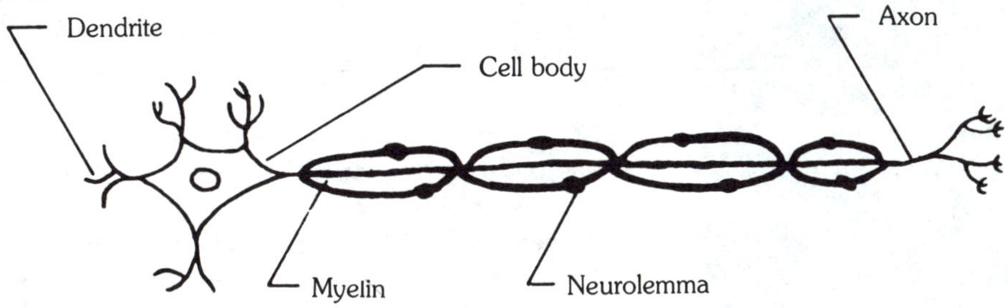

The cellular axon insulation is due to the (1) ________________, or Schwann sheath, and the (2) ________________ sheath.

---

1. neurolemma
2. myelin

8–7.

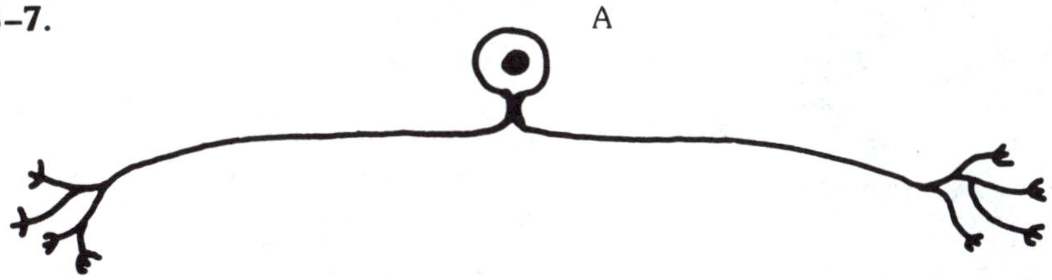

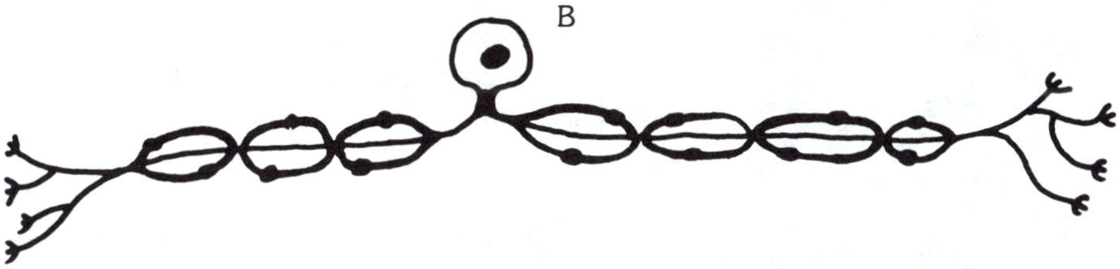

Which sensory fiber is unmyelinated (not insulated) ________________ (A or B)?

A

8–8. A nerve fiber encased in a myelin sheath is classified as a myelinated nerve fiber. If no myelin sheath is present, the nerve fiber is classified as an ________________ nerve fiber.

---

unmyelinated

8–9. The neurolemma is always (outside/inside) ________________ the myelin sheath.

---

outside

8–10.

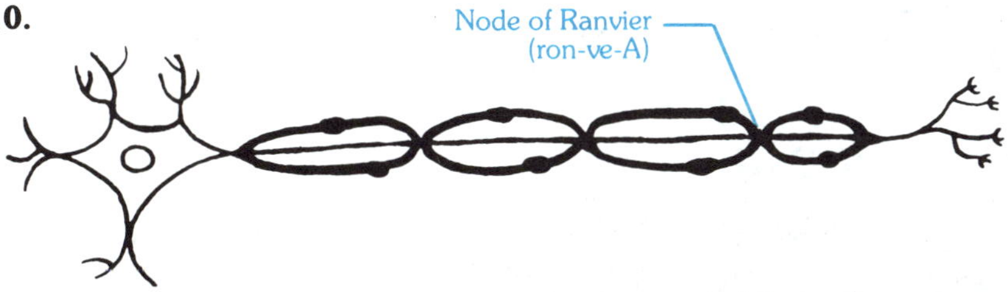

The myelin sheath is not continuous and uninterrupted, but rather a segmented sheath. The constrictions, where the axon is not insulated, occur at more or less regular intervals. Where the neurolemma dips down to touch the axon, a node of ________________ occurs.

Ranvier

8–11. A nerve is many nerve fibers combined in a single bundle. This bundle of fibers, a ________________, may contain both sensory and motor fibers.

nerve

8–12. Fibers that receive and then conduct impulses from a receptor, or sense organ, to the central nervous system are called AFFERENT (AF—er—ent) or sensory neurons or fibers.

Fibers that conduct impulses from the central nervous sytem to the effector, an organ or a muscle for example, and incite motor activity are called EFFERENT (EF—er—ent) or ________________ neurons or fibers.

motor

8–13. In Chapter 3 of this book, reference was made to fibers that transmitted sensory impulses as (1) ________________ fibers and to fibers that transmitted motor impulses as (2) ________________ fibers.

1. afferent
2. efferent

8–14. Afferent fibers that transmit impulses from the periphery of the body to the brain or spinal cord have their cell bodies located outside the central nervous system. These fibers transmit (sensory/motor) ________________ impulses.

sensory

8–15. Efferent fibers that transmit impulses to the periphery of the body from the brain or spinal cord have their cell bodies located within the central nervous system. These fibers transmit (sensory/motor) ________________ impulses.

motor

8–16. A NERVE NUCLEUS is a group of neuron or nerve cell bodies within the central nervous system. The cell bodies of (1) ________________ fibers are part of a nerve (2) ________________.

1. motor (efferent)
2. nucleus

8–17. A GANGLION (GANG—gli—un) is a group of nerve cell bodies outside the central nervous system. Cell bodies of (1) ________________ fibers can combine to form a (2) ________________.

1. sensory (afferent)
2. ganglion

8–18. A group of nerve cell bodies within the central nervous system is:

a. a ganglion (move to frame 19)
b. a nucleus (move to frame 20)

a.

8–19. No, a ganglion is a group of neuron cell bodies outside the central nervous system. The cell bodies of afferent fibers combine to form a ganglion.

(Move to frame 20.)

b.

8–20. Yes, a nucleus is a group of cell bodies within the central nervous system. The cell bodies of (motor/sensory) ________________ neurons combine to form a nerve nucleus.

motor

8–21. When a stimulus, applied to a (1) __________ neuron, is transmitted to the central nervous system and translated into a specific action impulse, the impulse is then sent by way of a motor neuron to a muscle. This produces an involuntary movement called a reflex.

A reflex would depend upon the reception of a sensory impulse and transmission of an action impulse from the central nervous system to the muscle by one or more (2) __________ neurons.

1. sensory
2. motor

8–22. You recall from Chapter 3, the elements of a reflex pathway are

1. a stimulus received by a (1) __________ neuron; then
2. an impulse transmitted to the (2) __________; and, then
3. an impulse conducted from the spinal cord to a (3) __________ neuron.

1. sensory
2. spinal cord
3. motor

8–23. When the afferent impulse from the skin or tendon travels to the central nervous system, it crosses a (1) __________. The impulse then travels to an extensor muscle, inducing involuntary contraction. This is an example of a (2) __________.

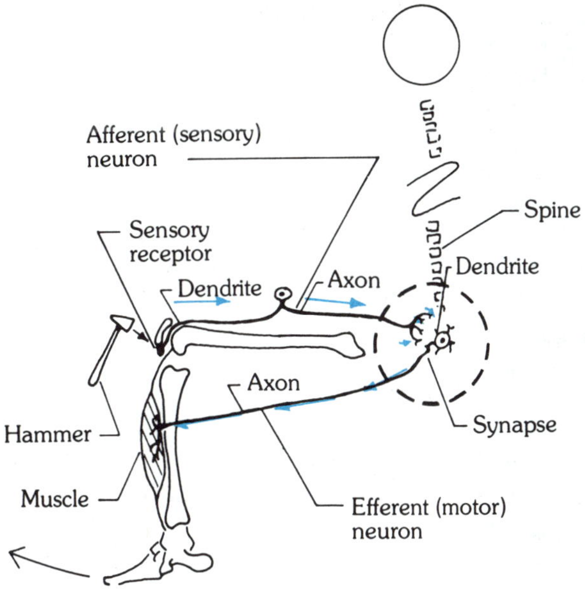

1. synapse
2. reflex

8–24. A reflex arc is an established pattern of stimulus and response. Sometimes a new stimulus is presented, and you are slow to respond; this is because there is no reflex arc established for this impulse, and it must be sent to the __________ for analysis.

brain

8–25. The reflex example used in this program has been simplified to emphasize the route of impulse transmission. Many reflexes are not a single impulse transmission but rather a complex combination of many (1) __________ and (2) __________ impulse transmissions.

1. afferent (sensory) 2. efferent (motor) (either order)	**8–26.** REFER TO THE ILLUSTRATION IN FRAME 23 The afferent, or presynaptic fiber, transmits the impulse to the central nervous system. The impulse is then transmitted to an efferent, or postsynaptic fiber, by way of a ________________ junction.
synaptic	**8–27.** You recall from Chapter 3 that a synapse is a functional connection between the axon of one neuron and the dendrites or cell body of another neuron. There is a functional, but not a ________________ connection.
structural	**8–28.** The functional connection between neurons is established by chemical substances present in the ends of the axons of the ________________ or presynaptic fiber. These chemicals are called NEUROTRANSMITTERS because they are responsible for the passage of the neural impulse to the postsynaptic fiber.
afferent	**8–29.** Several chemical substances have been identified as (1) ________________, and they fall into two groups. One type of chemical substance is released from the axon of the afferent fiber and facilitates transmission of the impulse to the (2) ________________ fiber. The other type acts on either the pre- or postsynaptic fiber to prevent the (3) ________________ of the impulse across the synapse or the stimulation of the efferent neuron.
1. neurotransmitters 2. efferent 3. transmission	**8–30.** When an impulse reaches the end of an axon, it may or may not be transmitted to the next neuron, depending on the type of ________________, or chemical substance, released by the axon.
neurotransmitter	**8–31.** Because several axon ends may be functionally connected to a single postsynaptic neuron, impulse transmission also depends on the sum of the action of the types of ________________ that each individual axon releases. (More will be said about neurotransmitters in Part 2 of this chapter.)
neurotransmitters	**8–32.** Most of the nerve cells of your body are within the central nervous system. A group of neuron cell bodies within the central nervous system is a nerve ________________.
nucleus	**8–33.** A bundle of many afferent and efferent fibers is a (1) ________________. A group of sensory cell bodies outside the central nervous system is called a (2) ________________.
1. nerve 2. ganglion	**8–34.** A presynaptic fiber of a simple reflex arc is an (1) ________________ fiber. A postsynaptic fiber of a simple reflex arc is an (2) ________________ fiber.

1. afferent
2. efferent

8–35. For convenience and clarity, nerve tissue is divided grossly into two systems:

The Central Nervous System (CNS)

The Peripheral Nervous System

Although this is a classic gross description, all nerve tissue is one functionally integrated system, the Nervous System.

(No response required—move to frame 36.)

The Brain and the Spinal Cord

8–36. The central nervous system consists of the BRAIN and SPINAL CORD, which are housed in the skull—cranial vault—and vertebral or spinal column, respectively. Afferent impulses, which are transmitted first to the

(1) ____________ ____________, may then either go to the

(2) ____________ or directly to an effector organ (muscle) by

(3) ____________ ____________.

1. spinal cord
2. brain
3. efferent fibers

8–37. The central nervous system, CNS, consists of both nerve cells and nerve fibers—afferent and efferent—that conduct impulses between the principle parts of the CNS, the (1) ____________ and (2) ____________ ____________, before the impulses arrive at the end of their travels.

1. brain
2. spinal cord

8–38. The brain and spinal cord are separated from the bony elements of the cranial vault and vertebral column by three fibrovascular membranes, the MENINGES (men—IN—jes). The central nervous system receives protection and support from the fibrovascular membranes, the (1) ____________, the cranial (2) ____________, and the vertebral column.

1. meninges
2. vault

8–39. A fresh brain removed from the cranial vault is soft and tends to lose its shape because of the forces exerted by its own weight. Within the cranial vault, the brain receives support from the ____________ and a watery fluid between these fibrovascular membranes.

meninges

8–40. The outer MENINX (singular of meninges), the DURA MATER (DU—ra—MA—ter), lies immediately next to the inner surface of the skull and vertebral column. The brain and spinal cord are separated from the bony inner surfaces of the skull and vertebral column by the outer meninx, the

____________ ____________.

dura mater

8–41. The space within your skull—cranial vault—is not occupied by one large unsegmented "glob" of brain. The brain is divided into communicating segments; each segment is protected and supported by the folds of the outer meninx, the ____________ mater.

dura

8–42. A second protective membrane, the ARACHNOID (a—RAK—noid), is separated from the dura mater by a watery fluid. The middle of the three protective meninges is the ________________.

arachnoid

8–43.

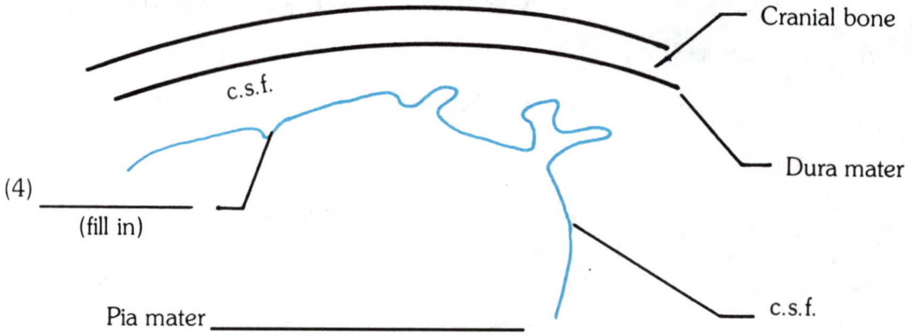

Your brain is supported and protected by three membranes. The outer meninx, the (1) ________________ mater; the mid-meninx, the (2) ________________; and the inner meninx the (3) ________________ ________________.

1. **dura**
2. **arachnoid**
3. **pia mater**
4. **arachnoid**

8–44. A watery fluid containing water and electrolytes—the CEREBROSPINAL (SER—e—bro—SPI—nal) fluid—is located between the outer fibrovascular membrane, the (1) ________________ mater, and the arachnoid. Cerebrospinal (2) ________________ is also located between the arachnoid and the inner fibrovascular membrane, the (3) ________________ mater.

1. **dura**
2. **fluid**
3. **pia**

8–45. The three membranes—the dura, arachnoid, and pia mater—and the (1) ________________ fluid provide (2) ________________ and (3) ________________ for the brain and spinal cord.

1. **cerebrospinal**
2. **protection**
3. **support**

8–46. The pia mater is a highly vascular membrane that closely conforms to the irregular brain surface. Blood vessels extend into the substance of the brain and spinal cord from the (dura/pia) ________________ mater.

pia

8–47. At the tip of the spinal cord, the pia mater extends as a threadlike structure through the arachnoid and dura mater and attaches to the coccyx of the spine.

The area between the pia mater and arachnoid is occupied by (1) ________________ ________________, which can circulate from the brain to the tip of the (2) ________________ ________________.

1. **cerebrospinal fluid**
2. **spinal cord**

8–48.

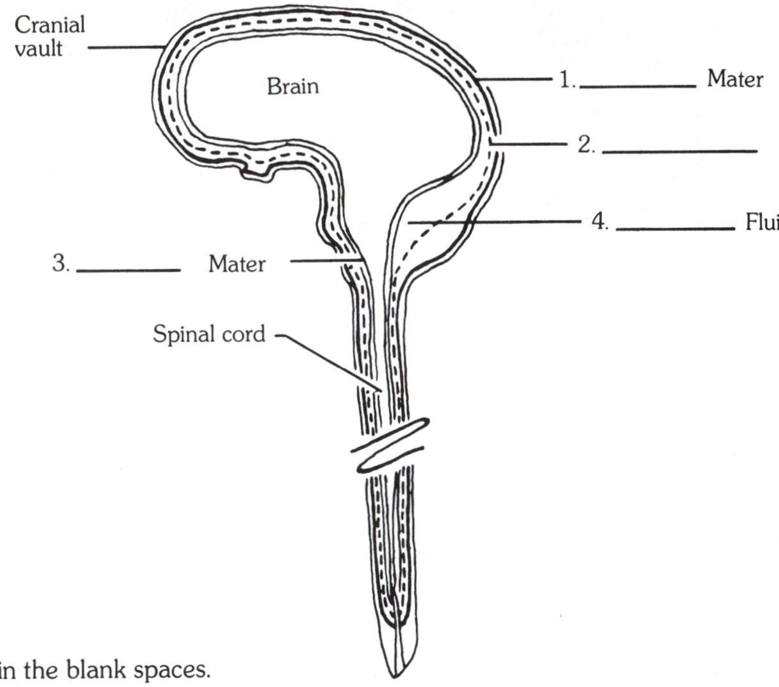

Fill in the blank spaces.

---

1. **dura**
2. **arachnoid**
3. **pia**
4. **cerebrospinal**

8–49. For convenience the following terms will be abbreviated in this program, as they are in medicine:

Central Nervous System .... CNS
Peripheral Nervous System .... PNS
Cerebrospinal fluid .... CSF
(No response required—move to frame 50.)

8–50. The space between the arachnoid membrane and the pia mater is referred to as the subarachnoid space.

Cerebrospinal fluid, (1) _______ _______ _______, is located in the space between the arachnoid and the pia mater, the (2) _______ space, and in the space between the dura mater and the arachnoid, the (3) _______ dural space.

---

1. **CSF**
2. **subarachnoid**
3. **sub**

8–51. Located within the brain and spinal cord are cavities that also contain CSF. The CSF within these cavities of the brain and spinal cord (balance/unbalance) _______ the effect of the CSF on the outer surface of the brain.

---

balance

8–52. The cavities within the brain that contain CSF are called VENTRICLES (VEN—tri—kls), and the cavity within the spinal cord that contains CSF is called the CENTRAL CANAL. The CSF is produced in the cavities within the brain called _______.

ventricles

8–53. CSF is located in the subarachnoid and subdural spaces and also within the brain and the spinal cord in the (1) ____________________ and the (2) ____________________ ____________________, respectively.

1. ventricles
2. central canal

8–54. If the brain and spinal cord were dissected as illustrated in the drawing below, you could locate the CSF, brain cavities, and the spinal canal.

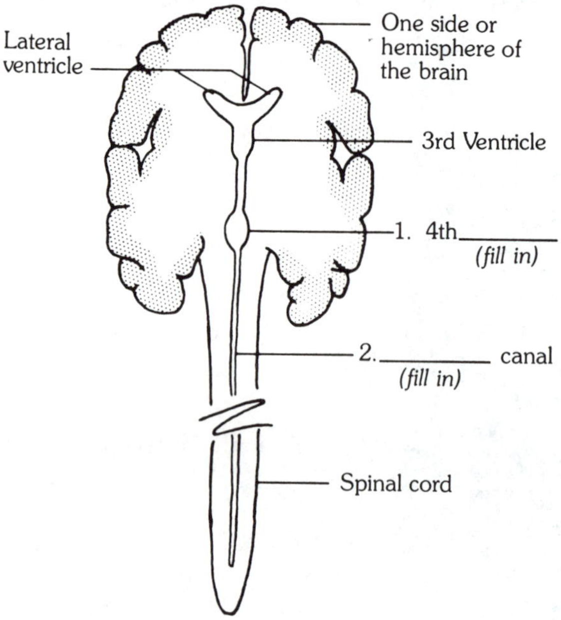

Fill in the blanks.

1. ventricle
2. central

8–55. The CSF that is formed in the ____________________ flows to the subarachnoid space.

ventricles

8–56. Some of the CSF flows upward over the surface of the brain, and some of the CSF flows down in the (1) ____________________ space and the (2) ____________________ ____________________ of the spinal cord.

1. subarachnoid
2. central canal

8–57. The fluid pressure created by the CSF on the outer surface of the brain is balanced by the CSF fluid pressure in the (1) ____________________ of the brain and (2) ____________________ ____________________ of the spinal cord.

1. ventricles
2. central canal

8–58. If the ventricles of the brain continued to produce CSF and if the body had no functional means of reabsorbing the excess, increased pressure would result.

There are within the cranial vault three major venous sinuses that function in part by absorbing CSF, thereby preventing any increase in ____________________.

pressure

8–59. Note that the CSF is passing from the subarachnoid space through (1) ________________ ________________ into a (2) ________________ ________________.

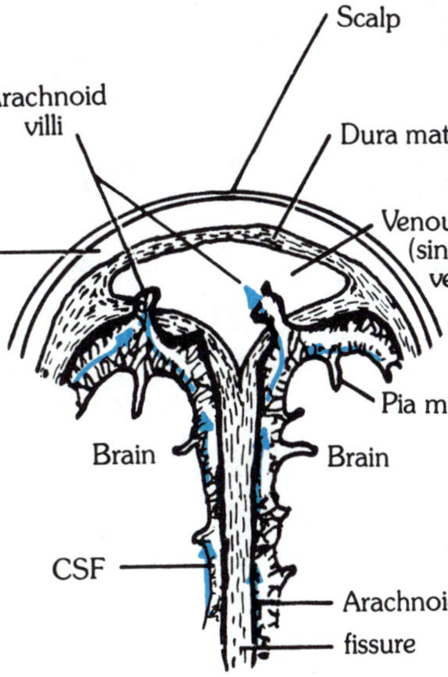

1. arachnoid villi
2. venous sinus

8–60. After the CSF has been circulated over the surfaces of the brain and the spinal cord, the fluid is then absorbed through the (1) ________________ ________________ into the (2) ________________ ________________.

1. arachnoid villi
2. venous sinus

8–61. The CSF within and on the surface of the brain and spinal cord provides a balanced amount of (1) ________________ and (2) ________________ for these organs.

1. support
2. protection

8–62. You may want to take a break before you continue. The next series of frames will deal with the brain.

The Brain

8–63. The human brain is divided into two hemispheres called CEREBRAL (SER—e—bral) hemispheres. The combined weight of the two ________________ hemispheres is about 1350 gm or approximately 3 lb for the average adult human.

cerebral

8–64. The surface of each cerebral hemisphere is irregular or convoluted. A deep furrow is called a (1) ________________ and a shallow indentation, a (2) ________________.

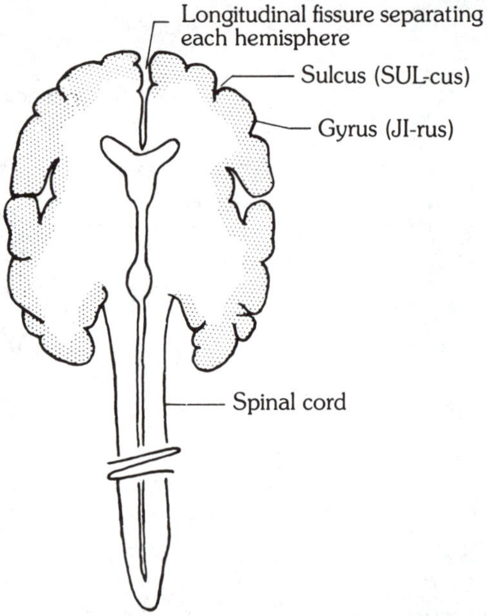

1. fissure
2. sulcus

8–65. Refer to the illustration in frame 64.

A prominent rounded elevation on the surface of a cerebral hemisphere is called a ________________.

gyrus

8–66. Note: The plural of sulcus is <u>SULCI</u>, and the plural of gyrus is <u>GYRI</u>.

(No response required—move to frame 67.)

8–67.

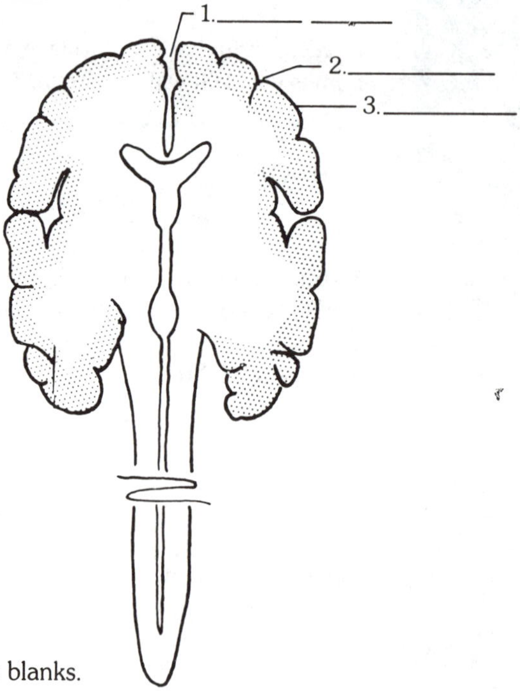

Fill in the three blanks.

1. **longitudinal fissure**
2. **sulcus**
3. **gyrus**

8–68. Each cerebral hemisphere has four main subdivisions or LOBES. In the illustration below locate each lobe.

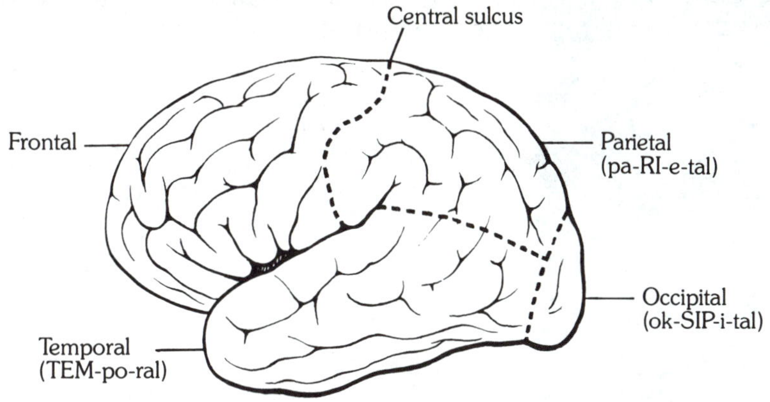

No answer required. Move to frame 69.

8–69. The front and temple sections of your skull are familiar to you. Label the two subdivisions or (1) ____________ of the cerebral hemisphere illustrated below.

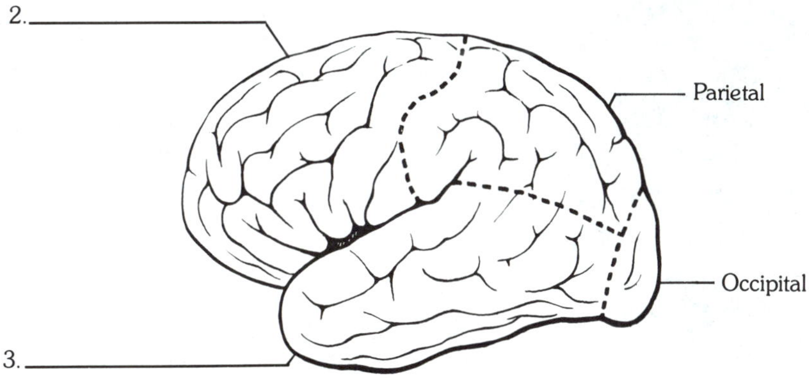

1. **lobes**
2. **frontal**
3. **temporal**

8–70. Label the four lobes of the cerebral hemisphere illustrated below.

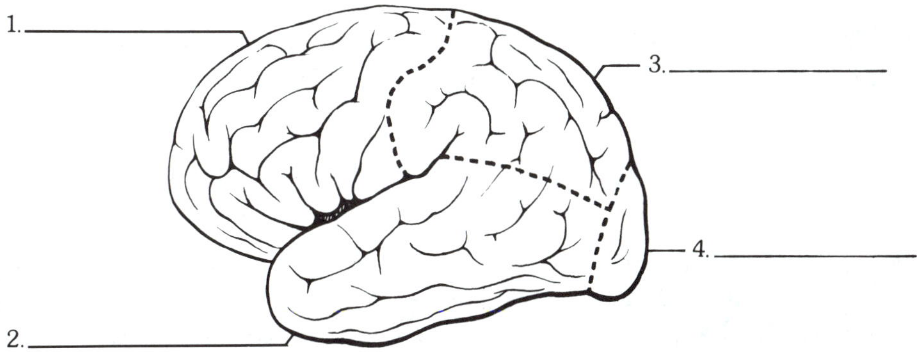

1. **frontal**
2. **temporal**
3. **parietal**
4. **occipital**

8–71. Each cerebral lobe has a different functional activity. For example, the frontal lobe is a coordinating center for moral activities, the temporal lobe—hearing, the parietal (pa—RI—e—tal) lobe—touch, and the occipital (oc—SIP—i—tal) lobe—sight.

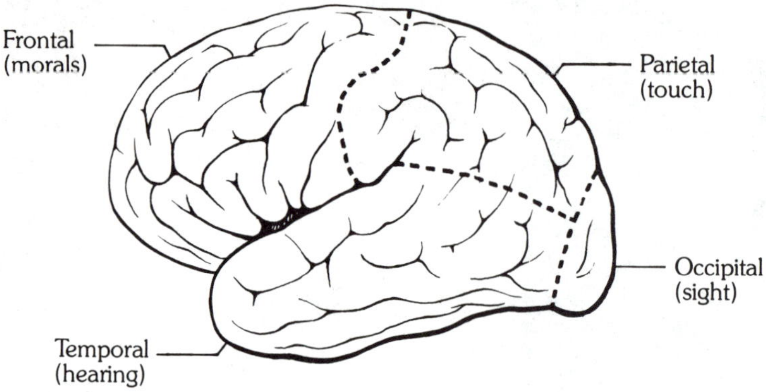

8–72.

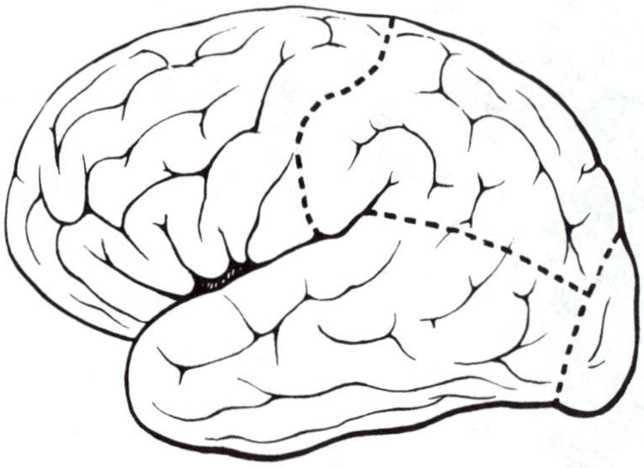

Match the cerebral lobe areas in the left hand column with the functional activities in the right hand column.

a. Frontal	1. Touch
b. Occipital	2. Sight
c. Temporal	3. Morals Coordinating Center
d. Parietal	4. Hearing

a—3
b—2
c—4
d—1

8–73. When you become concerned about the ethical and moral standards of your community, the (1) ________________ lobe of your cerebrum is activated.

Hearing is centered in the (2) ________________ lobe and sight in the (3) ________________ lobe of your cerebral hemispheres.

Your sense of touch is located in the (4) ________________ lobe.

1. **frontal**
2. **temporal**
3. **occipital**
4. **parietal**

8–74. The activities of the cerebral lobes are not as specific as the examples used in this program. Each lobe has additional activities not included here. You have been exposed to the more commonly related ________________ of each of the cerebral lobes.

activities

8–75. A MESIAL (MES—i—al)—middle—view of one hemisphere of the brain reveals another lobe, the ____________________, and the brain stem.

NOTE: You should remember that this illustration is representative of but one of the two cerebral hemispheres.

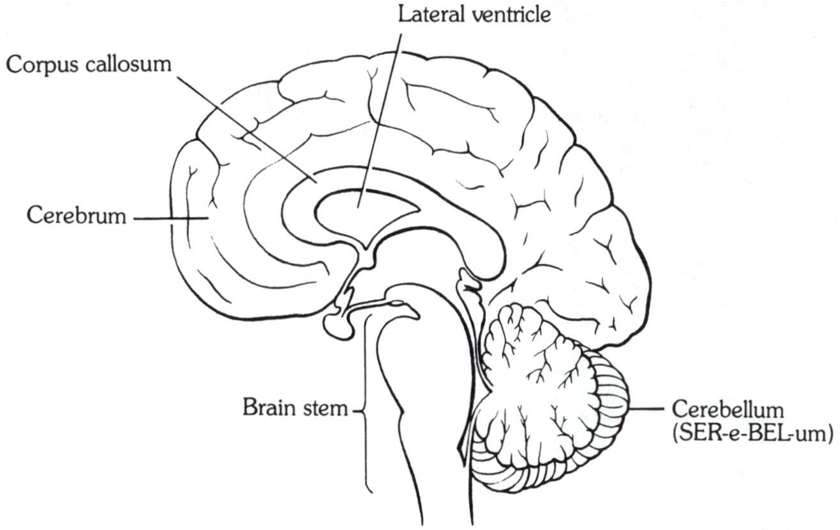

cerebellum

8–76. A human brain has, including the two lobes of the cerebellum, a total of how many lobes.

a. five lobes
b. ten lobes
c. eight lobes
(Select either a, b, or c and move to frame 77.)

b.

8–77. There are ten lobes or subdivisions of the human brain—
(1) ____________________ (number) lobes in each cerebral hemisphere and the two
(2) ____________________ lobes.

1. four
2. cerebellar

8–78. The cerebellum, like the cerebral lobes, has a specific function—that of maintaining coordination and balance. If you are unable to walk in a straight line, the (cerebrum/cerebellum) ____________________ may not be functioning properly.

cerebellum

8–79.

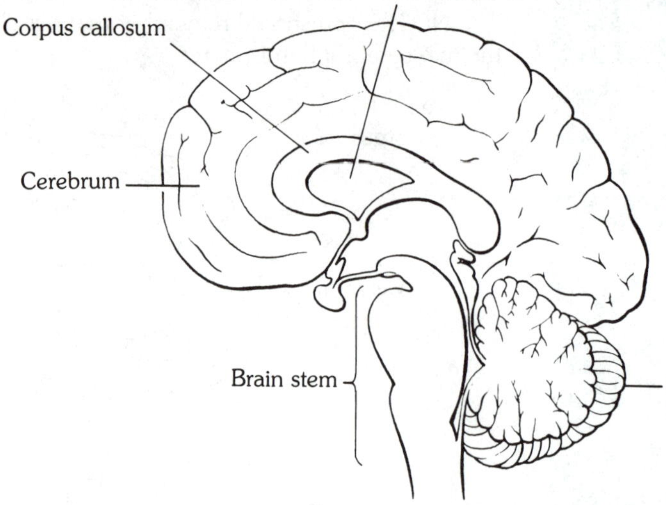

An anatomical structure, the CORPUS CALLOSUM (KOR—pus ka—LO—sum), lies superior to the lateral ventricle. This structure joins together the cerebral hemispheres. Locate the corpus callosum in the illustration above.

(No response required—move to frame 80.)

8–80. The corpus callosum is also a communication bridge between the cerebral hemispheres. An impulse from the left side of your brain may be transferred to the right cerebral hemisphere by way of the ________________ ________________.

---

corpus callosum

8–81. A mesial surface illustration of the human brain usually contains shaded and nonshaded areas. The nonshaded, or white, areas of the brain are concentrations of myelinated nerve fibers called WHITE MATTER. The corpus callosum contains mostly myelinated nerve fibers and is therefore a large mass of ________________ ________________.

---

white matter

8–82. GRAY MATTER is composed of cell bodies; white matter is composed of concentrations of myelinated nerve ________________.

---

fibers

8–83. Nerve cell fibers in the brain and spinal cord are primaily located in the (1) ________________ (white/gray) matter.

Nuclei in the brain and spinal cord are primarily located in the (2) ________________ (white/gray) matter.

white
gray

8–84. Locate and label the gray and white matter in the illustration below.

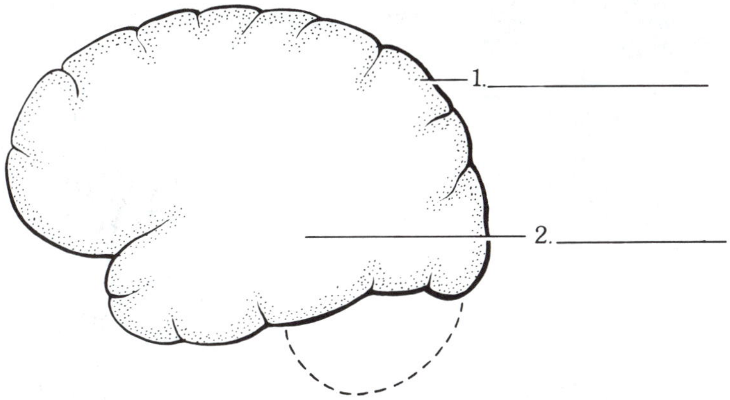

1. gray matter
2. white mater

8–85.

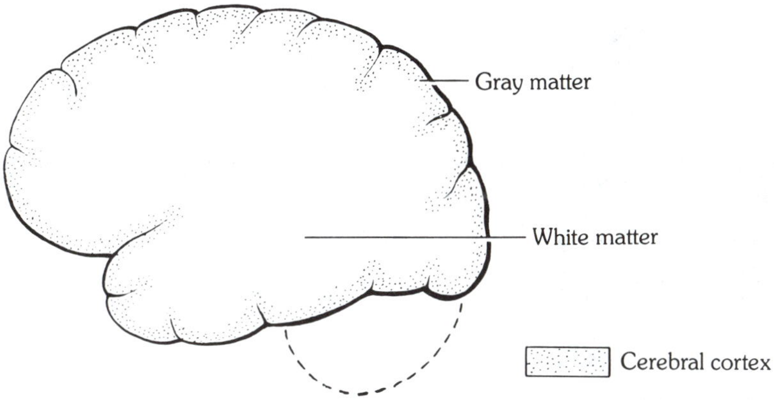

The outer surface of the cerebral hemispheres—the cerebrum—is covered with gray matter called the cerebral ________________.

cortex

8–86. If you refer to the drawing in frame 85, you will find centers of gray matter buried within, or mixed with, the white matter. These areas of gray matter are collections of (nuclei/cell processes) ________________ of the CNS.

nuclei

8–87. The outer portion of the cerebrum or (1) ____________ ____________ is composed primarily of cell bodies or (2) ____________ matter.

A large mass of myelinated nerve fibers or (3) ________________ matter, the (4)____________ ____________, is one communicating link between the two cerebral hemispheres.

1. cerebral cortex
2. gray
3. white
4. corpus callosum

8–88. There are several bridges of white matter that connect the cerebellum to the BRAIN STEM. These bridges would, like the corpus callosum, be composed of (nerve fibers/cell bodies) ____________ ____________.

nerve fibers

8-89. The connective bridges, peduncles, are composed of (1) ________________ matter. These peduncles connect the cerebellum to the brain (2) ________________.

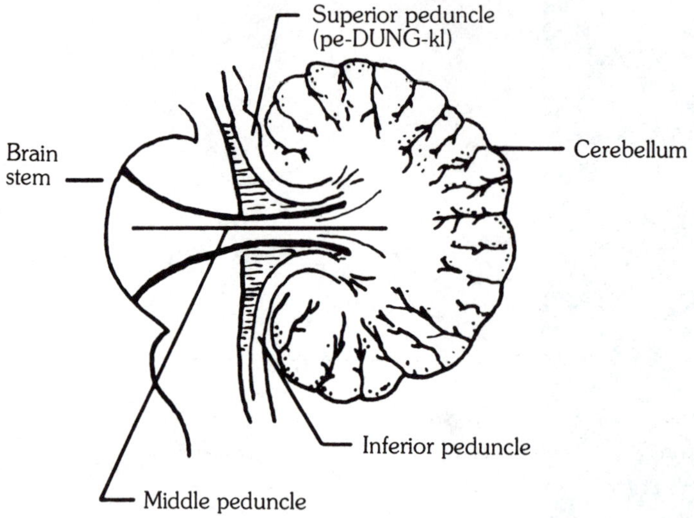

1. white
2. stem

8-90. The three bridges that connect the cerebellum to the brain stem are called ________________.

peduncles

8-91. Impulse communication between the cerebellum and brain stem is maintained by way of the masses of white matter, the (peduncles/corpus callosum) ________________.

peduncles

8-92. The cervical—neck—region of the spinal cord is connected to the brain proper by the ________________ ________________.

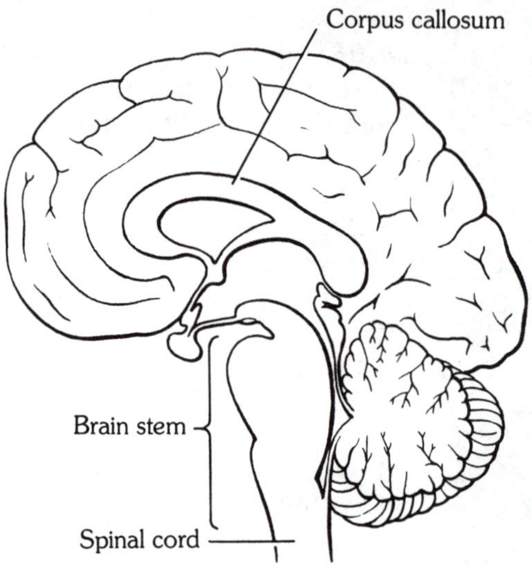

brain stem

8–93.

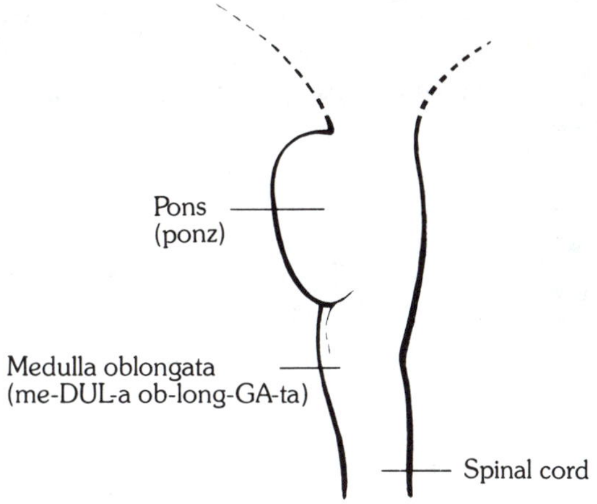

The brain stem is composed of two parts, the (1) ________________ and the (2) ________________ ________________.

---

1. pons
2. medulla oblongata

8–94. The area of the brain stem that is in direct contact with the spinal cord is the (pons/medulla oblongata) ________________

---

medulla oblongata

8–95. The medulla oblongata is the connecting link between the spinal cord and the ________________ of the brain stem.

---

pons

8–96. **Brain Stem**

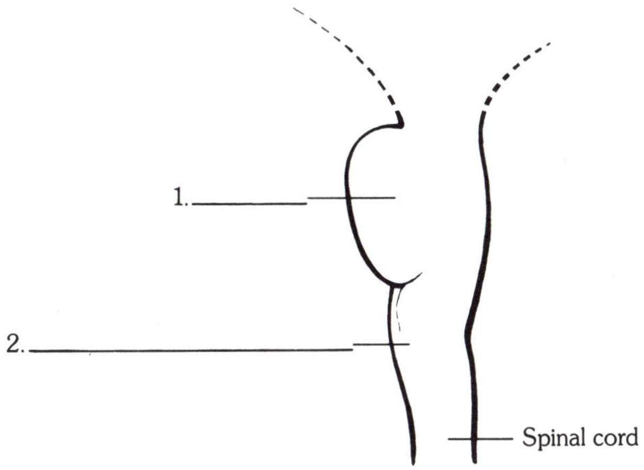

Fill in the numbered blanks.

---

1. pons
2. medulla oblongata

8–97. The surface of the pons is a mass of white matter; it is therefore composed of nerve (1) ________________.

There is a gray matter or nerve cell (2) ________________ within the pons.

---

1. fibers
2. bodies

8–98. The medulla oblongata contains nuclei for transmitting impulses from the spinal cord. This section of the brain stem must, therefore, contain (gray/white) ________________ matter.

gray

8–99. Sensory impulses are transmitted by (1) ______________ nerve fibers to the spinal cord. These impulses then may be communicated to the brain by passing up the spinal cord and through the (2) ______________ ______________ and (3) ______________ of the brain stem.

1. afferent
2. medulla oblongata
3. pons

8–100. It is not difficult to determine one function of the brain stem because of its position. The brain stem is the ______________ "link" between the brain and the spinal cord.

communications

8–101. In addition to its function as a conductor of impulses, the medulla oblongata is a reflex center for the circulatory and respiratory systems. Your rate of (1) ______________ beat and your rate of (2) ______________ are influenced by the medulla oblongata.

1. heart
2. breathing

8–102. The pons has two functions: conducting impulses from the medulla oblongata to the (1) ______________ and aiding in the (2) ______________ of your breathing rate.

1. brain
2. regulation

8–103. Impulses that originate in the nuclei of the medulla oblongata are transmitted to the brain by the ______________.

pons

8–104. Superior and anterior to the brain stem, on either side of the 3rd ventricle, are the THALAMUS (THAL—a—mus) and HYPOTHALAMUS (HI—po—THAL—a—mus), areas that are principally masses of gray matter.

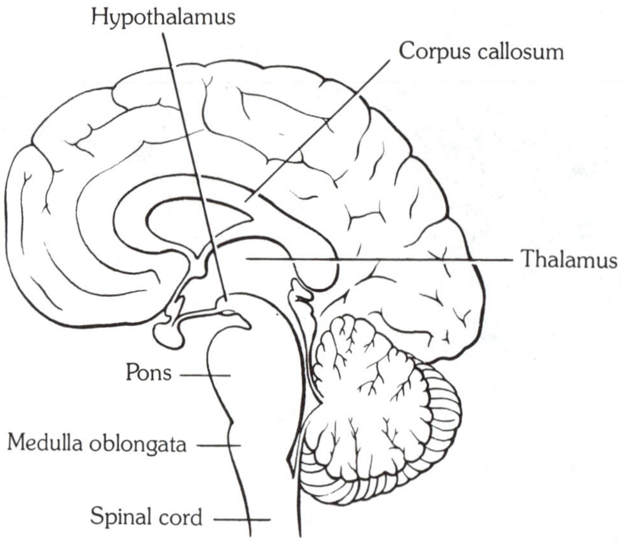

(No response required—move to frame 105.)

8–105. Locate and label the thalamus and hypothalamus. After you have done so, check your labeling with the illustration in frame 104.

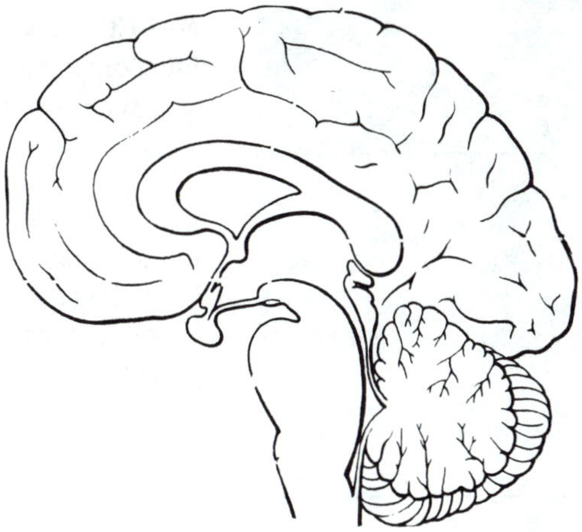

---

check illustration in frame 104.

8–106. In between the corpus callosum and the brain stem, on either side of the 3rd ventricle, are masses of gray matter, the (1) ________________ and the (2) ________________.

---

1. thalamus
2. hypothalamus (either order)

8–107. A mass of gray matter, the (thalamus/hypothalamus) ________________, is directly below the corpus callosum.

Refer to the illustration in frame 104 if you are unsure of the location of either the thalamus or hypothalamus.

---

thalamus

8–108. Below the thalamus is another mass of gray matter, the ________________.

---

hypothalamus

8–109. Although the thalamus and hypothalamus are covered by a thin layer of white matter, they contain a number of cell bodies and therefore consist largely of ________________ ________________.

---

gray matter

8–110. The hypothalamus contains nuclei that relay messages from the cerebral cortex.

Special hypothalamic nuclei control certain involuntary activities, especially water balance.

The hypothalamus also contains reflex centers for the regulation of body temperature and blood pressure.

The thalamus contains nuclei that transmit afferent ________________ impulses to the sensory areas of the cerebral cortex.

---

sensory

8–111. Elevation or lowering of the body temperature is influenced by the:

a. thalamus (move to frame 113)
b. hypothalamus (move to frame 112)

---

b.

8–112. Right! The hypothalamus does contain nuclei that influence, but not entirely control, the temperature of the body.

(No response required—move to frame 114.)

a.

8–113. The thalamus does not influence the temperature of the body; this function is one of the activities of the hypothalamus.

(No response required—move to frame 114.)

8–114. The thalamus can influence the sleep and appetite control centers located in the nuclei of the hypothalamus.

Thalamic functions:

1. relay station for (1) ________________ impulses being transmitted to the cerebral cortex.
2. influence the (2) ________________ and (3) ________________ control centers.

1. sensory
2. sleep
3. appetite

8–115. Hypothalamic functions:

1. relay messages from the ________________ cortex.
2. control certain ________________ activities, such as water balance in the body.
3. contain ________________ centers that influence body temperature and blood pressure.
4. act as ________________ centers for sleep and appetite.

1. cerebral
2. involuntary
3. reflex
4. control

8–116. The thalamus and hypothalamus are located on either side of the 3rd (1) ________________, between the corpus callosum and the brain (2) ________________.

1. ventricle
2. stem

8–117. The two cerebral hemispheres of the human brain are convoluted with deep furrows—(1) ________________, shallow indentations—(2) ________________, and rounded elevations—gyri.

1. fissures
2. sulci

8–118. Each cerebral hemisphere has (1) ________________ (number) main subdivisions or (2) ________________.

1. four
2. lobes

8–119. In the illustration below, label the lobes of the cerebral hemisphere, and below each label identify the functional activity of each lobe.

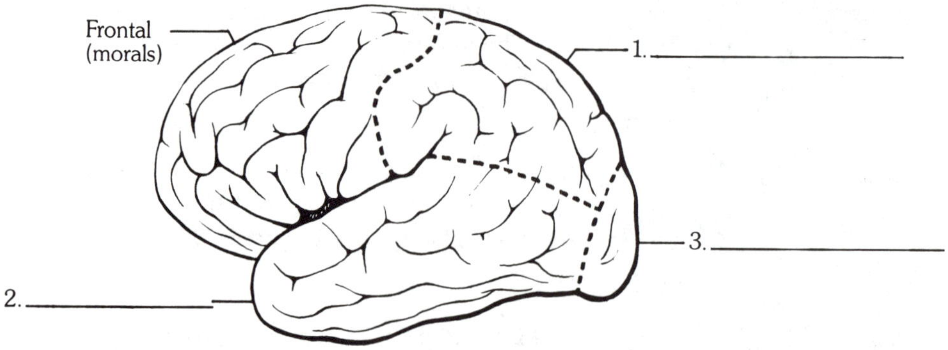

1. parietal touch **2. temporal hearing** **3. occipital sight**	**8–120.** There are ten main lobes of the human brain: the four lobes found in each of the two cerebral hemispheres and the two lobes of the ________.
cerebellum	**8–121.** The cerebellum functions as a coordinating center of: a. touch b. sight c. balance d. ethics (Select either a, b, c, or d and move to frame 122.)
c.	**8–122.** A large band of white matter that joins each cerebral hemisphere is the: a. superior peduncles b. corpus callosum (Select either a or b, and move to frame 123.)
b.	**8–123.** The corpus callosum may act as an impulse ________ between the two cerebral hemispheres.
conductor (communicator)	**8–124.** The corpus callosum is composed principally of nerve fibers, which are (gray/white) ________ matter.
white	**8–125.** A group of neuron cell bodies within the CNS is called a (1) ________. Such an aggregate of neuron cell bodies is (gray/white) (2) ________ matter.
1. nucleus **2. gray**	**8–126.** The cerebral cortex is the outer limit of the: a. cerebrum b. cerebellum (Select either a or b, and move to frame 127.)
a.	**8–127.** The cerebral cortex is composed primarily of cell bodies, or ________ matter.
gray	**8–128.** The bridges of white matter that connect the cerebellum with the brain stem are called: a. corpus callosum b. peduncles
b.	**8–129.** Communication of impulses between the cerebellum and the brain stem travels by way of the ________.

peduncles

8–130. The cervical section of the spinal cord is connected to the brain proper by the __________ __________.

brain stem

8–131. The upper portion of the brain stem is the (1) __________ and the lower section is the (2) __________ __________.

1. pons
2. medulla oblongata

8–132. Because of the nuclei present within the medulla oblongata, this section of the brain stem contains __________ __________.

gray matter

8–133. Of the following, what two functions can be related to the functions of the medulla oblongata?

a. impulse communicator
b. sight center
c. touch center
d. reflex center for respiration and circulation

a. and d.

8–134.

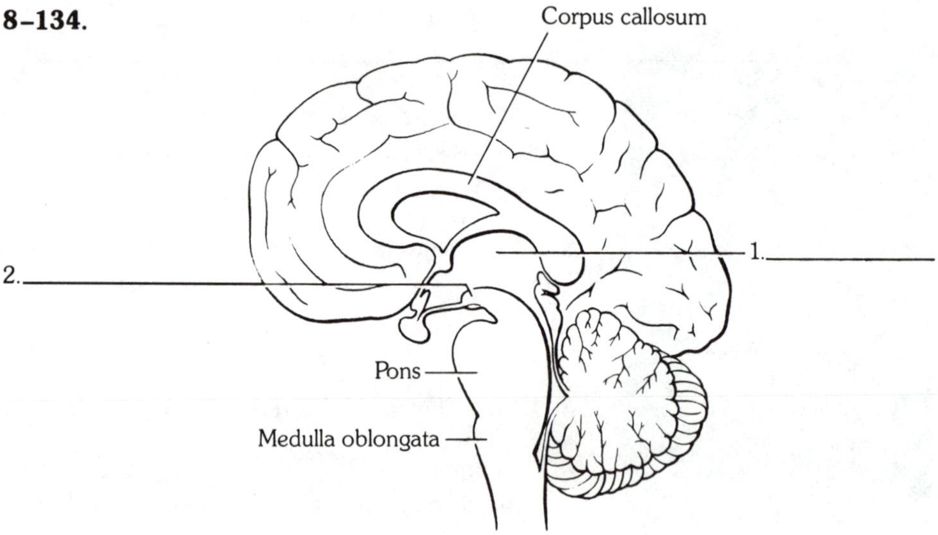

Fill in the unlabeled sections of this drawing.

1. thalamus
2. hypothalamus

8–135. The two masses of gray matter on either side of the 3rd ventricle are the (1) __________ and (2) __________.

1. hypothalamus
2. thalamus
(either order)

8–136. BRAIN AREA

a. Thalamus
b. Hypothalamus

FUNCTIONS

1. Relay messages from cerebral cortex
2. Control body water balance
3. Reflex center for the regulation of temperature
4. Transmits sensory impulses to the cerebral cortex
5. Influences the sleep and appetite control centers

Match the brain areas with the related functions.

The Spinal Cord

a. 4 and 5
b. 1, 2, and 3

8–137. Below the medulla oblongata and continuous with the brain stem is the second major part of the CNS, the ________ ________.

spinal cord

8–138. The spinal cord is approximately 18 inches long and ¾ of an inch in diameter, and because the vertebral column outgrows the spinal cord, the spinal cord in adults (does/does not) ________ extend to the coccyx.

does not

8–139. The spinal cord, like the brain, is protected by bone. The skeletal structure that protects the spinal cord is the ________ ________.

vertebral column

8–140. You will recall, from the discussion of the vertebral column in Chapter 5, that a typical vertebra consists of a rounded (1) ________ anteriorly or ventrally and a (2) ________ process at the posterior portion or dorsally.

Thoracic Vertebra

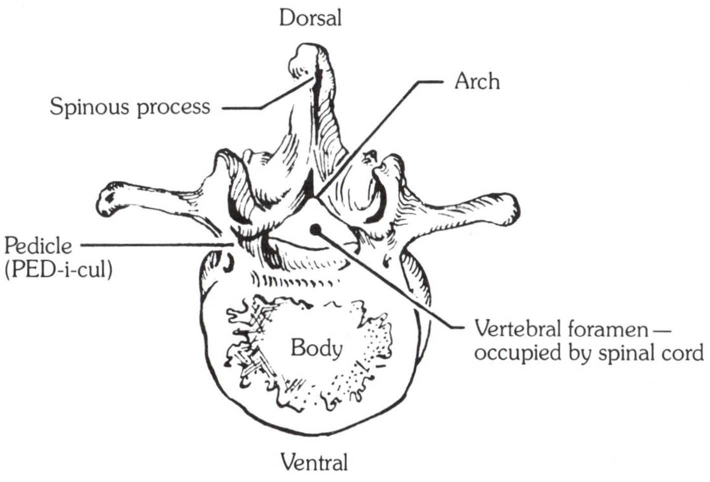

1. body
2. spinous

8–141. The PEDICLE (PED—i—cul), which is located between the body and spinous process of the vertebra, has a notch on the inferior (lower) and superior (upper) surface. When the surfaces of this notch are opposite a corresponding notch, in vertebrae either above or below, a foramen (small opening) is formed through which ________________ nerves pass to and from the spinal cord.

spinal

8–142.

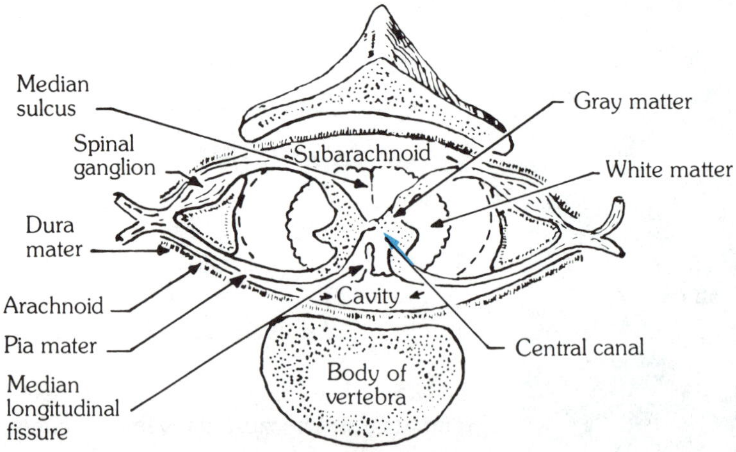

The anterior or ventral surface of the spinal cord has a deep furrow extending throughout its entire length, the (1) ________________ longitudinal (2) ________________.

The posterior surface of the spinal cord has a shallow furrow, the (3) ________________ ________________, which also extends the entire length of the spinal cord.

1. median
2. fissure
3. median sulcus

8–143.

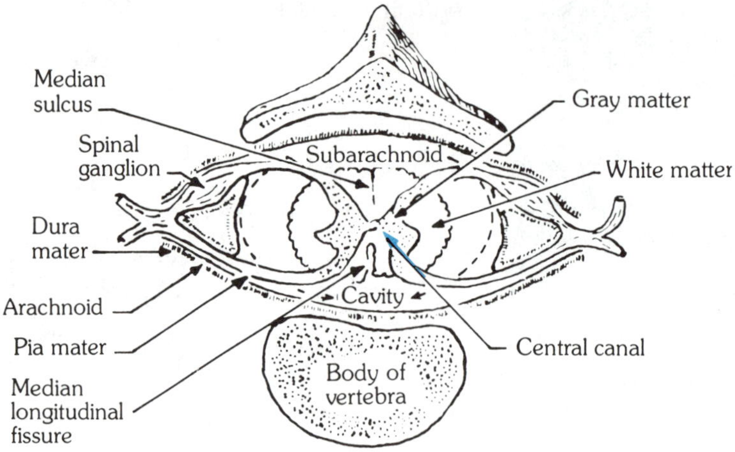

A cross section of the spinal cord reveals (1) ________________ matter surrounded by (2) ________________ matter. This arrangement of white and gray matter in the spinal cord is (the same as/opposite from) (3) ________________ the arrangement of white and gray matter in the human brain.

(1) gray
(2) white
(3) opposite from

8–144. Refer to the illustration in frame 143.

This cross section of the spinal cord also reveals that the spinal cord receives protection and support from the (1) ____________ mater, (2) ____________ mater, arachnoid and (3) ______ ______ ______.

1. pia
2. dura
(either order)
3. CSF

8–145. The gray matter of the spinal cord conforms roughly to the shape of the letter "H" (refer to the illustration in frame 143), with a central cavity in the crossbar), the ____________ ____________, which contains CSF.

central canal

8–146. The dorsal and ventral portions of each lateral (arms of the "H") half of the gray matter are commonly called "HORNS." However, because the gray matter extends throughout the entire ____________ of the spinal cord, these areas are also commoniy referred to as COLUMNS.

length

8–147. The horns of gray matter or (1) ____________ of the spinal cord contain, as does all gray matter, nerve (2) ____________ ____________.

1. columns
2. cell bodies

8–148.

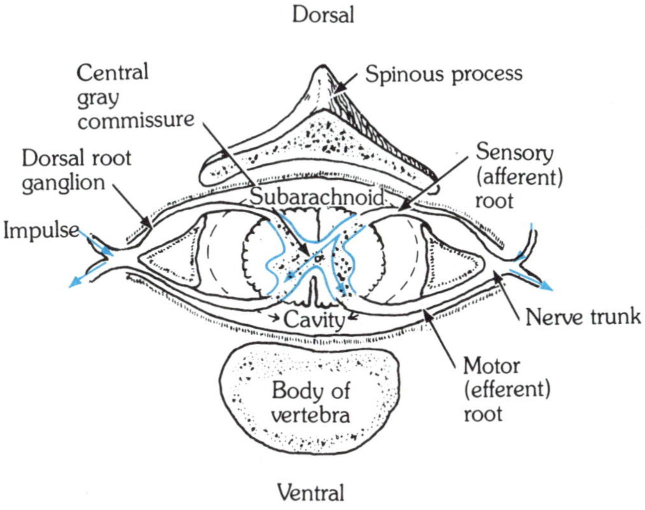

Cell bodies of ventral columns are connected to (1) ____________ nerve fibers or roots.

Cell bodies of dorsal columns are connected to (2) ____________ nerve fibers or roots.

1. motor (efferent)
2. sensory (afferent)

8–149. A sensory impulse enters the dorsal spinal column by way of a ganglion known as the ____________ ____________ ____________. This impulse may then:

1.

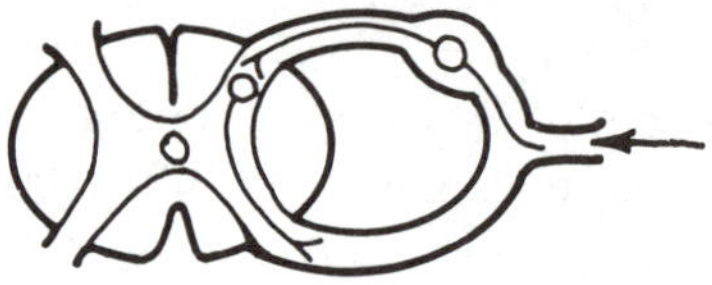

travel to a ventral column on the same side

3.

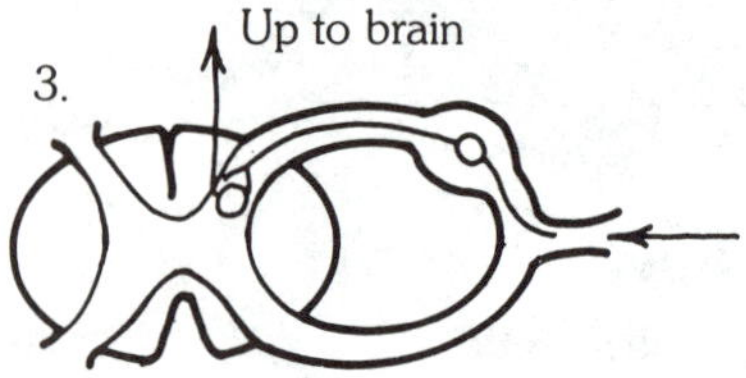

travel up to the brain

2.

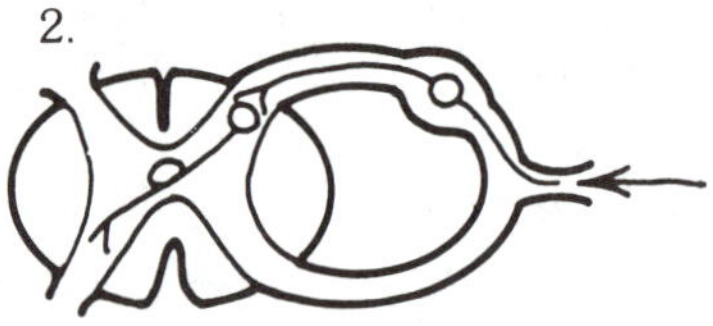

cross over to a ventral column on the opposite side

4.

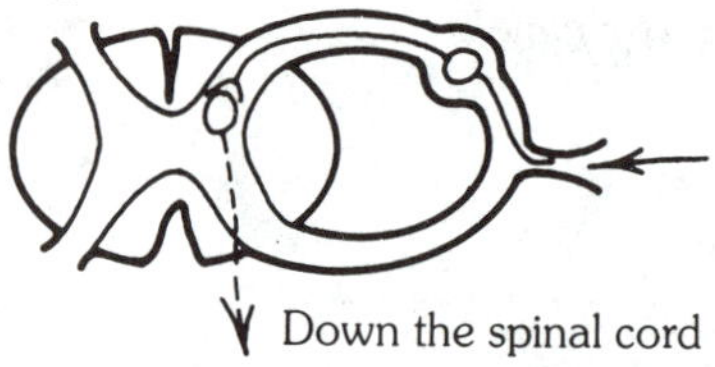

travel down the spinal column

Or all four possible routes may occur prior to a motor impulse leaving the spinal cord on a ventral root.

---

dorsal root ganglion

8–150. 1. Dorsal column
2. Ventral column

a. Sensory fiber
b. Motor fiber
c. efferent fiber
d. afferent fiber

Match items in the left-hand column with the correct items in the right-hand column.

---

1. a. and d.
2. b. and c.

8–151. Refer to the illustration in frame 149 to help you respond to the next few frames.

An impulse may synapse from a sensory to a motor root on the same side of the spinal cord, or a sensory impulse may cross over to a motor root on the opposite side through the CENTRAL COMMISSURE (KOM—i—SHOOR).

8–152. Impulses are conducted from one neuron within the spinal cord to another neuron within the spinal cord by INTERNUNCIAL (in—tur—NUN—si—al) neurons.

A sensory impulse entering by way of the (ventral/dorsal)
(1) ____________ spinal root may pass through an
(2) ____________ neuron before reaching an efferent or ventral root.

---

1. dorsal
2. internuncial

8–153. The cell bodies of the ventral spinal columns belong primarily to ____________ root fibers, which are distributed to skeletal muscles.

motor
(efferent)

8–154.

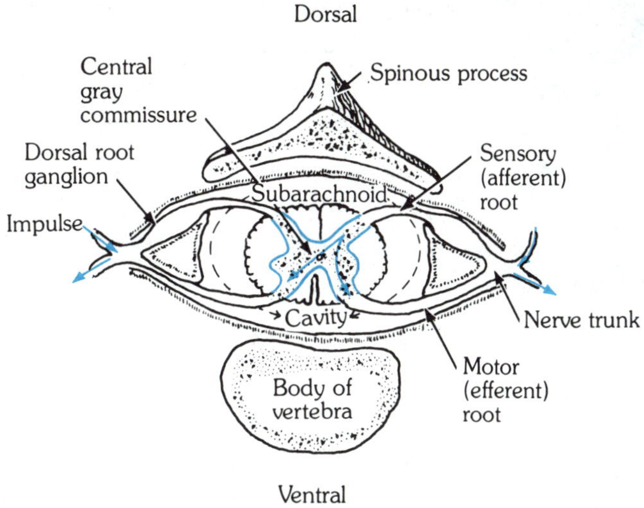

The cell bodies of all sensory neurons of spinal nerves are lodged in a (1) __________ __________ in a (2) __________ (dorsal/ventral) root.

---

1. spinal ganglion
2. dorsal

8–155. The spinal cord is a reflex center. Sensory impulses can pass from a dorsal root ganglion by an (1) __________ neuron to a ventral (2) __________ root and leave the cord without going to the (3) __________.

---

1. internuncial
2. motor
3. brain

8–156. Impulses may enter and leave the spinal cord on the same side, or an impulse may cross over to the opposite side of the spinal cord by the central __________ __________.

---

gray commissure

8–157. The white matter of the spinal cord, which is located __________ (within/around) the gray matter, consists of nerve fibers coursing in all directions but mostly longitudinally.

---

around

8–158. Impulses that are transmitted to or from the brain travel along the longitudinal TRACTS of the __________ matter.

---

white

8–159. Large collections of spinal cord white matter are referred to as the FUNICULI (fu—NIK—u—li).

The nerve pathways to or from the brain called (1) __________ are located in the (2) __________ of the spinal cord.

NOTE: plural-funiculi—singular-funiculus

---

1. tracts
2. funiculi

8–160. Spinal "horns" or columns refer to the (1) __________ (gray/white) matter. The large collections of white matter within the spinal cord, the (2) __________, contain (3) __________ that carry impulses to and from the brain.

1. **gray** 2. **funiculi** 3. **tracts**	**8–161.** Tracts that ascend to the brain would be called (1) ________________ (sensory/motor) tracts, and tracts that descend from the brain would be called (2) ________________ (sensory/motor) tracts.
1. **sensory** 2. **motor**	**8–162.** The SPINOTHALAMIC tract is a sensory tract because it conducts impulses from the (1) ________________ ________________ to the (2) ________________ portion of the brain.
1. **spinal cord** 2. **thalamic**	**8–163.** The CORTICOSPINAL tract conducts impulses from the brain down the spinal cord. This tract is a (sensory/motor) ________________ tract.
motor	**8–164.** Your spinal cord: a. extends the entire length of your vertebral column. b. does not extend the entire length of your vertebral column. (Select either a or b, and then move to frame 165.)
b.	**8–165.** 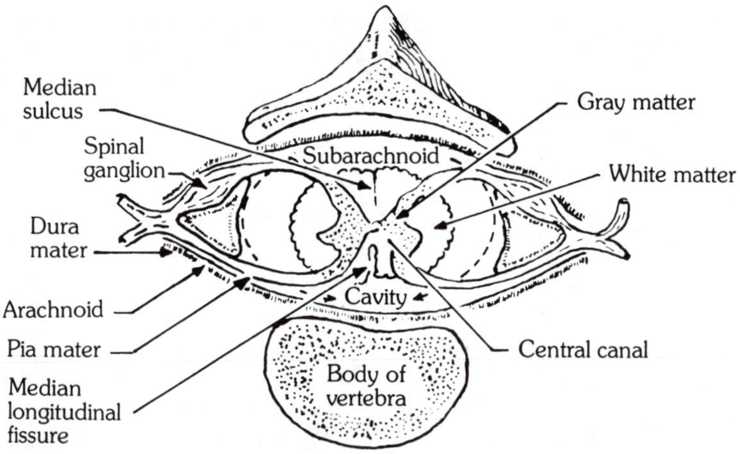 Spinal nerves enter and leave the vertebral column by passing through notches in the (pedicles/body) ________________ of the vertebra.
pedicles	**8–166.** The mid-posterior surface of the spinal cord has a shallow furrow, the median (1) ________________, which extends the entire length of the cord. The mid-anterior surface of the spinal cord has a deep furrow, the (2) ________________ longitudinal (3) ________________, which also extends the entire length of the cord.
1. **sulcus** 2. **median** 3. **fissure**	**8–167.** The gray matter of the spinal cord is located (inside/outside) ________________ a quantity of white matter.

inside	**8–168.** The three meninges, the (1) ________ mater, the (2) ________ mater, and the (3) ________, and the water-electrolyte liquid, the (4) ____ ____ ____, protect and support the spinal cord.
1. dura **2. pia** **(either order)** **3. arachnoid** **4. CSF**	**8–169.** Located in the center of the spinal cord gray matter is the ________ ________, which also contains CSF.
central canal	**8–170.** The dorsal and ventral portions of the gray component of the spinal cord are called: a. "horns" b. tracts c. columns (Select either a, b, or c, and then move to frame 171.)
a. or c.	**8–171.** The dorsal spinal columns contain the (1) ________ (motor/sensory) cell bodies. FIbers of these cell bodies synapse with internuncial neurons, which in turn (2) ________ with cell bodies of the (3) ________ root. Ventral spinal columns contain cell bodies of the (4) ________ (motor/sensory) root.
1. sensory **2. synapse** **3. motor (ventral)** **4. motor**	**8–172.** Impulses that enter the spinal cord can pass, by an internuncial neuron to: a. efferent fibers on the same side of the cord b. efferent fibers on the opposite side of the cord c. both of the above (Select either a, b, or c, and then move to frame 173.)
c.	**8–173.** A sensory impulse crossing over from a dorsal column to the ventral column of the opposite side of the cord must pass through the central gray ________ (commissure/anterior root).
commissure	**8–174.** Spinal (1) ________, which carry impulses to or from the brain, are located in the (2) ________ (gray/white) matter.
1. tracts **2. white**	**8–175.** The fibrous tracts are located in the large areas of the spinal cord white matter the: a. arachnoid cavity b. dorsal root c. funiculi (Select either a, b, or c, and then move to frame 176.)

c.	**8–176.** A corticospinal tract is a: a. sensory tract to the brain b. motor tract from the brain (Select either a or b, and then move to frame 177.)
b.	**8–177.** You have reached the end of Chapter 8, Part 1.

Part 2

The Peripheral Nervous System

	8–178. Although the nervous system is functionally integrated, nerve tissue is structurally divided into two systems, the central nervous system and the ____________ nervous system.
peripheral	**8–179.** The central nervous system consists of the (1) ____________ and the (2) ____________ ____________. The peripheral nervous system consists of the cranial nerves and the spinal nerves.
1. brain 2. spinal cord	**8–180.** The peripheral nervous system, then, includes nerves that connect the body parts to the ____________ ____________ ____________.
central nervous system	**8–181.** One division of the peripheral nervous system connects the CNS with the body wall, or soma. This division is called the SOMATIC SYSTEM and carries impulses to and from the skin and skeletal muscle (the soma) and the CNS. (No response required, move to frame 182.)
	8–182. You must remember that a nerve is composed of a bundle of nerve (1) ____________. The (2) ____________ ____________ that comprise a nerve can be (3) a. sensory only. b. motor only. c. both sensory and motor in a nerve. (Select either a, b, or c, and move to frame 183.)
1. fibers 2. nerve fibers 3. c.	**8–183.** Peripheral nerves contain both sensory and motor fibers because they must conduct impulses from the receptor organs to the spinal cord and from the spinal cord by way of ____________ fibers to effector organs.

motor (efferent)	**8–184.** The two groups of nerves that comprise the peripheral nervous system are the (1) ____________ nerves and the (2) ____________ nerves.
1. cranial **2. spinal**	**8–185.** The CRANIAL NERVES are so named because they orginate in the ____________ and brain stem.
brain	**8–186.** There are 12 pairs of cranial nerves. The nuclei of ____________ ____________ 1–3 originate in the cerebrum. The nuclei of cranial nerves 4–8 arise from the pons.
cranial nerves	**8–187.** Nuclei of cranial nerves 9 and 10 are located in the medulla oblongata, and the (1) ____________ of cranial nerve 11 are found in the medulla oblongata and the spinal cord. Nuclei of (2) ____________ ____________ 12 are located in the medulla oblongata.
1. nuclei **2. cranial nerve**	**8–188.** The cranial nerves are arranged in ____________ so that each side of the body can be independently innervated.
pairs	**8–189.** Efferent impulses that influence circulation or respiration may be transmitted <u>from</u> nuclei of the medulla oblongata, along the fibers of ____________ ____________ pairs 9–11, to the effector organ.
cranial nerve	**8–190.** In addition to the 12 pairs of cranial nerves, the peripheral nervous system consists of 31 pairs of ____________ nerves.
spinal	**8–191.** Cranial nerves <u>are</u> similar <u>functionally</u> to the spinal nerves. Cranial nerves <u>are not exactly</u> similar <u>structurally</u> to the spinal nerves. You (1) ____________ (would/would not) expect all the (2) ____________ (number) pairs of spinal nerves to be anatomically like the cranial nerves.
1. would not **2. 31**	**8–192.** Because a nerve is a bundle of nerve fibers, in some nerves there are both (1) ____________ and (2) ____________ fibers that carry impulses to or away from the CNS.

1. sensory (afferent)
2. motor (efferent)

8–193.

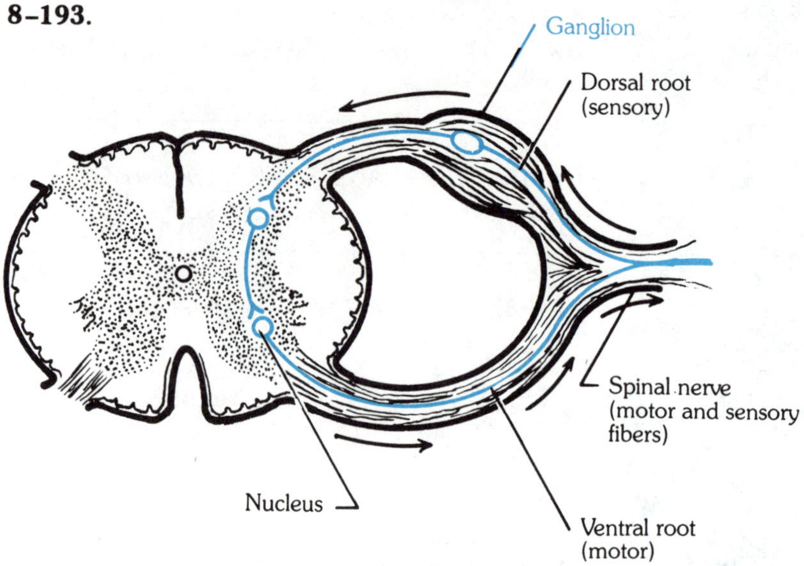

Spinal nerves and some cranial nerves contain both (1) ________ and (2) ________ fibers.

1. sensory
2. motor

8–194. Spinal roots (are/are not) ________ composed of mixed fibers.

are not

8–195. Where an efferent and afferent root unite, a ________ ________ is formed.

spinal nerve

8–196. If you are slapped on the back, the sensory impulse passes by way of a spinal nerve to a (1) ________ root, through a dorsal root (2) ________ to the dorsal spinal columns. The impulse is then conducted to the brain or by an (3) ________ neuron to the (4) ________ motor root.

(Refer to figure in frame 193 if you need help.)

1. dorsal
2. ganglion
3. internuncial
4. ventral

8–197. Fibers from various spinal nerves, after leaving the spinal cord, intermingle and form networks of nerves. This intermingling of nerves and fibers is called a PLEXUS (PLEX—sus).

Spinal nerve fibers from one level in the neck intermingle with other spinal nerve fibers from different levels in the neck to form the CERVICAL ________.

plexus	**8–198.** A plexus is an intermingling of (1) ________ and (2) ________. At the lower end of the spinal cord, spinal nerves leave the cord and intermingle with nerves and fibers from other levels of the spinal cord to form the LUMBOSACRAL (3) ________.
1. nerves 2. fibers 3. plexus	**8–199.** Such an intermingling of nerves and fibers allows for a diffuse reaction. For example, if you step on a nail, the impulse returns to the spinal cord by way of sensory fibers. Upon reaching the cord, the impulse can be conducted to many levels of the spinal cord by way of several of the intermingling sensory fibers located <u>outside</u> the cord, in the lumbosacral plexus. You react by lifting your foot, but you may also ________ (use your own words to describe additional reactions) because the original stimulus has been conducted simultaneously to many areas of the spinal cord and brain.
yell, jump, hold your foot, etc.	**8–200.** The PERIPHERAL NERVOUS SYSTEM includes nerves that connect body parts to the CNS. The (1) ________ division of this system connects the body wall, skin, and (2) ________ muscle to the (3) ________.
1. somatic 2. skeletal 3. CNS	**8–201.** The craniospinal nerves that comprise the PNS include (1) ________ (number) pairs of cranial nerves and (2) ________ (number) pairs of spinal nerves.
1. 12 2. 31	**8–202.** Spinal nerves and some (1) ________ nerves contain both (2) ________ and (3) ________ fibers.
1. cranial 2. sensory 3. motor	**8–203.** A network of intermingling nerves and fibers outside the spinal cord is called a a. root b. plexus (Select either a or b, and move to frame 204.)
b.	**8–204.** An impulse travels from the CNS by way of a nerve plexus to the wall of the body, inducing a reaction in the (1) ________ organs usually made up of (2) ________ muscle.
1. effector 2. skeletal	**8–205.** Because the (1) ________ nervous system must conduct impulses to and from the wall of the body, the nerves of this system or division must contain (2) ________ and (3) ________ fibers and be functionally integrated with other nervous tissue of the body.

1. somatic
2. sensory
3. motor

8–206.

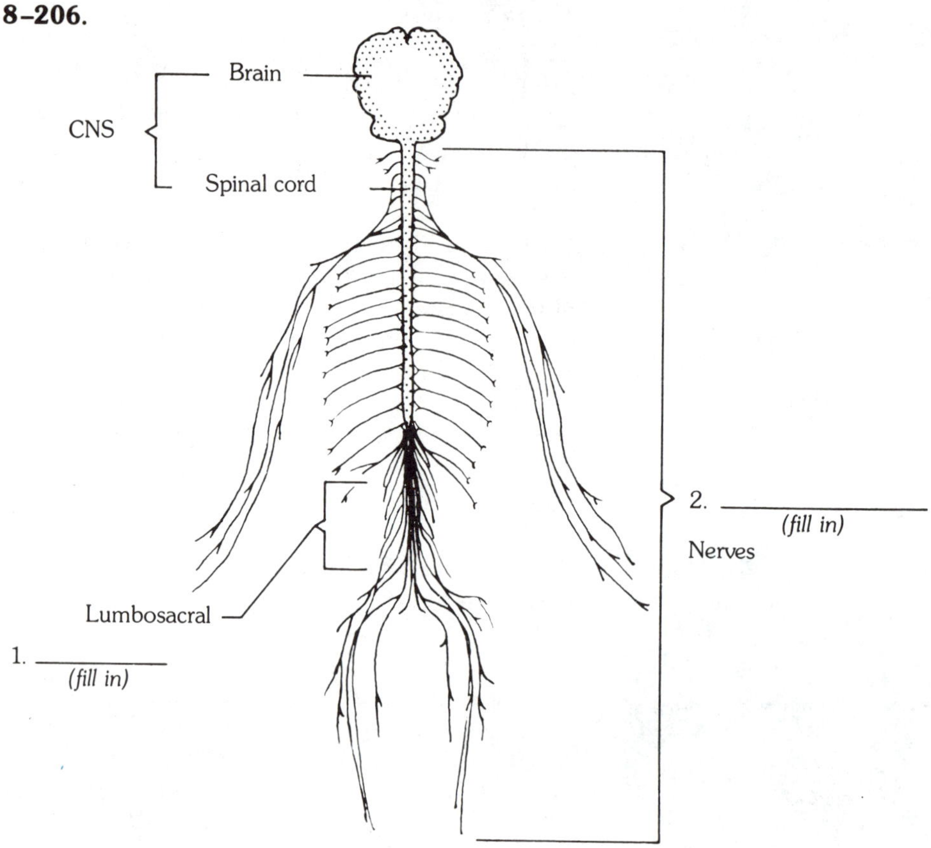

Fill in the unlabeled blanks.

1. plexus
2. peripheral

8–207. So far, you have learned how the peripheral nervous system conducts impulses to and from the skin and skeletal muscle. A ______________ arc is an established pattern of stimulus and response.

reflex

8–208. A second part of the (1) ______________ nervous system is called the AUTONOMIC DIVISION. The function of this division usually called the autonomic nervous system, abbreviated ANS, is to conduct impulses from smooth muscle, cardiac muscle, and glands to and from the (2) ______________ nervous system.

1. peripheral
2. central

8–209. The ANS is often referred to as the visceral nervous system because it conducts impulses to and from (1) ______________ muscle, cardiac muscle, and (2) ______________ that influence the viscera, or internal body structures.

1. smooth
2. glands

8–210. The term, Autonomic Nervous System, indicates that this system is autonomous, or that it functions independently of the CNS; this is not so. The ANS is ______________ ______________ with other nervous tissue of the body.

functionally integrated	**8–211.** The system that controls the functions of the viscera (organs) that are generally not under voluntary control is the _____ _____ _____.
ANS (Autonomic Nervous System)	**8–212.** Voluntary actions—movement of skeletal muscles—are regulated by the (1) _____ and (2) _____ nervous system. Involuntary actions—contractions or relaxations of smooth and cardiac muscles and secretions of many glands—are regulated by the (3) _____ nervous system.
1. central 2. somatic (either order) 3. autonomic	**8–213.** There are two divisions of the ANS with distinct connections to the CNS: the SYMPATHETIC (sim—pa—THET—ik) and the PARASYMPATHETIC (par—a—sim—pa—THET—ik). The two divisions of the CNS are the brain and spinal cord. The two divisions of the ANS are the (1) _____ and (2) _____.
1. sympathetic 2. parasympathetic (either order)	**8–214.** Fill in the four unlabeled blanks SYSTEM: CNS — DIVISIONS: 1. _____ 2. _____ PNS — Somatic; Autonomic — Divisions: 3. _____ 4. _____
1. brain 2. spinal cord (either order) 3. sympathetic 4. parasympathetic (either order)	**8–215.** The sympathetic preganglionic fibers originate from cell bodies in the lateral columns of the spinal cord gray matter, from approximately the level of the eighth cervical to the third lumbar vertebrae. This division of the ANS is, therefore, called the: a. Thoracolumbar outflow b. Craniosacral outflow (Select either a or b, and then move to frame 216.)
a.	**8–216.** The sympathetic system of the ANS is called the THORACOLUMBAR (tho—RA—ko—LUM—ber) outflow because it is from these areas of the vertebral column that the axons of the sympathetic system leave the _____ _____.
spinal cord	**8–217.** After the preganglionic fibers of the sympathetic neurons leave the spinal cord, they mingle within fibers of the ventral roots of the spinal nerves and may then _____ with autonomic ganglia outside the CNS. (Recall that a ganglion is a group of neuron cell bodies outside the CNS.)

synapse

8–218.

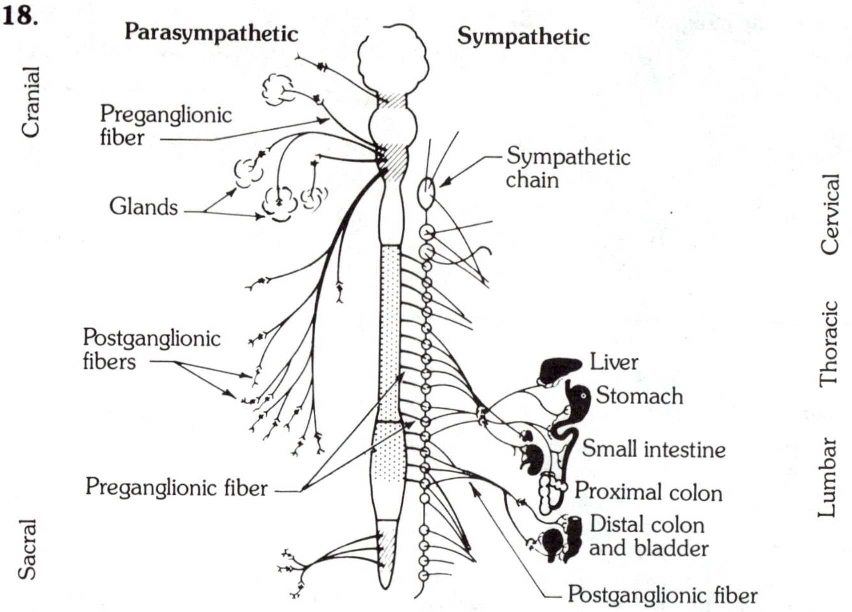

The small round circles to the right of the spinal cord in this illustration represent the autonomic ________________ of the sympathetic division.

---

ganglia

8–219. The ANS fibers that connect the ganglia to the spinal column are called:

a. preganglionic fibers
b. postganglionic fibers

(Select either a, or b, and then move to frame 220.)

---

a.

8–220. Preganglionic fibers of the ANS are myelinated and are referred to as the ________________ (white/gray) RAMI (RA—mi; the plural of Ramus) COMMUNICANTES (co—mu—ni—CAN—tes).

---

white

8–221. The preganglionic fibers or white (1) ________________ ________________ are insulated with a white, protein-lipid material knows as (2) ________________.

---

1. rami communicantes
2. myelin

8–222. There are three routes that the sympathetic preganglionic fibers can follow. After leaving the spinal column by the (1) ________________ ________________ communicantes the fibers can separate from the (2) ________________ nerves and synapse immediately with the neurons of the nearest ganglion in the sympathetic division.

1. white rami
2. peripheral

8–223.

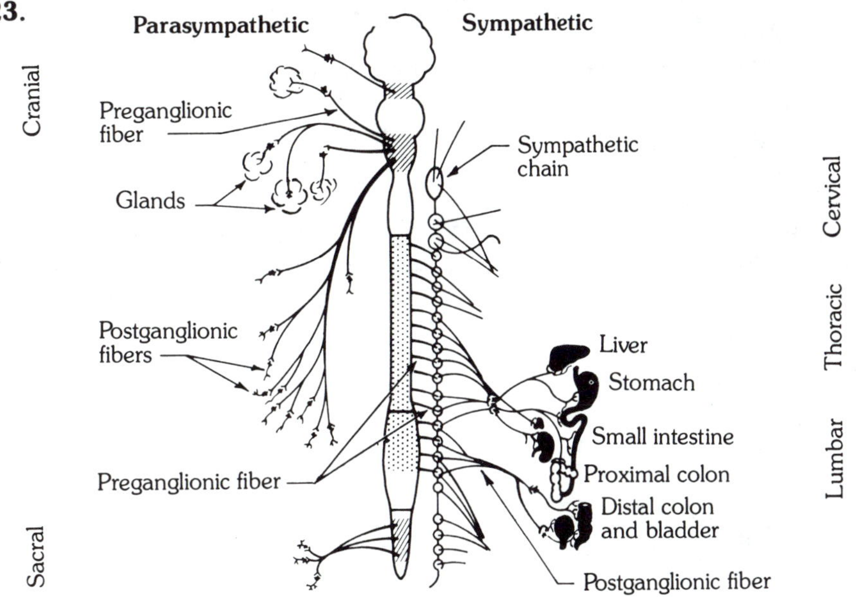

In the above drawing, you can see one set of 22 sympathetic ganglia, which are interconnected to form a chain. There are two sympathetic chains, each with 22 ____________ ____________, one chain on each side of the vertebral column.

---

sympathetic ganglia

8–224. A second route of impulse transmission for the sympathetic preganglionic fibers would be to enter the sympathetic chain and travel up or down the chain, ____________ with cells in distant ganglia of the chain.

---

synapsing

8–225. Finally the white (1) ____________ ____________ can pass through the sympathetic ganglia, without synapsing, leave the sympathetic chain and travel on the (2) ____________ nerves to the periphery of the body where they then synapse with peripheral ganglion cells (cardiac ganglion, superior mesenteric ganglion, etc.).

---

1. rami communicantes
2. peripheral

8–226. Impulses may be transmitted by the sympathetic preganglionic fibers by three routes.

First, an impulse may leave the spinal cord by way of the (1) ____________ (white/gray) rami communicantes, separate from the peripheral nerves and synapse (2) ____________ (immediately/later) with the neurons of a ganglion of the sympathetic division.

---

1. white
2. immediately

8–227. Second, because the (1) ____________ (number) sympathetic ganglia on each side of the vertebral column are interconnected, an impulse may be transmitted on preganglionic fibers that travel up or down the sympathetic chain (2) ____________ (with/without) synapsing until arriving at a distant ganglion.

1. 22 2. without	**8–228.** Third, an impulse can pass on a preganglionic fiber, white (1) ________ communicantes, without synapsing through a sympathetic chain neuron, by way of (autonomic/peripheral) (2) ________ nerves, to distant peripheral ganglia of the body where they then synapse.
1. rami 2. peripheral	**8–229.** The preganglionic fibers of the sympathetic division are: (1) ________ (afferent/efferent) with respect to CNS, (2) ________ (afferent/efferent) with respect to the sympathetic ganglia.
1. efferent 2. afferent	**8–230.** It is important to remember that the afferent preganglionic fibers of the sympathetic division arise from the lateral columns in the spinal cord. (Few/None) ________ of the preganglionic fibers arise from sympathetic ganglia.
None	**8–231.** Preganglionic fibers of the ANS are also called ________ ________ ________.
white rami communicantes	**8–232.** Preganglionic fibers of the ANS sympathetic division arise from the lateral columns of the (1) ________ ________ and then synapse with a sympathetic ganglia and are therefore (2) ________ (afferent/efferent) in function.
1. spinal column 2. afferent	**8–233.** Fibers that conduct impulses from a sympathetic ganglion to the effector organ are called: a. preganglionic white rami b. postganglionic fibers (Select either a or b, and then move to frame 234.)
b.	**8–234.** The afferent preganglionic fibers of the ANS arise from the (1) ________ ________ of the spinal cord. Most fibers arising from the sympathetic ganglia (axons of the sympathetic ganglia) are (2) ________ (motor/sensory) in function and are called (3) ________ fibers.
1. lateral columns 2. motor 3. postganglionic	**8–235.** A good example of the functional integration of the ANS with other nervous tissue is a VISCERAL REFLEX ARC. The peripheral nervous system sensory fibers from an innervated organ conduct an impulse to the ________ ________ of the spinal cord.
lateral columns	**8–236.** The impulse then travels from the lateral columns on a (1) ________ fiber—white ramus communicantes—that synapses with a (2) ________ ________.

1. preganglionic
2. postganglionic fiber

8–237. The axons of neurons in the sympathetic ganglion that are (1) ________________ (pre/post) ganglionic fibers, carry the impulse back to the innervated organ completing the formation of the Visceral (2) ________________ ________________.

1. post
2. Reflex Arc

8–238. A myelinated ANS fiber that arises in the lateral columns of the spinal cord is a (1) ________________ fiber.

An unmyelinated ANS fiber that is a continuation of a sympathetic ganglia cell is a (2) ________________ fiber.

1. preganglionic
2. postganglionic

8–239. Preganglionic sympathetic fibers are (1) ________________ (myelinated/unmyelinated).

Postganglionic sympathetic fibers are (2) ________________ (myelinated/unmyelinated).

1. myelinated
2. unmyelinated

8–240.

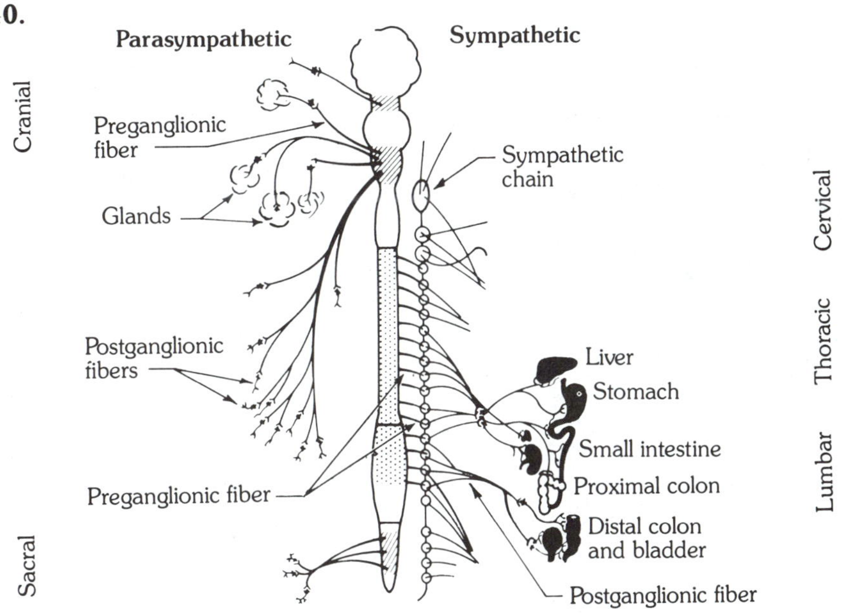

The second division of the ANS, the (1) ________________ division, is sometimes called (2) ________________ (thoracolumbar/craniosacral) outflow.

1. parasympathetic
2. craniosacral

8–241. Refer to the drawing in frame 240.

The ganglia of the (1) ________________ division are located close to the spinal cord, and the ganglia of the (2) ________________ division are located in or near the organ innervated by the efferent fibers.

1. sympathetic
2. parasympathetic

8–242. The preganglionic fibers of the parasympathetic division have their origin in the nuclei of the (1) ________________ and the (2) ________________ region of the spinal cord.

1. brain
2. sacral

8–243. Preganglionic parasympathetic fibers carry impulses to or near the innervated (1) ________________. The preganglionic fibers then synapse with a parasympathetic (2) ________________.

1. organ
2. ganglion

8–244. Relatively, sympathetic ganglia are located (1) ________________ (close to/far away from) the innervated organ.

Parasympathetic ganglia are located (2) ________________ (close to/far away from) the innervated organ.

1. far away from
2. close to

8–245. Sympathetic postganglionic fibers "fan out" in many directions before they reach the innervated organs. You would therefore expect a wide spread of diffuse body reaction from impulses conducted by the (sympathetic/parasympathetic) ________________ division of the ANS.

sympathetic

8–246. The preganglionic fibers of the (1) ________________ division synapse with a ganglion that is (2) ________________ to or within the innervated organ. The postganglionic fibers of this division are short and more direct. Stimulation of this division results in a (3) ________________ (diffuse/local) effect.

1. parasympathetic
2. close
3. local

8–247. The scope of the ANS functions can be better appreciated if you consider the wide distribution of smooth muscle throughout the body. For example, smooth muscle is found in the blood vessels from head to foot, in the digestive organs from the mouth to the anus, and in the urinary tract organs, reproductive organs, and special sensory organs such as the eye.

(No response required, move to frame 248.)

8–248. The overall function of the ANS is to maintain a constant environment for the cells of the body. This includes constant body temperature, regulation of the heart rate and respiratory rate, digestive functions, all the processes vital to cell integrity, and a constant fluid ________________.

environment

8–249. In order to live in outer space, an astronaut must be provided with an environment that closely approximates conditions on earth. The atmosphere within the space capsule is such an environment.

Your body, like an astronaut's, must maintain a constant environmental state. The ________________ nervous system helps your body when internal and external environmental changes occur.

autonomic

8–250.

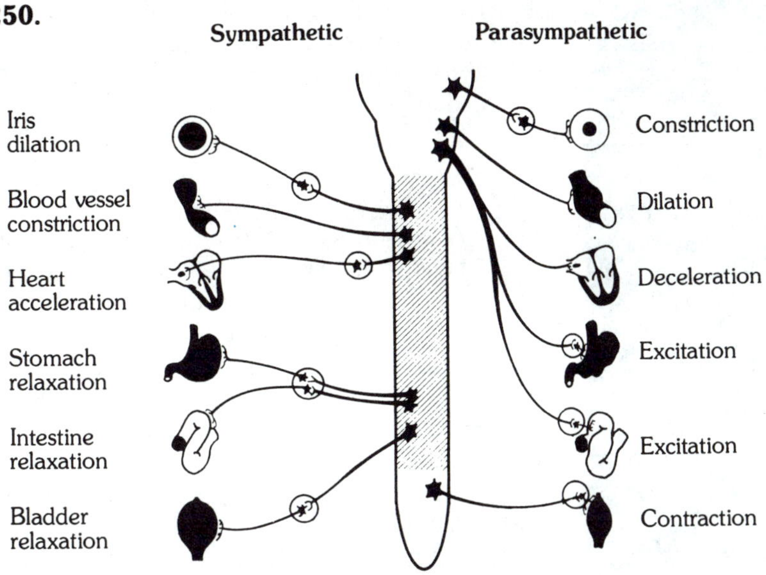

You will note by this simplified drawing that most organs innervated by the sympathetic division of the ANS are also innervated by the ______________________ ______________________.

parasympathetic division

8–251. You should also note, from the illustration in frame 250, that the postganglionic fibers of one division are usually antagonistic to the fibers of the other division.

For example, impulses from the sympathetic division cause the bladder to (1) ______________________, and impulses to this organ from the parasympathetic division induce bladder (2) ______________________.

1. relax
2. contraction

8–252.

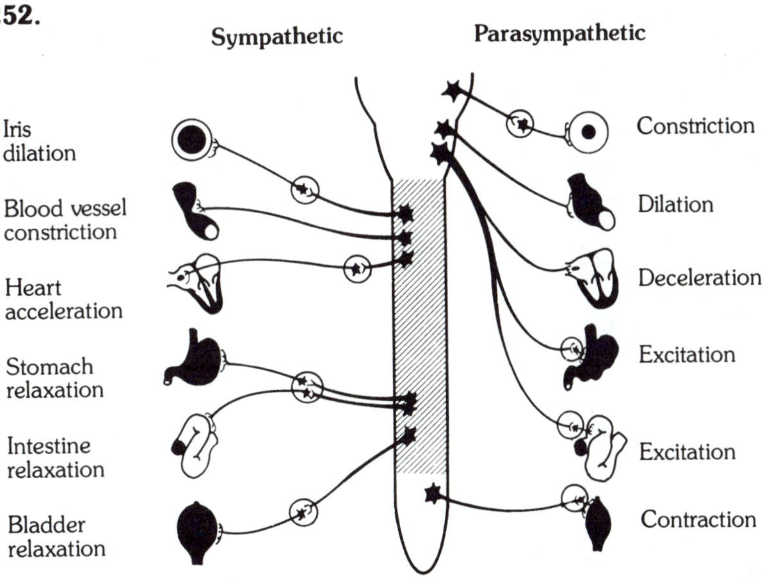

The sympathetic division tends to increase your body defenses. Starvation, temperature changes, and physical attack are situations that would require assistance from the ______________________ division of the ANS.

sympathetic	**8–253.** The parasympathetic division is directed toward the conservation and restoration of the body's resources. When you become frightened and your heart begins to beat excessively, it is restored to a more normal pace by the __________ division of the ANS.
parasympathetic	**8–254.** You are not expected to remember all the functions of both divisions of the ANS; however, you should remember that when your body is called upon to fight and your heart beats faster to provide more energy to the cells, it is the (1) __________ division that induces this increased (2) __________ action.
1. sympathetic 2. defensive	**8–255.** After your body has met the challenge, it is the parasympathetic division that induces the (1) __________ of used body energy and the (2) __________ of the body's resources in preparation for other emergencies.
1. restoration 2. conservation	**8–256.** The 12 pairs of cranial nerves and 31 pairs of spinal nerves constitute the __________ nervous system.
peripheral	**8–257.** Voluntary actions, such as movement of __________ muscles, are controlled by the CNS.
skeletal	**8–258.** Involuntary actions (excluding reflexes of skeletal muscles), such as movement of smooth and cardiac muscles, are regulated by the __________ nervous system.
ANS	**8–259.** The ANS __________ (is/is not) functionally integrated with other nerve tissue.
is	**8–260.** The ANS is divided into two main divisions the thoracolumbar outflow or (1) __________ division and the craniosacral outflow or (2) __________ division.
1. sympathetic 2. parasympathetic	**8–261.** Sympathetic preganglionic fibers originate from cell bodies in the (1) __________ __________ of the spinal cord (2) __________ matter.
1. lateral columns 2. gray	**8–262.** Another name for the sympathetic division is the thoracolumbar outflow. This name refers to origin of the sympathetic preganglionic fibers, the (1) __________, (2) __________, and (3) __________ areas of the spinal cord.

1. cervical 2. thoracic 3. lumbar	**8–263.** The ganglia of the sympathetic division of the ANS are located close to the ______________ (vertebral column/innervated organ).
vertebral column	**8–264.** Sympathetic preganglionic fibers are (1) ______________ (myelinated/unmyelinated). These fibers are called (2) ______________ ______________ ______________.
1. myelinated 2. white rami communicantes	**8–265.** Sympathetic preganglionic fibers can, after leaving the spinal cord: 1. Separate from the spinal nerves and immediately synapse with one of the 22 ______________ ______________ on each side of the vertebral column,
sympathetic ganglia	**8–266.** 2. travel up or down the chain of interconnected sympathetic ganglia and synapse with distant ______________ of the sympathetic chain,
ganglia	**8–267.** 3. pass without synapsing with a (1) ______________ ______________ and be conducted by way of (2) ______________ nerves to the effector organ.
1. sympathetic ganglion 2. peripheral	**8–268.** Sympathetic preganglionic fibers that are afferent in function arise from the lateral columns of the spinal cord and then synapse with a ______________ ______________.
sympathetic ganglion	**8–269.** All fibers arising from the sympathetic ganglia are (1) ______________ in function and are called (2) ______________ (post-ganglionic/preganglionic) fibers. Because these fibers are unmyelinated, they are called (3) ______________ (white/gray) rami communicantes.
1. motor (efferent) 2. post-ganglionic 3. gray	**8–270.** A visceral reflex arc is formed when the sensory fibers of the (1) ______________ nervous system transmit an impulse from an innervated organ to the (2) ______________ columns of the spinal cord.
1. peripheral 2. lateral	**8–271.** The impulse is then conducted by (1) ______________ (white/gray) rami communicantes to a sympathetic (2) ______________.
1. white 2. ganglion	**8–272.** From the sympathetic chain, the impulse is conducted away from the sympathetic ganglia on (1) ______________ (dendrites/axons) of the sympathetic ganglion cells, (2) ______________ (white/gray) rami communicantes, or on (3) ______________ nerves to the innervated organ.

1. axons 2. gray 3. peripheral	**8–273.** A visceral reflex arc is a good example of the ____________ ____________ of all nervous tissues.
functional integration	**8–274.** Preganglionic fibers of the parasympathetic division of the ANS—craniosacral outflow—arise from the gray matter of the (1) ____________ or sacral region of the (2) ____________ ____________.
1. brain 2. spinal cord	**8–275.** Preganglionic parasympathetic fibers of the ANS conduct impulses to or near the innervated organ where the fibers synapse with a ____________.
ganglion	**8–276.** The ganglia of the ____________ (sympathetic/parasympathetic) division are located close to or within the innervated organ.
parasympathetic	**8–277.** Before reaching the innervated organ, an impulse conducted on a ____________ ANS fiber spreads out in many directions.
sympathetic	**8–278.** Impulses conducted on ____________ ANS fibers would induce a local effect.
parasympathetic	**8–279.** The ANS acts to maintain a constant body (1) ____________ when affected by either (2) ____________ or (3) ____________ body changes. The ANS functions help to maintain homeostasis.
1. environment 2. internal (either 3. external order)	**8–280.** When the sympathetic and parasympathetic divisions of the ANS respond to internal or external body changes, they usually act (synergistically/antagonistically) ____________.
antagonistically	**8–281.** Your body defense system would be activated by the ____________ division of the ANS if you are physically attacked.
sympathetic	**8–282.** The parasympathetic division of the ANS helps your body (1) ____________ and (2) ____________ its resources.
1. restore 2. conserve (either order)	**8–283.** Your heart rate can be accelerated by the (1) ____________ division of the ANS and decelerated by the (2) ____________ division of the ANS.
1. sympathetic 2. parasympathetic	**8–284.** These two divisions, working together, enable your body to respond to stressors, internal or external events that produce stress. The stress response that helps the body to make adjustments and restore homeostasis is ____________.

adaptation	**8–285.** Your nervous system is, in an operational sense, like a high-speed computer. Data—impulses—are fed into the computer and are then analyzed, and a response, "feedback," is provided. (No response required; move to frame 286.)
	8–286. The cerebral cortex, the highest governing level of the nervous sytem, is, however, much more complex than our most sophisticated computers. You will recall that the cerebral cortex is connected to the brain stem and spinal cord by ascending and descending ________________.
tracts	**8–287.** Some impulses travel directly from the (1) ________________ ________________ to the cerebral cortex on direct high speed (2) ________________.
1. **spinal cord** 2. **tracts**	**8–288.** There is another connective network between the spinal cord and cerebral cortex. This network consists of small fibers in the spinal cord and brain stem and is called the RETICULAR (re—TIK—u—ler) SYSTEM. An indirect tract to the cerebral cortex would pass through the ________________ system.
reticular	**8–289.** Impulses that pass through the indirect route of small ________________ fibers are carried to the cerebral cortex on thalamic fibers.
reticular	**8–290.** If a sensation is to be experienced, you must be awake. The reticular system keeps the cerebral cortex awake and is therefore referred to as the ________________ activating system.
reticular	**8–291.** Your cerebral cortex is kept alert to conscious sensations by impulses that originate in the reticular (1) ________________ system and are transmitted to the cortex by (2) ________________ fibers.
1. **activating** 2. **thalamic**	**8–292.** Impulses enter the CNS on (1) ________________ (sensory/motor) neurons. They pass from the spinal cord directly to the brain on highspeed (2) ________________ or through the small fibers of the reticular activating system, which is located in the spinal cord and (3) ________________ ________________.
1. **sensory** 2. **tracts** 3. **brain stem**	**8–293.** To experience a sensation, your cerebral cortex must be (1) ________________. The reticular (2) ________________ system transmits impulses by (3) ________________ fibers to keep the cortex alerted to external stimuli.

1. awake
2. activating
3. thalamic

8–294. PERIPHERAL MOTOR ACTIVITIES are controlled by the cerebral cortex and nuclei in the brain stem.

Actions such as walking and talking are examples of ________________ motor activities.

peripheral

8–295. The PERIPHERAL AUTONOMIC MOTOR ACTIVITY, like other peripheral motor activities, is also controlled by nuclei in the upper ________________ ________________ and other sections of the brain.

brain stem

8–296.

Brain Surface-One Hemisphere
Medial View

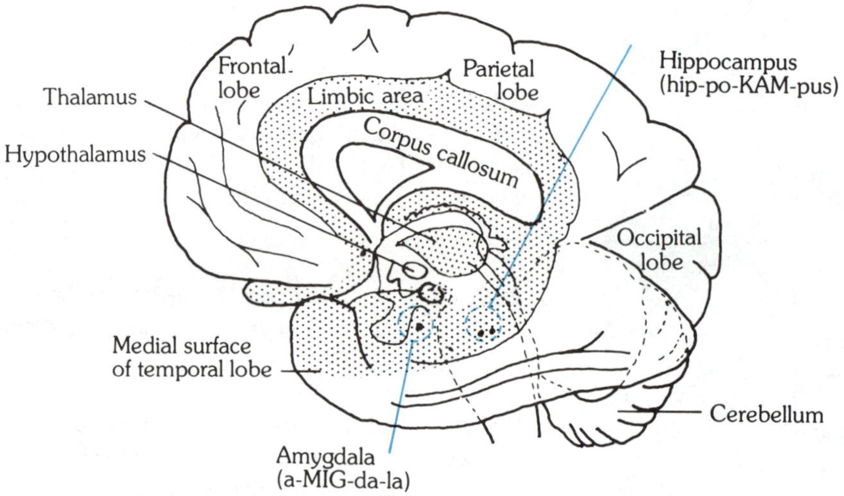

The ring of gray matter in the above drawing is referred to as the autonomic or visceral brain. This area of the brain would control the peripheral ________________ motor activities.

autonomic

8–297. Refer to the illustration in frame 296.

The LIMBIC (LIM—bik) area of the ANS is composed of three main sections, the (1) ________________, the (2) ________________ and the (3) ________________. The limbic area controls the peripheral (4) ________________ ________________ activities of the body.

1. amygdala
2. hippocampus
3. hypothalamus
(any order)
4. autonomic motor

8–298. The limbic area consisting of the (1) ________________, (2) ________________ and (3) ________________, is referred to as the limbic system. This system influences the emotional and intellectual aspects of conscious behavior.

1. amygdala
2. hypothalamus
3. hippocampus
(any order)

8–299. Sexual activity—reproduction—and securing of food—survival—are activities that have the most profound influence on your behavior. These activities are under the control of the ________________ area of the brain.

limbic	**8–300.** The amygdala influences our survival patterns. The hippocampus influences our sexual activities. Match the areas of the brain in the column at the left with the activities in the right-hand column: a. amygdala — 1. sexual activities b. hippocampus — 2. emotion and intellect c. limbic system — 3. survival
a. 3 **b. 1** **c. 2**	**8–301.** Peripheral motor activity is controlled by the cerebral cortex of the brain and the (1) ______ ______. Peripheral autonomic motor activity is controlled by the visceral brain—(2) ______ ______ and the hypothalamus.
1. brain stem **2. limbic area**	**8–302.** Peripheral autonomic motor activity is regulated and correlated by two sections of the brain, the (1) ______ and the (2) ______ areas.
1. hypothalamus **2. limbic**	**8–303.** It is interesting to note that efferent impulses involving either survival or sexual activities leave the limbic system and travel to peripheral motor and peripheral autonomic motor centers resulting in behavior patterns that involve both the (1) ______ and (2) ______ nervous systems.
1. central **2. autonomic** **(either order)**	**8–304.** The cerebral cortex is kept awake by impulses originating in the (1) ______ ______ system. These impulses are then transmitted to the cortex by (2) ______ fibers.
1. reticular activating **2. thalamic**	**8–305.** The ANS peripheral motor activities are regulated and correlated by the (1) ______ and (2) ______ areas of the brain.
1. limbic **2. hypothalamic**	**8–306.** Another example of the functional integration of nerve tissues is the manner in which afferent impulses involving sexual activity or survival are transmitted from the limbic area on peripheral motor and peripheral autonomic motor nerves of both the (1) ______ and (2) ______ nervous systems.
1. central **2. autonomic**	**8–307.** Match the brain structures at the left with the functions listed at the right: STRUCTURE — FUNCTION a. Limbic system — 1. Conscious thinking b. Thalamus — 2. Alerts cerebral cortex c. Amygdala — 3. Correlates emotion and intellect d. Hippocampus — 4. Sexual activities e. Cerebral cortex — 5. Survival f. Reticular activating System — 6. Transmits alerting impulses

a. 3
b. 6
c. 5
d. 4
e. 1
f. 2

8–308. Impulses leave the brain and brain stem by direct high speed tracts or through the reticular system and pass to either the peripheral motor or peripheral __________________ motor fibers.

autonomic

8–309. You know that impulse transmission along the axon and dendrite of a neuron is a(n) __________________ (electrical/mechanical) phenomenon.

electrical

8–310. In a resting state—no impulse transmission, nerve fibers are POLARIZED (PO—ler—iz—ed). A polarized fiber has positive ions (charged atoms) located on the outer surface of the membrane that surrounds the fiber and an equal number of __________________ ions located on the inner surface of the nerve fiber membrane.

negative

8–311. Nerve fibers are surrounded by a semipermeable membrane. When an impulse is transmitted along a nerve fiber, the permeability of the fiber is increased, and the charged surface becomes DEPOLARIZED or neutralized. An impulse permits the (1) __________________ charges on the outer surface and (2) __________________ charges on the inner surface of a neuron to unite.

1. positive (+)
2. negative (−)

8–312. Impulses increase the (1) __________________ of the membrane surrounding a nerve fiber resulting in a (2) __________________ or neutralized fiber.

1. permeability
2. depolarized

8–313. The depolarization action "runs" along the neuron until the impulse reaches the nerve endings. You should remember that the transmission of an impulse along a nerve fiber is an __________________ phenomenon; the transmission of an impulse from one neuron ending to the next neuron, or organ, is a chemical phenomenon.

electrical

8–314. At the nerve endings of:
-All parasympathetic and sympathetic preganglionic fibers,
-All parasympathetic postganglionic fibers, sympathetic postganglionic fibers to sweat glands,
-Motor fibers to skeletal muscles,
a chemical substance known as ACETYLCHOLINE (a—SE—til—KOL—een) is produced. Acetylcholine is a neurotransmitter.

(No response required, move to frame 315.)

8–315. A chemical such as (1) __________________, which is released at (2) __________________ (all/most) parasympathetic postganglionic endings, induces a (3) __________________ reaction, permitting an impulse to pass from one neuron to the next. This is a neurotransmitter that facilitates the movement of the impulse across the (4) __________________ to the next neuron or organ.

Answers	Frames
1. acetylcholine 2. all 3. chemical 4. synapse	**8–316.** Acetylcholine, which induces impulse transmission between neurons, is produced by the nerve endings of: -all (1) ____________ sympathetic and parasympathetic fibers -all parasympathetic (2) ____________ fibers -sympathetic postganglionic fibers going to (3) ____________ ____________ -motor fibers to (4) ____________ muscles.
1. preganglionic 2. postganglionic 3. sweat glands 4. skeletal	**8–317.** Impulse transmission between neurons is a ____________ phenomenon. Impulse transmission along a neuron is an ____________ phenomenon.
1. chemical 2. electrical	**8–318.** An all-inclusive statement about nerve impulse transmission would be one indicating that impulse transmission is a combined ____________ phenomenon.
electrochemical	**8–319.** The acetylcholine produced at the nerve endings is rapidly destroyed (in one thousandth of a second) by an enzyme called CHOLINESTERASE (KOL—in—ES—ter—ase). The enzyme—____________—destroys acetylcholine.
cholinesterase	**8–320.** After a nerve impulse is transmitted from one neuron to the next, the (1) ____________ that induced the transmission is destroyed by the enzyme (2) ____________.
1. acetylcholine 2. cholinesterase	**8–321.** Nerve fibers that produce acetylcholine are called CHOLINERGIC (KOL—en—ER—gic). Postganglionic sympathetic sweat gland nerves would be classified as ____________ nerve fibers.
cholinergic	**8–322.** All parasympathetic and sympathetic preganglionic fibers are also ____________.
cholinergic	**8–323.** With the exception of the postganglionic sympathetic sweat gland fibers, all sympathetic postganglionic fibers produce a chemical known as NOREPINEPHRINE (nor—EP—i—NEF—rin). Efferent sympathetic fibers to the heart would produce (1) ____________ at their nerve endings. Norepinephrine is a chemical substance, (2) ____________, that facilitates transmission of an impulse across the synapse.
1. norepinephrine 2. neurotransmitter	**8–324.** Epinephrine = ADRENALINE (ad—REN—al—in) Norepinephrine = Noradrenaline Epinephrine and norepinephrine are related compounds, and are many times referred to as (1) ____________ and (2) ____________, respectively.

1. adrenaline 2. noradrenaline	**8–325.** All (1) ________________ (preganglionic/postganglionic) sympathetic nerve endings, except those afferent sympathetic fibers that innervate sweat glands, produce (2) ________________ at their nerve endings.
1. postganglionic 2. norepinephrine	**8–326.** Nerve fibers that produce norepinephrine are called ADRENERGIC (ad—REN—er—gic) fibers. Most postganglionic sympathetic fibers are ________________ (cholinergic/adrenergic).
adrenergic	**8–327.** All preganglionic fibers of both the parasympathetic and sympathetic divisions produce (1) ________________ at their nerve endings and are therefore referred to as (2) ________________ nerve fibers.
1. acetylcholine 2. cholinergic	**8–328.** Acetylcholine is destroyed by the enzyme, ________________, which is present at the cholinergic nerve endings.
cholinesterase	**8–329.** Most (1) ________________ sympathetic nerve fibers produce the chemical (2) ________________. These nerves are called (3) ________________ fibers.
1. postganglionic 2. norepinephrine 3. adrenergic	**8–330.** A nerve in a resting state has (1) ________________ ions on the outside of the membrane that surrounds the nerve fiber. This fiber is polarized if there are an equal number of (2) ________________ ________________ on the inner surface of the neuron membrane.
1. positive 2. negative ions	**8–331.** The (1) ________________ of a nerve fiber increases when stimulated, and the nerve fiber becomes (2) ________________.
1. permeability 2. depolarized	**8–332.** The electrical depolarization action "runs" along the neuron until the impulse reaches the nerve (1) ________________, where the impulse is transmitted to the next neuron or organ by a (2) ________________ reaction.
1. endings 2. chemical	**8–333.** An impulse from a motor fiber is transmitted to a skeletal muscle by way of ________________, which is produced at the nerve endings.
acetylcholine	**8–334.** A stimulated nerve returns to a resting state when the (1) ________________ produced at the nerve endings is destroyed by the (2) ________________ cholinesterase.

1. acetycholine 2. enzyme	**8–335.** The neurotransmitter substances that interfere with the transmission of impulses across synapses are not as clearly understood. One identified substance suspected to act in this way is GAMMA-AMINOBUTYRIC ACID, usually called GABA. (No response required.)
	8–336. Epinephrine and norepinephrine are also produced by the MEDULLA (me—DUL—a) of the ADRENAL (ad—RE—nal) glands of the body. There are two adrenal glands—one above and near each kidney. The release of these substances from the (1) ________ ________ is due to stimulation by the (2) ________ (sympathetic/parasympathetic) division of the ANS.
1. adrenal medulla 2. sympathetic	**8–337.** Postganglionic sympathetic fibers of the ANS tend to increase the body (1) ________ against external or internal changes by producing (2) ________.
1. defenses 2. norepinephrine	**8–338.** The SYMPATHOADRENAL (SIM—pa—tho—ad—RE—nal) AXIS refers to the combined effect of the (1) ________ nerves and (2) ________ glands to produce norepinephrine for body defenses.
1. sympathetic 2. adrenal	**8–339.** Norepinephrine is produced by the sympathetic nerve endings and the adrenal glands, which are known collectively as the ________ axis.
sympathoadrenal	**8–340.** The production of norepinephrine by these two major sources is achieved if the ________ ________ is stimulated.
sympathoadrenal axis	**8–341.**

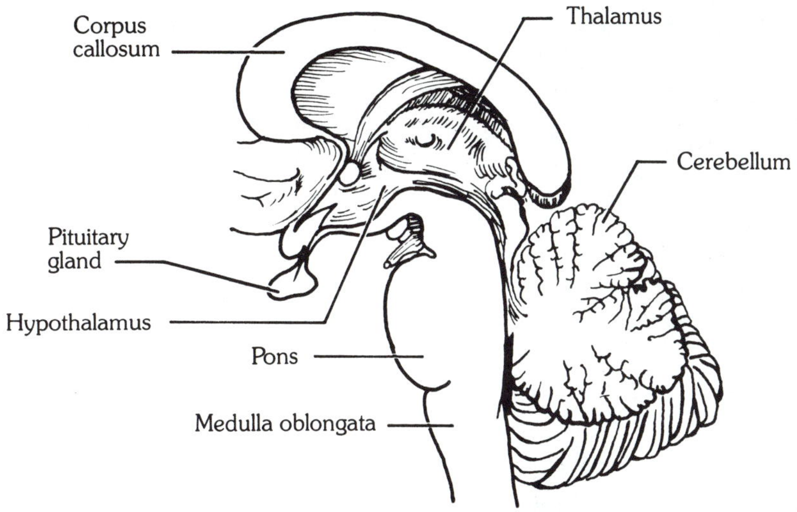

Another example of the functional integration of the nervous system with other body systems is the Hypothalamic-PITUITARY (pi—TU—i—ter—i) Axis. You will recall that the hypothalamus is a group of nuclei located beneath the ________.

thalamus

8–342. The hypothalamus is involved with several regulatory activities and as a structure of the limbic system, correlates and regulates the peripheral motor activity of the________________ (CNS/ANS).

ANS

8–343.

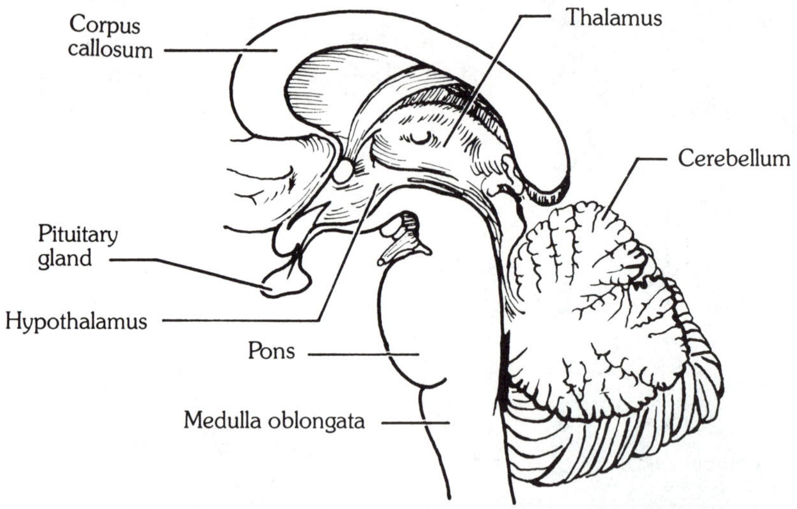

The hypothalamic nuclei also influence the ________________ gland, which is located below the hypothalamic region of the brain.

pituitary

8–344. Stimulation of the hypothalamus, by changes internal or external to the body, results in release of chemical messages from the (1) ________________. These messages are sent to the (2) ________________ gland.

1. **hypothalamus**
2. **pituitary**

8–345. The sympathoadrenal axis and the hypothalamic-pituitary axis are illustrations of the ________________ ________________ of the nervous system with other body systems.

functional integration

8–346. The pituitary gland and the adrenal glands are parts of the ENDOCRINE (EN—do—krin) system. The endocrine system will be the next chapter, Chapter 9, of this course that you will study.

9

The Endocrine System

9–1. In past units, you have read about systems composed of various organs of the human body; the system that is the subject of this unit is composed of glands. The glands of the body are divided into two categories: the EXOCRINE (EX—o—krin) glands and the ENDOCRINE (END—o—krin) ________________.

glands

9–2. Glands, you remember, are body organs that produce and secrete important substances. Glands are classified as "endocrine" or "exocrine" according to where they ________________ the particular substances they produce.

secrete

9–3. The characteristics of exocrine glands are that they have tubes called ducts that carry their secretions to the outside of the body or into hollow organs. Endocrine glands, on the other hand, do not have ducts, but secrete their substances into the bloodstream. To express this distinction, the ________________ glands are also referred to as the "ductless" glands.

endocrine

9–4. The glands that you have encountered as part of the integumentary system and elsewhere—sweat, mammary, sebaceous glands, and so on—all secrete through ducts and are therefore exocrine glands.

This unit covers the other category, the ductless (1) ________________ glands that compose the (2) ________________ system.

1. endocrine
2. endocrine

9–5. The secretions of endocrine glands are called HORMONES (HOR—moans). Hormones are first formed and secreted by the endocrine glands and then transported to various parts of the body by means of (ducts/the bloodstream) ________________.

the bloodstream

9–6. Hormones can be defined as chemical substances that are produced by gland cells in one part of the body and travel through the bloodstream to affect, in most cases, the function of cells in (another /the same) ________________ part of the body.

another

9–7. The part of the body whose cells are affected by the hormones from a given endocrine gland is called the TARGET ORGAN. Through the agency of the hormones it produces, then, an endocrine gland exercises a kind of chemical remote control over its particular ________________ organ or organs.

target

9–8. It is this remote control function exercised by the various (1) ________________ glands over their respective (2) ________________ ________________ that accounts for the importance of the endocrine system.

Endocrine Glands

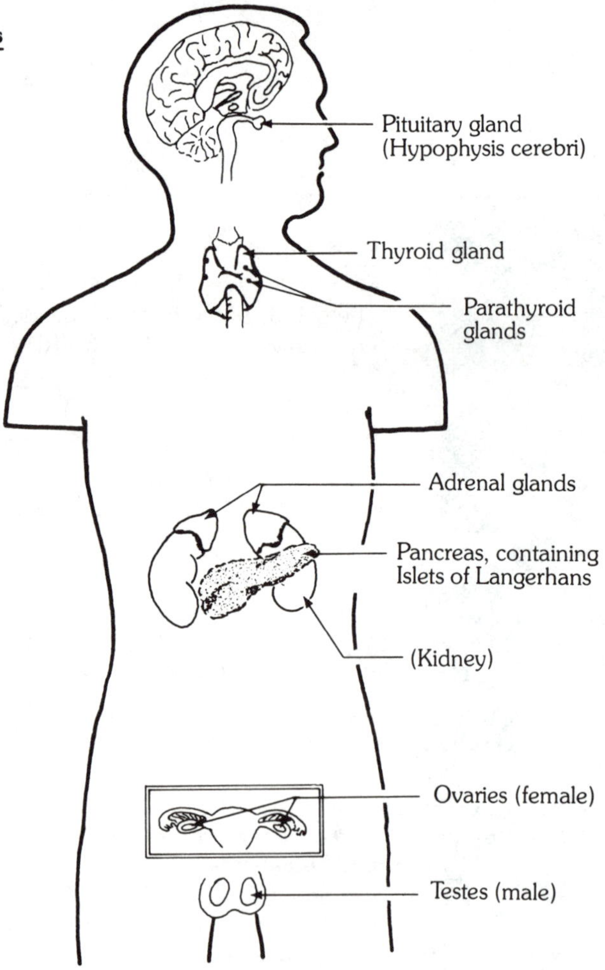

1. endocrine
2. target organs

9–9. Look at the drawing of the endocrine system that shows the locations of the glands in the endocrine system.

As you can see, these glands are (compactly arranged/widely separated) ____________ in the body.

widely separated

9–10. The glands in the endocrine system are not attached to each other in any way; they are related not in structure but in function. The endocrine system consists of a group of glands all of which exercise the function of chemical remote ____________.

control

9–11. The endocrine system, along with the nervous system, is one of the two major regulatory or ____________ systems in the human body.

control

9–12. In general, the hormones of the endocrine glands act to regulate the various metabolic processes of the body. It should be understood that "metabolism" simply means "change;" in its broadest sense, the term metabolism is used to include all the chemical and physical ____________ involved in the development and functioning of the living body.

changes

9–13. Thus, for example, the electrolytic changes undergone in the body by the essential minerals such as calcium or sodium, the changes undergone by the nutrients such as the carbohydrates, the gross changes undergone by the tissues themselves—all these are termed ____________ processes.

metabolic

9–14. The maintenance of normal growth, development, and function of the adult human body is therefore dependent on the presence and proper functioning of all the endocrine (1) ____________ in the endocrine (2) ____________.

The Thyroid Gland

1. glands
2. system

9–15. Now let's examine the individual endocrine glands in detail.

Look at the drawing again; the large single gland located in the neck is called the ____________ gland.

thyroid

9–16. The THYROID (THI—roid) gland is located in the front part of the neck; it consists of two lateral portions or lobes joined in front by a narrow band of tissue.

The arrows on the diagram point to the ________________ of the thyroid gland.

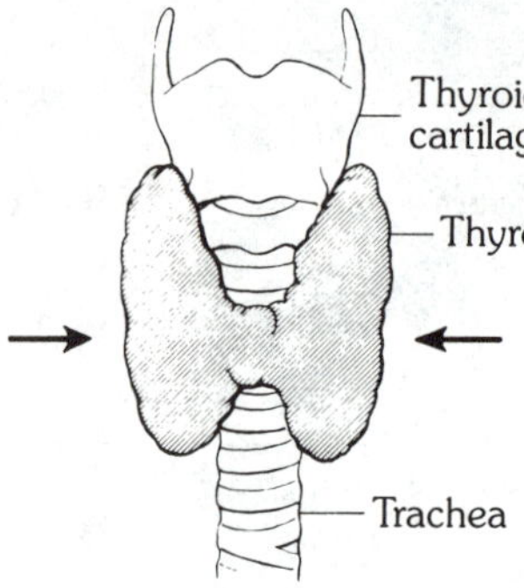

Front view

lobes

9–17. The function of the thyroid gland is to secrete a hormone that regulates the rate of metabolism of human body cells. The building up and, to a somewhat lesser extent, the breaking down of protoplasm throughout the entire body is influenced by the hormone secreted by the ________________ gland.

thyroid

9–18. A normal rate of metabolism is an important part of normal body function, most particularly growth. The thyroid hormone, because it regulates the (1) ________________ of metabolism of cells throughout the body, is thus essential to maintaining (2) ________________ body function.

1. rate
2. normal

9–19. The body, however, can continue to function without the thyroid hormone even if it does not function normally. In other words, although the thyroid hormone is essential to normal daily activity by virtue of its influence on the rate of (1) ________________, it is not essential to (2) ________________.

1. metabolism
2. life

9–20. The hormone secreted by the thyroid gland is called THYROXINE, each molecule of which contains four atoms of iodine. The atoms of iodine are the most important components of ________________.

thyroxine

9–21. Iodine is ingested with the food we eat, is circulated in the bloodstream, and is absorbed by epithelial cells in the thyroid gland, where it is then used to manufacture the thyroid hormone ________________.

thyroxine

9–22. The importance of iodine to thyroxine is reflected by several special characteristics of the thyroid gland. The epithelial cells in the thyroid, for instance, are more efficient at extracting iodine from the system than other tissue in the body. The result of this is that practically all of the (1) ________________ in the body is concentrated in the (2) ________________ gland.

1. iodine
2. thyroid

9–23. Again, if the system is deficient in iodine for any reason, the thyroid gland will enlarge itself in an effort to trap the available iodine more effectively. When the thyroid is thus ________________, it is called a goiter.

---

enlarged

9–24. When the thyroid gland is functioning as it should be, it is producing just the amount of thyroxine necessary to maintain a normal rate of metabolism for the rest of the body. Because the body's metabolic rate can be measured, it is relatively easy to determine whether or not the thyroid gland is ________________ properly.

---

functioning

9–25. The rate of metabolism is measured as follows:

The chemical activity involved in cell metabolism produces heat. An individual's metabolic rate can be determined by measuring the amount of ________________ being produced by metabolic activity within his body.

---

heat

9–26. However, since other factors besides metabolism also produce heat—mental and physical activity, the ingestion of food—the individual is asked to avoid these activities as far as possible, so that the heat in his body is being produced almost solely by ________________ activity.

---

metabolic

9–27. An individual who is refraining from mental and physical activity and the ingestion of food is described as being in a resting or BASAL (BAY—sl) condition, and the measurement of his heat production in that condition is called his ________________ metabolic rate or BMR.

---

basal

9–28. An individual who has avoided (1) ________________ and (2) ________________ activity and the ingestion of (3) ________________ is described technically as being in a (4) ________________ condition.

---

1. mental
2. physical
   (either order)
3. food
4. basal

9–29. The BMR or (1) ________________ ________________ ________________ can give a good indication of whether or not the (2) ________________ gland is functioning properly.

---

1. Basal Metabolic Rate
2. thyroid

9–30. When the thyroid gland is secreting an abnormal amount of thyroxine, the BMR rises or falls accordingly. An abnormally high BMR, for instance, would result from an excessive amount of metabolic activity in the body and would indicate that the thyroid was producing (too much/too little) ________________ thyroxine.

---

too much

9–31. An excess of thyroid hormone in the body is called HYPERTHYROIDISM. "Hyper" means "too much," as in "hyperactive" or "hypersensitive." An abnormally high metabolic rate is a reliable sign of ________________ thyroidism.

hyper	**9–32.** When metabolic activity is intensified by the presence of too much thyroxine, the body may be compared to an automobile engine with the idling speed set too high; the body, like the engine, produces more waste, burns more fuel, and produces more heat than is normal for it. All of these are characteristics of the bodily condition called ________________.
hyperthyroidism	**9–33.** The thyroid gland may also produce too little thyroid hormone, with a resultant slowing down of the metabolic rate. The term for this condition is HYPOthyroidism. Just as the prefix "hyper" means "too much," the prefix "________________" means "too little."
hypo	**9–34.** When the metabolic rate slows down as a result of insufficient thyroid hormone (caused by a diseased thyroid or a lack of iodine in the system), the body has a tendency to produce less heat and consume less fuel; these signs are characteristics of the condition called ________________.
hypothyroidism	**9–35.** An individual suffering from hypothyroidism is likely to feel (1) ________________ (cold/hot); also, because his appetite may be unaffected but he is burning (2) ________________ (more/less) fuel, he would tend to (3) ________________ (gain/lose) weight.
1. cold 2. less 3. gain	**9–36.** An individual with hyperthyroidism, on the other hand, would have a feeling of (1) ________________ in his body and would tend to (2) ________________ weight.
1. hotness, heat 2. lose	**9–37.** Hyper- and hypothyroidism are not usually fatal. If they are not treated, they can be damaging to ________________ growth and body function.
normal	**9–38.** For this reason, the thyroid hormone thyroxine is particularly important in the periods of rapid ________________ during infancy and youth.
growth	**9–39.** Let's review briefly what you've learned so far. The thyroid gland is part of the (1) ________________ system. This system, with the nervous system, is one of the two major (2) ________________ systems of the body. In general, the glands of this system act to regulate the (3)________________ processes of the body.
1. endocrine 2. control 3. metabolic	**9–40.** All endocrine glands secrete into the (1) ________________. The secretions of all endocrine glands are called (2) ________________.

1. bloodstream
2. hormones

9–41. The thyroid hormone is called (1) ________________.

The most important components of the thyroid hormone are its atoms of (2) ________________.

1. thyroxine
2. iodine

9–42. Thyroxine regulates the (1) ________________ of ________________ of the cells throughout the body.

For his thyroid activity to be measured, an individual should be in a (2) ________________ condition.

The measurement made when he is in that condition is the (3) ______ ______ ______ (initials).

1. rate (of) metabolism
2. basal
3. BMR

9–43. An abnormally high BMR indicates a condition of (1) ________________, or (2) ________________ (too much/too little) thyroid hormone in the body.

An abnormally low BMR indicates a condition of (3) ________________.

1. hyperthyroidism
2. too much
3. hypothyroidism

9–44. Thyroid hormone is essential to (1) ________________ (your own words), but not essential to (2) ________________.

The Parathyroid Glands

1. normal growth and body function
2. life

9–45. Now refer to the endocrine system drawing again.

The small glands that appear to be imbedded in the larger thyroid gland (but may be either embedded in the thyroid or simply very near to it, depending on the individual) are the ________________ glands.

parathyroid

9–46. The parathyroid glands are each about the size of a pea; they are usually four in number, two associated with each lobe of the thyroid.

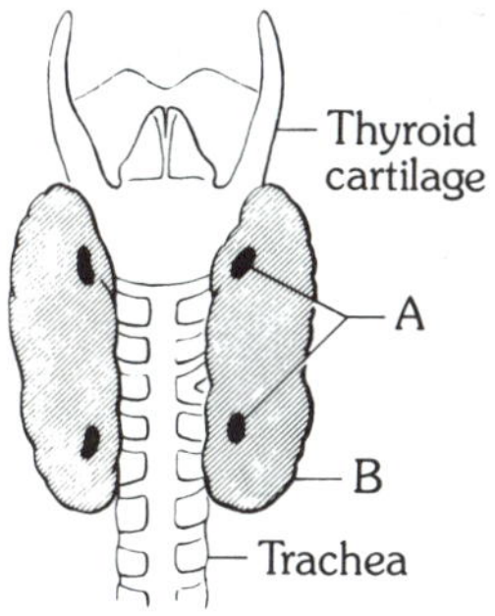

Back view

Arrow A points to two (1) ________________ glands.

Arrow B points to one (2) ________________ of the thyroid.

1. parathyroid
2. lobe

9–47. The product of the parathyroid is called PARATHORMONE. You should have no difficulty remembering that the (1) ________________-thyroid glands produce an endocrine secretion or (2) ________________ called (3) ________________.

1. para 2. hormone 3. parathormone	**9–48.** Unlike the thyroid hormone thyroxine, the parathyroid hormone, ________________, is not only essential to normal body function, but also essential to life.
parathormone	**9–49.** The most important metabolic effect of parathormone is on the essential mineral calcium; the electrolytic changes undergone by ________________ in the body are profoundly influenced by the parathyroid hormone.
calcium	**9–50.** The primary function of the parathyroid glands is to maintain a normal concentration of calcium in the blood and tissue fluids. Despite wide variations in calcium ingestion, the parathyroids control the levels of (1) ________________ in the (2) ________________ and tissue (3) ________________ within very narrow limits.
1. calcium 2. blood 3. fluids	**9–51.** In order to perform this function, parathormone acts principally in two areas. Most of the calcium in the body is stored in the bones; parathormone mobilizes calcium out of the (1) ________________ and into the (2) ________________ and (3) ________________ fluids.
1. bones 2. blood 3. tissue	**9–52.** The second way that parathormone acts to maintain calcium levels in the blood and tissue fluids is to promote the reabsorption of calcium by the kidneys. Thus, the two principal sites of parathormone's action are the bones from which calcium is mobilized and the ________________ in which calcium is reabsorbed.
kidneys	**9–53.** The effects of parathormone are therefore very widespread in the body, as may be seen particularly when the supply is abnormal. An oversupply of parathormone results in: Increased decalcification of the (1) ________________; Increased reabsorption of (2) ________________ by the (3) ________________.
1. bones 2. calcium 3. kidneys	**9–54.** The most important result of an oversupply of parathormone, however, is the (1) ________________ (increased/decreased) concentration of calcium in the blood and in the fluid of (2) ________________ throughout the body.
1. increased 2. tissues	**9–55.** Calcium plays a role not only in the mineralization of bones but also in many other body functions, including blood coagulation, the irritability of skeletal muscles, and the rhythm of the heart; and changes in blood and tissue fluid calcium concentration may affect all these functions. (No answer required.)
	9–56. Thus an overproduction of parathormone may not only demineralize and weaken the bones but also can lead ultimately to kidney and circulatory collapse; this condition would be called (hyper/hypo) ________________ parathyroidism and can be fatal if untreated.

hyper

9–57. The opposite situation—an undersupply of parathormone and resultant drop in the level of calcium in the blood and tissue fluids—is called ______ and can also be fatal.

hypoparathyroidism

9–58. Hypoparathyroidism can produce increased muscular and nervous irritability, leading to twitching, cramps, convulsions, and ultimate paralysis of the respiratory system if untreated. These symptoms are the result of an (oversupply/undersupply) ______ of parathormone.

undersupply

9–59. Furthermore, in addition to its effect on the metabolism of calcium, parathormone also influences the metabolism of phosphorus. Most of the phosphorus in the body (about 80%) is found in combination with calcium in the bones and teeth, and the metabolism of ______ is closely connected with that of calcium.

phosphorus

9–60. Thus although the princile metabolic effect of parathormone is on calcium, it also affects the metabolism of ______.

phosphorus

9–61. To review:

The vital function of the parathyroid hormone, (1) ______, is to regulate the level of (2) ______ in the (3) ______ and (4) ______ fluids.

1. parathormone
2. calcium
3. blood
4. tissue

9–62. It performs this function primarily by mobilizing calcium out of the (1) ______ and into the blood and tissue fluids, and at the same time by promoting the reabsorption of calcium in the (2) ______.

1. bones
2. kidneys

9–63. In addition to its effect on the metabolism of calcium, the hormone secreted by the (1) ______ glands also influences the metabolism of (2) ______.

The Adrenal Glands

1. parathyroid
2. phosphorus

9–64. The next endocrine glands to be studied are the ADRENAL (ad—REEN—al) glands.

Look at the endocrine system diagram. There are ______ (number) adrenal glands in the body.

two

9–65. The prefix "ad" means "near" and the word "renal" means "kidney." The location of the two ______ glands is "near the kidneys," one above each kidney.

adrenal	**9–66.** Each adrenal gland has two distinct parts, the MEDULLA (meh—DULL—ah) and the CORTEX (KOR—tex). The cortex is a pale pink envelope of tissue encasing the dark brown inner core, the ________________.
medulla	**9–67.** Both the medulla and the cortex of an adrenal gland secrete hormones; however, the hormones secreted by the inner core, the (1) ________________, are quite different from those secreted by the outer envelope, the (2) ________________.
1. medulla 2. cortex	**9–68.** The adrenal medulla secretes two hormones that you have already encountered in the preceding unit. The first is called EPINEPHRINE or ADRENALINE; "epinephrine" is the preferred term. You should however, remember that epinephrine and ________________ are the same hormone.
adrenaline	**9–69.** The second adrenal medullary hormone is called NOREPINEPHRINE (or noradrenaline). Although norepinephrine differs somewhat from epinephrine both in nature and in degree of activity, their effects are so closely related that epinephrine and ________________ are commonly referred to together as "the epinephrines."
norepinephrine	**9–70.** For the sake of simplicity, the following frames will use the term epinephrines when describing the activities of both hormones—epinephrine and (1) ________________—secreted by the adrenal (2) ________________.
1. norepinephrine 2. medulla	**9–71.** In the preceding unit you learned that the sympathetic nervous system releases an epinephrinelike substance used as a transmitting agent. The effects of the (1) ________________ nervous system on the body—increase of heart rate, blood pressure, mental alertness, and blood sugar content—can also be described as the effects of the adrenal medullary hormones, the (2) ________________.
1. sympathetic 2. epinephrines	**9–72.** The epinephrines are often called the "emergency" hormones because they mobilize what is termed the "fight or flight mechanism." The production of the epinephrines is stimulated by such strong emotions as fear, rage, and so on for the purpose of helping the body meet—by (1) ________________ or fleeing—whatever (2) ________________ caused these emotions.
1. fighting 2. emergency	**9–73.** The epinephrines put the whole body in a state of readiness for an emergency by increasing mental alertness, raising the level of sugar in the blood (as fuel for muscles), and increasing (1) ________________ pressure and the rate at which the (2) ________________ beats, for maximum efficiency.
1. blood 2. heart	**9–74.** Thus the epinephrines play an important role in helping the body adjust to its environment. The body can, however, survive without the epinephrines (in a protected environment)—in other words, they are not ________________ to life.

essential

9–75. To review: You should remember that the two hormones secreted by the adrenal (1) ______________—the inner core of the adrenal gland—are (2) ______________ and (3) ______________ also called adrenaline and noradrenaline.

1. medulla
2. epinephrine
3. norepinephrine

9–76. Epinephrine and norepinephrine are secreted by the adrenal medulla in response to strong emotions caused by an (1) ______________, and they act to mobilize the (2) ______________ or (3) ______________ mechanism.

1. emergency
2. fight (or) flight

9–77. The epinephrines prepare the body to meet an emergency by increasing:

1. mental ______________;
2. the ______________ content of the blood;
3. blood ______________; and
4. the rate at which the ______________ ______________.

1. alertness
2. sugar
3. pressure
4. heart beats

9–78. The adrenal medullary hormones (are/are not) ______________ essential to life.

are not

9–79. On the other hand, the hormones secreted by the outer envelope, the ______________, of the adrenal glands are essential to life.

cortex

9–80. The adrenal cortex secretes a number of hormones called, collectively, the ADRENAL CORTICOSTEROIDS. The part of the name that refers to their origin in the cortex is (1) "______________." This part is often omitted for brevity, making them simply the (2) "______________ ______________."

1. cortico
2. adrenal steroids

9–81. "Steroids" are chemical compounds of the lipid class and are related to each other in that they all have the same chemical nucleus. As a result, the different adrenal (1) ______________ (or adrenal (2) ______________) have many functions in common.

1. steroids
2. corticosteroids

9–82. The most important functions common to the majority of adrenal corticosteroids are the regulation of salt metabolism and of carbohydrate (sugar) metabolism. The adrenal steroids vary first of all according to whether their influence is exerted chiefly over the metabolism of (1) ______________ or the metabolism of (2) ______________.

1. salt
2. carbohydrate (either order)

9–83. Those adrenal steroids whose primary effect is on carbohydrate metabolism are termed GLUCOCORTICOIDS. Cells in various parts of the body—notably liver, muscle, and fat cells—are acted on most particularly by the (1) ______________ to convert fat and protein into (2) ______________.

1. glucocorticoids 2. carbohydrate (or sugar)	**9–84.** The single adrenal steroid hormone that shows the greatest effect on carbohydate metabolism is HYDROCORTISONE. As the glucocorticoids are the most potent carbohydrate metabolizers among the adrenal steroids, so ________________ is the most potent glucocorticoid.
hydrocortisone	**9–85.** The levels of blood sugar in the body, the depositions of fat, the life span of white blood cells, and a number of other vital body activities are affected to an important degree by the (1) ________________ of the adrenal cortex and most particularly by the hormone (2) ________________.
1. glucocorticoids 2. hydrocortisone	**9–86.** Another adrenal steroid of the glucocorticoid category that should be singled out for attention is CORTISONE. You have probably heard of the treatment of arthritis and also various skin disorders similarly typified by inflammation by the use of ________________.
cortisone	**9–87.** Although it exerts an important influence on carbohydrate metabolism like all the glucocorticoids, it is particularly for its anti-inflammatory powers that (1) ________________ is noted. Likewise, hydrocortisone, which has a major role in ________________ metabolism, acts as an anti-inflammatory agent.
1. cortisone 2. carbohydrate	**9–88.** The second category of adrenal steroid hormones includes those that exert their effect primarily on salt metabolism; these hormones are termed MINERALOCORTICOIDS. The absorption and retention of (1) ________________ by cells throughout the body are profoundly influenced by the (2) ________________ of the adrenal cortex.
1. salt 2. mineralocorticoids	**9–89.** One hormone in this category, called ALDOSTERONE, is estimated to be at least twenty-five times as effective as any other adrenal steroid in the retention of salt. Thus as hydrocortisone is the most potent glucocorticoid, so (1) ________________ is the most potent (2) ________________.
1. aldosterone 2. mineralocorticoid	**9–90.** As a result of their influence on (1) ________________ metabolism, the mineralocorticoids also exert far-reaching effects on electrolyte distribution, water balance, and blood pressure; in all of these areas, (2) ________________ is the adrenal steroid hormone with the greatest potency.
1. salt 2. aldosterone	**9–91.** Again, the mineralocorticoids and particularly aldosterone also promote the excretion of potassium from the kidneys. If, as a result of inadequate mineralocorticoid activity, the kidneys fail to excrete the normal quantity of ________________, the effects produced on the nervous and cardiovascular systems can be fatal.
potassium	**9–92.** It should be pointed out that <u>all</u> the adrenal corticosteroids act to promote salt retention and ________________ excretion to a greater or lesser degree.

potassium

9–93. The third and final category of adrenal steroids is a group of sex hormones that are similar in structure and action to the hormones secreted by the sex glands, which will be covered later in the text. The ________ hormones of the adrenal cortex differ from those of the sex glands in being substantially less potent.

---

sex

9–94. It is worth noting, however, that degree of effectiveness is the only major difference between the ________ ________ secreted by the adrenal cortex and those secreted by the sex glands.

---

sex hormones

9–95. Now a brief review.

The hormones secreted by the adrenal cortex—although they share most of their functions to a greater or lesser degree—are divided into 3 categories according to primary effect. These categories are:

1. the ________;
2. the ________;
3. the ________ ________.

---

1. **glucocorticoids**
2. **mineralocorticoids (either order)**
3. **sex hormones**

9–96. The glucocorticoids chiefly influence (1) ________ metabolism. Two glucocorticoids of particular importance are: (2) ________, because it is the most potent; and (3) ________, because of its anti-(4) ________ powers.

---

1. **carbohydrate (or sugar)**
2. **hydrocortisone**
3. **cortisone**
4. **inflammatory**

9–97. The mineralocorticoids chiefly influence (1) ________ matabolism; the most potent member of this group is (2) ________.

The sex hormones of the adrenal cortex differ from those of the sex glands chiefly in (3) ________ of effectiveness.

---

1. **salt**
2. **aldosterone**
3. **degree**

9–98. The hormones of the adrenal cortex are called, collectively, the (1) ________ ________, or more simply the (2) ________ ________.

The hormone epinephrine, also called adrenaline, is secreted by the adrenal (3) ________.

The Pancreas

---

1. **adrenal corticosteroids**
2. **adrenal steroids**
3. **medulla**

9–99. Refer to the endocrine system diagram. The PANCREAS (PAN—kree—as) is a gland located in the abdomen that secretes digestive juices into the small intestine through a duct. An integral part of the pancreas is a group of specialized cells that secrete hormones directly into the bloodstream. The pancreas as a whole, then performs (endocrine/exocrine/both endocrine and exocrine) ________ functions.

both endocrine and exocrine	**9–100.** The specialized cells within the pancreas that secrete directly into the bloodstream are called the ISLETS OF LANGERHANS (LANG—er—hans). The endocrine functions of the pancreas are performed by the ________________ ________________ ________________.
islets of Langerhans	**9–101.** The drawing shows a magnified section of the pancreas with an ________________ ________________ ________________, drawn in blue, embedded in it. 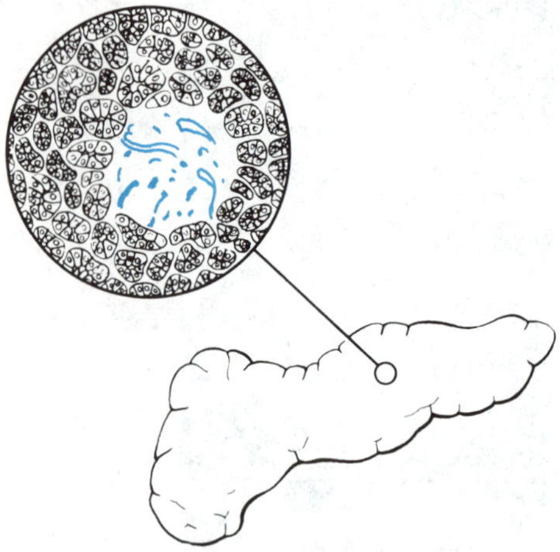
islets of Langerhans	**9–102.** The islets of Langerhans produce a hormone essential to life known as INSULIN. Insulin is indispensable to normal carbohydrate metabolism. Carbohydrate metabolism is regulated by the hormones of the adrenal cortex and medulla, by the thyroid, and by a hormone called ________________.
insulin	**9–103.** Insulin is the secretion of the (1) ________________ ________________ ________________ located in the (2) ________________.
1. islets of Langerhans 2. pancreas	**9–104.** Insulin is a complex protein that works at the cellular level to enable cells to take up carbohydrates in order to convert them for energy. Without insulin, (1) ________________ metabolism all but ceases at the (2) ________________ level.
1. carbohydrate 2. cellular	**9–105.** When they cannot be metabolized, the now useless carbohydrates accumulate in the bloodstream, raising the amount of sugar in the ________________ to a dangerous level.
blood, bloodstream	**9–106.** The excess sugar spills over through the kidneys into the urine, carrying with it a large amount of water. The loss of the (1) ________________ that is carried along by the excess (2) ________________ causes the body to become dehydrated.

1. water 2. sugar	**9–107.** The disease characterized by dehydration of the body, resulting in excessive thirst, and by excessive production of urine containing sugar, is familiar to you as diabetes mellitus. If untreated by means of insulin injections, ________ ________ can lead to coma, convulsions, and death.
diabetes mellitus	**9–108.** At the same time, if a lack of insulin prevents the cells from producing energy by converting carbohydrates, other substances—namely fat and protein—must be broken down to produce ________ for the body.
energy	**9–109.** The breaking down of fat and protein to produce energy also results in the production of certain acids that enter the bloodstream and raise the acidity of the blood. The production of these (1) ________, as a side-effect of a (2) ________ (lack/excess) of insulin, can be fatal if unchecked.
1. acids 2. lack	**9–110.** Finally, an excessive supply of insulin causes a (drop/rise) ________ in the level of blood sugar, which can lead also to convulsions and death.
drop	**9–111.** Another hormone secreted by the cells in the islets of Langerhans is GLUCAGON (GLUE—ka—gone). As its name implies, glucagon is a hormone that influences ________ metabolism.
sugar (glucose)	**9–112.** Whereas insulin acts to decrease the concentration of glucose or sugar in the bloodstream, ________ increases the glucose concentration in the blood
glucagon	**9–113.** The target organ for glucagon is the liver, where it enables the stored ________ to be released to the bloodstream.
glucose (sugar)	**9–114.** Insulin and glucagon must work together to maintain a normal level of glucose in the ________.
bloodstream	**9–115.** A third hormone that the cells of the islets of Langerhans secrete is SOMATOSTATIN (so—MA—toe—STA—tin). The exact action of this hormone is still being studied. However, research indicates that the important balance between insulin and glucagon secretion may be regulated by ________.
somatostatin	**9–116.** To review briefly: The endocrine functions of the pancreas are performed by the (1) ________ ________ ________, which secrete the hormones (2) ________, (3) ________, and somatostatin.

1. islets of Langerhans 2. insulin 3. glucagon	**9–117.** Insulin is indispensable to (1) ________________ metabolism at the cellular level and is therefore essential to maintaining a healthy level of (2) ________________ in the blood. Insulin is clearly essential to (3) ________________.
1. carbohydrate 2. sugar (glucose) 3. life	**9–118.** If the blood sugar level decreases, the cells in the islets of Langerhans respond by secreting (1) ________________, which acts on the liver to release (2) ________________ to the bloodstream.
1. glucagon 2. glucose (sugar)	**9–119.** The balance between the secretions of insulin and glucagon is thought to be regulated by the third hormone produced by the cells in the (1) ________________ ________ ________________. This hormone is called (2) ________________. **The Gonads**
1. islets of Langerhans 2. somatostatin	**9–120.** The GONADS (GO—nads) are the sex glands of the body. Both the male and female sex glands are called ________________.
gonads	**9–121.** In the male, the sex glands are the TESTES (TEST—eez), two oval glands contained in the scrotum, a pouch suspended between the legs. The singular of (1) ________________ is TESTIS. These two glands are the sex glands or (2) ________________ of the male.
1. testes 2. gonads	**9–122.** The male sex glands produce the male sex hormones, the most important of which is called TESTOSTERONE (tes—TOST—er—own), a steroid. The adrenal corticosteroids and the hormone (1) ________________, secreted by the male sex glands, the (2) ________________, are members of the same chemical family.
1. testosterone 2. testes	**9–123.** Embedded in the testes are a number of specialized interstitial cells, and these are the cells directly responsible for the production of the hormone ________________.
testosterone	**9–124.** Testosterone stimulates the development of male secondary sex characteristics. The primary sex characteristics of the male—presence of a penis and testicles—are not controlled by the testes. Testosterone is therefore essential only to the development of ________________ sex characteristics in the male.
secondary	**9–125.** Testosterone is responsible for such characteristics as the development of the beard, growth and distribution of body hair, and deepening of the voice. All these are classified as (1) ________________ sex (2) ________________.

1. secondary 2. characteristics	**9–126.** Testosterone is also responsible for the "sex drive" or LIBIDO (li—BEE—doe) of the male. A eunuch—a male deprived of testosterone, for instance, by the removal of the testes by castration—suffers the loss both of (1) ________ ________ ________ and of (2) ________.
1. secondary sex characteristics 2. libido	**9–127.** The testes are not only endocrine glands. They also perform an exocrine function, namely, in the production of SPERM cells. SPERM is the male name for the reproductive ________ of the male, which are, of course, carried to the outside of the body.
cells	**9–128.** Sperm is produced by the cells of a mass of minute tubes or tubules coiled within the testes. These tubules are connected to a system of EXCRETORY DUCTS. The (1) ________ cells secreted by the testes are carried to the outside of the body by the system of (2) ________ ________, that terminates in the penis.
1. sperm 2. excretory ducts	**9–129.** Thus the testes perform not only an endocrine function, the secretion of the hormone (1) ________, but also an exocrine function, the production of (2) ________ cells, which are the (3) ________ cells of the male.
1. testosterone 2. sperm 3. reproductive	**9–130.** The gonads of the female are the OVARIES (OH—va—reez). The ovaries are two small bean-shaped masses of tissue located in the lower part of the abdominal cavity. Like the testes, the ________ have both exocrine and endocrine functions.
ovaries	**9–131.** The ovaries produce two hormones. One is called "progesterone" and will be discussed later, in the section on reproduction. The other is called ESTROGEN (ES—tro—gen). Like testosterone in the male, estrogen is responsible for the (primary/secondary) ________ sex characteristics of the female.
secondary	**9–132.** Estrogen controls the female libido, and development of the breasts, widening of the hips, distribution of body fat, onset of menstruation, and other secondary sex characteristics of the (1) ________ body. Again, like testosterone, (2) ________ belongs to the chemical family of steroids.
1. female 2. estrogen	**9–133.** The exocrine function of the ovaries is the development and expulsion of egg or OVUM (O—vum), the female counterpart of the sperm. The female (1) ________ cell is the egg or (2) ________.
1. reproductive 2. ovum	**9–134.** The ova (plural of ovum) are produced by specialized cells. These epithelial cells group together into islands, each island producing an ________ at its center.

ovum	**9–135.** These islands are called FOLLICLES (FOL—ik—ls). The whole island, including epithelial cells and the (1) ________________ that they produce, is included in the term (2) ________________.
1. ovum 2. follicle	**9–136.** You should remember that the production of ova within the cell islands or (1) ________________ of the ovaries is an (2) ________________ (endocrine/exocrine) function comparable to the production of (3) ________________ by the testes.

The Pituitary Gland

1. follicles 2. exocrine 3. sperm	**9–137.** The last member of the endocrine system to be studied is the PITUITARY (pi—TU—i—terry) GLAND whose Latin name is "hypophysis cerebri." Look at the endocrine system diagram. The pituitary is a compact mass of tissue about the size of a pea, located in the ________________.
head or skull	**9–138.** The pituitary gland is contained in a small cavity in the floor of the skull, directly behind the bridge of the nose. The pituitary is a double gland. It consists of two lobes, like the thyroid, but each pituitary lobe is considered a separate ________________ gland.
pituitary	**9–139.** The lobe of the pituitary toward the back side or posterior of the skull is called the posterior pituitary gland. The lobe toward the front, or anterior, of the skull is called the ________________ ________________ ________________.
anterior pituitary gland	**9–140.** Both the lobe toward the back of the skull, the (1) ________________ pituitary gland, and the lobe toward the front of the skull, the (2) ________________ ________________ gland, secrete several distinct hormones.
1. posterior 2. anterior pituitary	**9–141.** The pituitary gland has the title of "master gland of the endocrine system." It owes this title to a group of hormones secreted by the anterior pituitary, whose function is to regulate the productivity of other glands of the ________________ system.
endocrine	**9–142.** These hormones are called the TROPIC hormones, the word "trcpic" meaning "affecting development." The healthy development and functioning of several of the endocrine glands that you have studied is under the direct control of the (1) ________________ hormones secreted by the (2) ________________ pituitary.

1. tropic 2. anterior	**9–143.** Thus the several GONADOTROPIC hormones of the anterior pituitary regulate the growth and proper functioning of the male and female ________________.
gonads	**9–144.** Both the endocrine and exocrine functions of the testes and ovaries are stimulated by the several ________________ hormones, as you will see in more detail in the next chapter.
gonadotropic	**9–145.** Another tropic hormone secreted by the anterior pituitary is called ADRENOCORTICOTROPIC HORMONE, or more simply ACTH. The secretion of hormones by the (1) ________________ ________________ is stimulated by (2) ______ ______ ______ ______ (initials).
1. adrenal cortex 2. ACTH	**9–146.** You should remember that not only the productivity but also the healthy growth and development of the adrenal cortex is dependent on ______ ______ ______ ______.
ACTH	**9–147.** The final member of the tropic hormone group is the THYROTROPIC hormone, which stimulates the growth of, and production of thyroxine by, the ________________ gland.
thyroid	**9–148.** Thus the tropic hormones of the anterior pituitary include: the several (1) ________________ hormones; (2) ______ ______ ______ ______; and the (3) ________________ hormone.
1. gonadotropic 2. ACTH 3. thyrotropic	**9–149.** The way in which the pituitary's own activity is regulated is called a <u>feedback mechanism</u>. You recall from Chapter 8 that there is a functional integration between the nervous system and the rest of the body. One way that integrated activity occurs is due to the relationship between the nervous system and the endocrine system. The hypothalamic-pituitary axis is one type of relationship. (No response required.)
	9–150. Whenever there is a drop in the blood level of a hormone, such as thyroxine, the hypothalamus detects the decrease. The hypothalamus sends hormonal messages to the anterior pituitary gland. The pituitary relays the message to the thyroid gland that more thyroxine is needed. The message from the pituitary is sent by the ________________ hormone.
thyrotropic	**9–151.** If the blood level of thyroxine increased, the pituitary would decrease its (1) ________________ hormone action. This signal would slow down the secretion of (2) ________________ into the bloodstream. The process of sending messages from the body to the glands involved in regulating metabolism is called a (3) ________________ ________________. This process is another example of how the body can respond to changes and maintain homeostasis.

1. **thyrotropic** 2. **thyroxine** 3. **feedback mechanism**	**9–152.** The anterior pituitary hormones discussed so far are those that control the activities of other glands and thus account for the pituitary's title of "master" gland. The anterior pituitary also produces an important hormone that is called the GROWTH HORMONE because of its direct influence on ________________.
growth	**9–153.** The proper name for the growth hormone is SOMATOTROPIC hormone. The word "somato" comes from the Greek word "soma", meaning "body"; the combination of (1) "________________" and "tropic" tells you that this hormone affects the growth or development of the (2) ________________.
1. **somato** 2. **body**	**9–154.** Although it is not the only growth-influencing hormone, its importance to normal body development can be seen in the fact that the rate of growth of an individual may be reduced by as much as 50% in the total absence of ________________ (proper name) hormone.
somatotropic	**9–155.** The somatotropic hormone stimulates body growth in general, especially the growth of long bones, and it exerts a particular influence on the metabolism of carbohydrates and proteins. Unlike the other tropic hormones of the anterior pituitary, somatotropic hormone produces these effects (directly/through its regulation of other endocrine glands) ________________.
directly	**9–156.** A lack of somatotropic hormone during the body's period of growth will result in a midget: person of normal proportions but unusually small size. Conversely, an excess of somatotropic hormone during the growth period will produce an individual of normal proportions but unusually ________________ size.
large	**9–157.** A circus giant is a good example of a person whose anterior pituitary gland produced (too much/too little) ________________ somatotropic hormone during his growing period.
too much	**9–158.** Giants are persons of abnormal size but normal proportions. If, on the other hand, an excessive production of somatotropic hormone occurs after the body's normal period of growth, the result will be a person of abnormal ________________; outsized hands, feet, and facial bones (diseases referred to as acromegaly) are the most common examples of this condition.
proportions	**9–159.** Thus, the anterior pituitary secretes both the gland-stimulating tropic hormones, which give the pituitary as a whole its title of (1) "________________ gland of the endocrine system;" it also secretes a hormone essential to proportionate growth and to the rate of growth, the (2) ________________ hormone.
1. **master** 2. **somatotropic**	**9–160.** Important hormones are also produced by the other lobe of the pituitary gland, the ________________ pituitary. The posterior pituitary is also closely related to the hypothalamus.

posterior	**9–161.** The actions of one posterior pituitary hormone, called OXYTOCIN, are restricted to the female body; it stimulates the secretion of (1) ________________ from the mammary glands during lactation and stimulates contraction of the female uterus. Obstetricians commonly make use of (2) ________________ to induce labor.
1. milk 2. oxytocin	**9–162.** By stimulating the (1) ________________ of the uterus during childbirth and the (2) ________________ of milk from the mammary glands afterwards, the posterior pituitary hormone (3) ________________ plays an important role in the total process of reproduction.
1. contraction 2. secretion 3. oxytocin	**9–163.** The second posterior pituitary hormone is called ANTIDIURETIC HORMONE, or more simply, ADH, and exerts its major effect on kidney function. The influence of (1) ______ ______ ______ (initials) will be covered in the second part of this chapter, which examines (2) ________________ function in detail.
1. ADH 2. kidney	**9–164.** For the time being, you need remember only that the second hormone secreted by the (1) ________________ pituitary is called (2) ______ ______ ______ and that this hormone has its major effect on the excretion of urine by the (3) ________________.
1. posterior 2. ADH (or antidiuretic hormone) 3. kidney	**9–165.** The following frames will review briefly the principal hormones of the pituitary gland—whose Latin name is "hypophysis cerebri" and whose title is (1) "________________ gland of the (2) ________________ system."
1. master 2. endocrine	**9–166.** The gland-stimulating hormones of the (1) ________________ pituitary are: The several (2) ________________ hormones; (3) ______ ______ ______ ______ (initials); and the (4) ________________ hormone. The growth-stimulating (5) ________________ hormone is also secreted by this lobe of the pituitary.
1. anterior 2. gonadotropic 3. ACTH 4. thyrotropic 5. somatotropic	**9–167.** The two hormones secreted by the (1) ________________ pituitary are (2) ________________ and (3) ______ ______ ______ (initials).
1. posterior 2. oxytocin 3. ADH	**9–168.** Chapter 10 will pay particular attention to the genitourinary system and its relationships to the endocrine system.

10 The Genitourinary System

10–1. The GENITOURINARY SYSTEM actually consists of two distinct groups of organs, the genital or reproductive organs and the urinary organs. These organs form two distinct systems of their own within the ________________ system.

genitourinary

10–2. It is important to remember that in spite of their different functions, all the organs of both groups are included under the collective title of the ________________ system.

genitourinary

The Reproductive System

10–3. The genital or reproductive organs, which will be considered first, are grouped together to form the REPRODUCTIVE SYSTEM. As its name indicates, the function of this system is to carry out the process of human ________________.

reproduction

10–4. For the purpose of this course, the human reproductive process will be divided into four successive stages, with which you should be familiar in order to understand the functions of the individual organs of the ________________ system.

reproductive	**10–5.** Human reproduction begins with the manufacture of male and female reproductive cells by the male and female sex glands or ________.
gonads	**10–6.** Thus the first stage of the human reproductive process would be the ________ of reproductive cells.
manufacture	**10–7.** Once the reproductive cells have been produced by their respective gonads, a male and a female cell must be brought together to form the first cell of a new human being. This stage is called FERTILIZATION; a wholly new cell is formed when a female cell is ________ by a male cell.
fertilized	**10–8.** The second stage of human reproduction is the ________ of a female reproductive cell by a male cell.
fertilization	**10–9.** Another term for fertilization is CONCEPTION. When the male and female cells unite to form a new cell, this may be described as either fertilization or ________.
conception	**10–10.** The terms (1) "________" and (2) "________" are used interchangeably for the (3) ________ stage of human reproduction.
1. fertilization 2. conception (either order) 3. second	**10–11.** The new cell formed at conception or fertilization develops within the female body during PREGNANCY. The whole period from conception to birth is termed ________.
pregnancy	**10–12.** The new member of the species develops from a single cell into a fully formed infant during (1) ________, the (2) ________ stage of the reproductive process.
1. pregnancy 2. third	**10–13.** The fourth stage of reproduction is ________ itself, when the child is delivered from the mother.
birth	**10–14.** Thus the four stages of the reproductive process (as defined for the purpose of this course) are: (1) ________ (your own words), (2) ________ or ________, (3) ________, and (4) ________.

1. manufacture of reproductive cells (or equivalent statement)
2. fertilization (or) conception
3. pregnancy
4. birth

10–15. It should be pointed out that lactation, the manufacture of milk for the new infant by the mother, is also standardly included in the reproductive process, but because the mammary glands, which are responsible for lactation, are not structurally a part of the reproductive system, they will not be studied along with the ________________ of that system.

organs

10–16. The organs of the male and female reproductive systems and their respective functions will be examined successively in the following frames.
(No answer required.)

The Male Reproductive System

10–17. The primary function of the male system is the production of male reproductive cells or sperm cells. Thus, the organs that produce sperm, the male gonads or (1) ________________, are referred to as the
(2) ________________ organs of the male reproductive system.

1. testes
2. primary

10–18. The secondary or accessory function of the male system is to provide the means by which the sperm cells can travel to and fertilize the female cell or ovum. The organs that, in their various ways, help the sperm cells
(1) ________________ to the ovum are referred to as the secondary or
(2) ________________ reproductive organs.

1. travel
2. accessory

10–19. The sperm cells, you remember, travel from the testes to the outside of the body through a system of ________________.

ducts

10–20. These ducts provide the route by which the sperm cells may leave the male body on their way to the female body and are thus ________________ organs of the male.

accessory

10–21. On their way through the ducts, the sperm cells are joined by fluid substances secreted by other accessory organs, notably the PROSTATE (PROSS—tate) GLAND and the SEMINAL VESICLES (VES—ik—ls). The sperm cells will need the secretions of the (1) ________________ gland and the
(2) ________________ vesicles as a medium in which to travel to and in the female body.

1. prostate
2. seminal

10–22. Sperm cells alone cannot reach the ovum without the fluid medium provided by the secretions of the (1) ________________ gland and the
(2) ________________ ________________.

1. prostate
2. seminal vesicles

10–23. The substance that is formed by the combination of the sperm cells and the secretions of the prostate gland and the seminal vesicles is called SEMEN (SEE—men). Since the sperm cells cannot reach the ovum without their fluid medium, it is as components of this substance, ________________, that they make the rest of the trip.

semen

10–24. The sperm cells leave the male body and enter the female body, during copulation, through the PENIS (PEE—nis). The copulatory organ of the male reproductive system is the (1) ________________, which provides the means by which the sperm cells—contained in the substance called (2) ________________—can enter the female body.

1. penis
2. semen

10–25. The final accessory organ of the male reproductive system is the copulatory organ, the ________________.

penis

10–26. Thus, the principle organs of the male reproductive system are the primary organs, the (1) ________________, and the accessory organs, including the (2) ________________ through which the sperm travels; the (3) ______________ gland and the (4) ______________ ______________, which provide the fluid medium for the sperm cells; and the copulatory organ, the (5) ________________.

1. testes
2. ducts
3. prostate
4. seminal vesicles
5. penis

10–27. This drawing shows the locations of the principal organs of the male reproductive system.

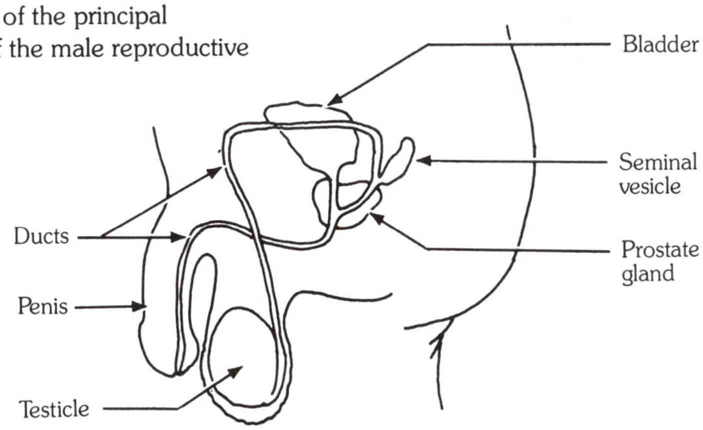

10–28. The primary function of the male reproductive system is to produce (1) ________________ cells, and its secondary function is to provide the means by which these cells can (2) ________________ to the ovum in order to fertilize it.

1. sperm
2. travel

10–29. The combination of sperm cells and their fluid medium leaves the male body and enters the female body in the form of ________________.

Female Reproductive System

(in section)

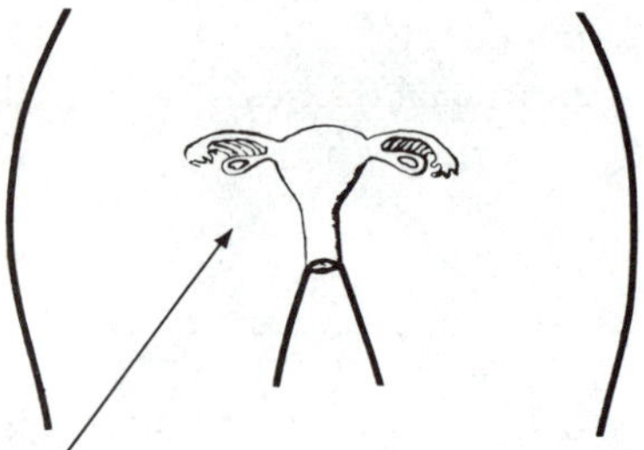

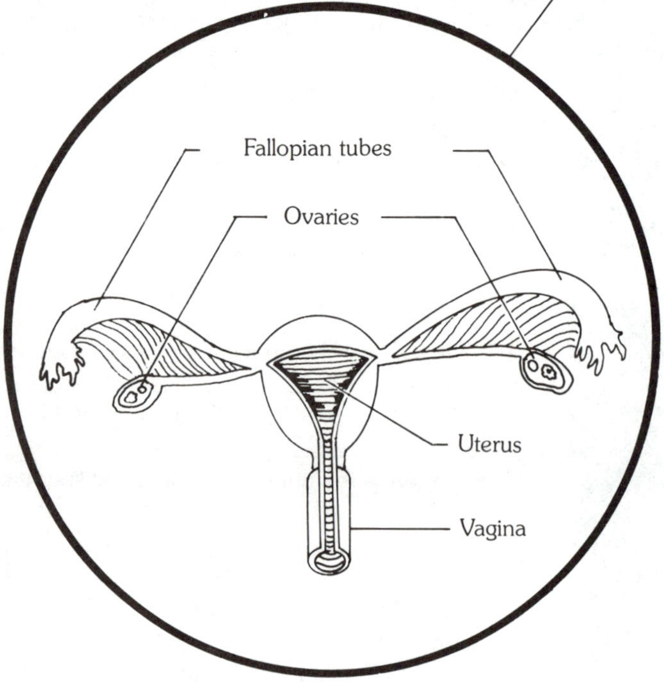

The Female Reproductive System

semen

10–30. Now refer to the drawing of the female system. The part of the female reproductive system that leads to the outside of the body is the tubular channel called the ___________________.

VAGINA
(va—JY—nah)

10–31. The sperm cells, contained in semen, enter the female body through the ___________________.

vagina

10–32. The vagina leads into the hollow pear-shaped organ called the UTERUS (YOU—ter—us). Sperm cells pass through the vagina and into the ___________________.

uterus

10–33. Look at the drawing again. Extending from the top of the uterus toward the two OVARIES (the organs that produce the ova) are two tubes called the FALLOPIAN (fal—LOPE—i—an) TUBES. Each ovary is located near the fringed funnel-shaped opening of one of the ___________________ tubes.

Each ovary is attached to the uterus by cords of dense connective tissue called ligaments.

fallopian

10–34. The ovum is produced by the ovary and expelled into the funnel-shaped opening of the ____________ ____________.

fallopian tubes

10–35. The sperm cells leave the uterus and travel up through the fallopian tubes; the ovum meanwhile travels down the tube until it meets the sperm, and ____________ takes place.

fertilization
or
conception

10–36. Thus, before the sperm cells can fertilize the ovum, they must travel much of the length of the female reproductive system, entering through the (1) ____________, passing through the (2) ____________, and traveling up the (3) ____________ ____________ until they meet the ovum produced by the (4) ____________.

1. vagina
2. uterus
3. fallopian tube (s)
4. ovary (ies)

10–37. Label the reproductive organs of the female in the drawing below.

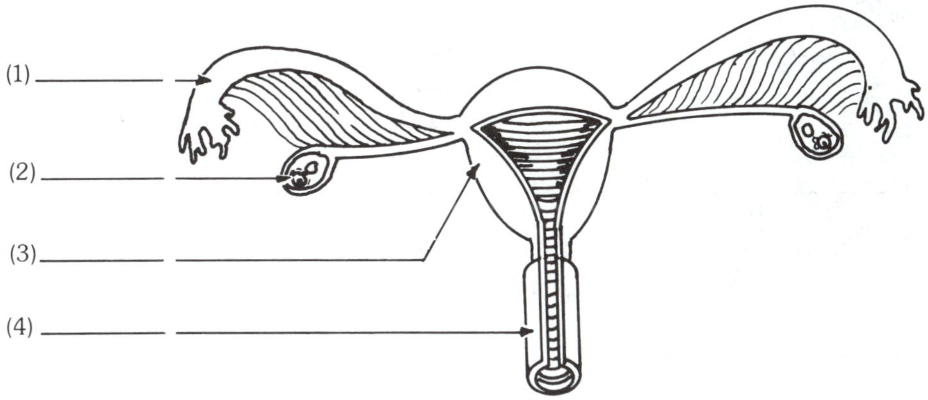

1. fallopian tube
2. ovary
3. uterus
4. vagina

10–38. Like the testes, which produce sperm, the ovaries, which produce ova, are the ____________ organs of the female reproductive system.

primary

10–39. The other functions of the female system, however, are of course completely different from those of the male system. The difference appears first of all in the way sperm and ova are produced, which will be discussed in the following frames.

(No answer required.)

10–40. The male and female gonads both begin producing their respective cells during the period of human growth called puberty. The beginning of sperm or ova production, in fact, is commonly used as the way of defining the onset of ____________ in the male or female.

puberty	**10–41.** Sperm production, once it has begun, is a continuous process; in other words, from the time the male reaches (1) ________________, his testes are secreting sperm cells in a more or less (2) ________________ flow.
1. puberty 2. continuous	**10–42.** The ovaries, on the other hand, produce ova only at regular intervals, over a definite period of time or cycle. Thus, ova production, unlike sperm production, is a (continuous/cyclical) ________________ process.
cyclical	**10–43.** The period of time during which ova are produced is called the MENSTRUAL CYCLE. As you probably know, the length of the average ________________ cycle is about 28 days.
menstrual	**10–44.** An ovary expels an ovum once every 28 days (on an average)—in other words, once during every ________________ cycle.
menstrual	**10–45.** The menstrual cycle, however, involves more than just the production of ova. Every time an ovum is expelled from an ovary, the organs of the female system must prepare for the possibility that the ovum will be ________________ by sperm cells.
fertilized	**10–46.** Every time the newly produced ovum fails to be fertilized, these preparations are abandoned, to begin all over again during the next ________________ ________________.
menstrual cycle	**10–47.** This process of producing an ovum, preparing for fertilization and the ensuing pregnancy, and abandoning these preparations when fertilization does not occur takes place in the female reproductive organs approximately every ________________ days, beginning at puberty.
28	**10–48.** The menstrual cycle is suspended only when fertilization does occur: during the ensuing ________________ and for (generally) a short time afterwards.
pregnancy	**10–49.** The following frames will examine in more detail the way in which the female reproductive organs (1) ________________ ova and (2) ________________ for the possibility that the ovum will be (3) ________________ by a sperm cell.
1. produce 2. prepare 3. fertilized	**10–50.** In the average 28-day menstrual cycle, the expulsion of the ovum from the follicle of the ovary takes place on the 14th day. The expulsion of the ovum thus marks the (middle/end) ________________ of the cycle.

middle	**10–51.** The first half of the cycle is taken up with three important processes that (precede/follow) ________________ the expulsion of the ovum.
precede	**10–52.** The first of these is the process by which the follicle-stimulating hormone of the anterior pituitary (that you encountered in Chapter 9) stimulates the (1) ________________ to grow to maturity. An ovum can be expelled only by a (2) ________________ follicle.
1. follicle 2. mature	**10–53.** The follicle requires about 14 days of pituitary stimulation to grow to ________________ .
maturity	**10–54.** During this same period, the follicle-stimulating hormone also acts to produce an increased secretion of estrogen from the ovary. As the first half of the cycle progresses, more and more ________________ is secreted by the ovary.
estrogen	**10–55.** The estrogen secreted by the ovary in turn stimulates an increased growth of the cells in the lining of the uterus or womb—thus preparing for the possibility that the ovum will be fertilized. If the ovum is fertilized, it will lodge ultimately in the ________________ of the uterus to continue its development.
lining	**10–56.** The second process taking place in the first half of the menstrual cycle leading up to the expulsion of the ovum is the preparation of the (1) ________________ of the (2) ________________ or womb for the possibility of pregnancy.
1. lining 2. uterus	**10–57.** The third process that must take place before an ovum is expelled is the formation of the ovum itself from its parent cell in the follicle. The follicle begins each cycle as a cluster of follicle cells enclosing a single ________________ cell.
parent	**10–58.** As the first half of the cycle progresses, the follicle cells and this enclosed ________________ ________________ mature simulataneously.
parent cell	**10–59.** About 2 or 3 days before the middle of the cycle, the parent cell begins to divide. Reproductive cell division, however, is quite different from the process of ________________ undergone by other cells.
division	**10–60.** The ovum that results from the division of the parent cell contains only half the usual number of chromosomes. When the ovum is ready for expulsion from the follicle, in other words, it contains only ________________ (number) chromosomes.

23	**10–61.** The ovum contains only 23 chromosomes because it is only one of the two parties to the new cell that will be formed if fertilization occurs. The sperm cells, as you may guess, have gone through a similar process of (1) ______ and also have only 23 (2) ______.
1. division 2. chromosomes	**10–62.** The cell that will be formed if the ovum unites with a sperm will thus have the full complement of ______ (number) human chromosomes.
46	**10–63.** The process of division undergone by the parent cell in the follicle is therefore another way in which the female reproductive organs are ______ for the possibility of fertilization.
preparing	**10–64.** By the 14th day of the cycle, the follicle is fully (1) ______, and the (2) ______ of the uterus is developed. The ovum—containing (3) ______ chromosomes—is expelled from the follicle into the funnel-shaped end of the (4) ______ tube.
1. matured 2. lining 3. 23 4. fallopian	**10–65.** If the ovum is not fertilized after it is expelled from the follicle, it disintegrates in the fallopian tube, and the cycle continues. As you know, the menstrual cycle runs out its 28-day period every time that fertilization (does/does not) ______ occur.
does not	**10–66.** The estrogen secretion of the ovaries begins to decrease with the expulsion of the ovum. As the second half of the cycle progresses, the cells in the lining of the uterus receive progressively (less/more) ______ estrogen stimulation.
less	**10–67.** By the 28th or last day of the cycle, the lining begins to deteriorate. The flow of blood through the vagina that punctuates the menstrual cycle is caused by the ______ and subsequent shedding of the lining of the uterus.
deterioration	**10–68.** You should note that the onset of bleeding is considered to mark the beginning of the menstrual cycle; in other words, the bleeding period occupies the (first/last) ______ few days of the cycle.
first	**10–69.** After an average bleeding period of about 4 days, the follicle begins to mature again; the production of estrogen by the ovaries increases, and a new menstrual ______ is under way.
cycle	**10–70.** The following frames will examine what happens when fertilization does occur, and the menstrual cycles are suspended during the ensuing pregnancy. (No answer required.)

Pregnancy

10–71. If sperm is present in the fallopian tube, the ovum will probably be fertilized. The newly fertilized ovum travels slowly down the tube. By about the tenth day after it was expelled from the follicle, it has become attached to the ________________ of the uterus.

lining

10–72. As estrogen secretion diminishes, the anterior pituitary stimulates the ovaries to produce the hormone progesterone (pro—JEST—er—own), which was mentioned in the first part of this unit. The job of maintaining the lining of the uterus is largely taken over from estrogen by ________________.

progesterone

10–73. As estrogen is the primary sex hormone of the female, so ________________ is the secondary sex hormone.

progesterone

10–74. Progesterone is secreted by the follicle from which the ovum was expelled. A follicle that has expelled its ovum is yellow in color and is called a CORPUS LUTEUM (LU—tee—um), "corpus" meaning "body" and "________________" meaning "yellow."

luteum

10–75. By stimulating the continued growth of the cells in the lining of the uterus where the fertilized ovum is housed, the ________________ ________________, which produces progesterone, plays an important role in early pregnancy.

corpus luteum

10–76. The pituitary hormone primarily responsible for stimulating the corpus luteum to produce progesterone is the LUTEINIZING (LOO—tee—in—izing) HORMONE, abbreviated LH. What the follicle-stimulating hormone (FSH) is to estrogen, the ________________ ________________ (LH) is to progesterone.

luteinizing hormone

10–77. Fill in the blanks below:

The ANTERIOR PITUITARY secretes (1) ________________ ________________ hormone, which promotes the maturing of the follicle and increases the production of (2) ________________, and (3) ________________ hormone, which promotes the growth of the corpus luteum and increases the production of (4) ________________.

1. follicle-stimulating
2. estrogen
3. luteinizing
4. progesterone

10–78. You should note, however, that although the luteinizing hormone stimulates a sharp rise in the production of progesterone after the expulsion of the ovum, the production of estrogen by no means ceases; and the levels of both progesterone and estrogen remain high throughout pregnancy.

10–79. The early stages of pregnancy are marked by the development of two organs essential to the growth of the child in the uterus. These organs, as well as the child itself, develop from the fertilized ________________.

ovum

10–80. After the fertilized ovum has become embedded in the lining of the uterus, a part of it develops into the pancakelike organ called the PLACENTA (pla—SENT—ah). It is through this organ, the ________________, that the developing child will receive its nourishment.

placenta

10–81. The placenta absorbs nutrients from the mother's blood stream and transmits them through the membranous UMBILICAL (um—BIH—li—kl) CORD to the developing child. At the same time, waste materials are returned from the child through the (1) ________________ cord to the (2) ________________ for disposal through the mother's bloodstream.

1. umbilical
2. placenta

10–82. The blood supply of the developing child is shared by the (1) ________________, which absorbs the child's nutrients from the mother, and the (2) ________________ ________________ through which these nutrients pass.

1. placenta
2. umbilical cord

10–83. It is important to note, however, that the blood supply of the mother is completely separate from the blood supply shared by the child, the umbilical cord, and the placenta. Although nutrients and waste materials pass across the placenta, the blood of the mother and the blood in the placenta (do/do not) ________________ intermix.

do not

10–84. By the time the placenta and the umbilical cord have been formed (during the second month of pregnancy), all the basic mechanisms are established that will support the developing child until ________________—approximately 9 months after conception.

birth

10–85. The processes by which a fully-formed child is developed in the uterus from a single fertilized ovum are too complex to be covered in the present course. You should, however, be familiar with the terminology that applies to the stages of the child's ________________ within the uterus.

development

10–86. The first stage of pregnancy is called the period of the ovum. The period of the ovum begins when the ovum is fertilized by the sperm cell and ends when it becomes embedded in the lining of the uterus. Thus, the period of the ________________ covers about the first 10 days of pregnancy.

ovum

10–87. The period of the ovum is followed by the period of the EMBRYO (EM—bree—o), which lasts until the end of the third month of pregnancy. The fertilized ovum is called an (1) ________________ from the time it becomes (2) ________________ in the lining of the uterus until the end of the (3) ________________ month in the uterus.

1. embryo 2. embedded 3. third	**10–88.** By the end of the third month, the placenta and the umbilical cord are fully formed, and most of the organs of the child have begun to take recognizable shape. These developments take place during the period of the ____________, the second stage of pregnancy.
embryo	**10–89.** The expulsion of an embryo from the uterus for any reason during this period is called an EMBRYONIC ABORTION. The term (1) ____________ ____________ applies to expulsion of the developing child by any cause, whether spontaneous or induced, during the period of the embryo, that is, up to the end of the (2) ____________ month of pregnancy.
1. embryonic abortion 2. third	**10–90.** The third and final stage of pregnancy is the period of the FETUS (FEE—tus). From the beginning of the fourth month until birth, the developing child is called a ____________.
fetus	**10–91.** After about seven full months in the uterus, the developing child becomes viable, or capable of surviving expulsion from the uterus. The developing child—called a (1) ____________ during this period—is (2) ____________ during the last two months of the normal term of pregnancy.
1. fetus 2. viable	**10–92.** The term "miscarriage" means the expulsion of the developing fetus from the uterus before it is capable of surviving, that is, before it becomes ____________.
viable	**10–93.** Miscarriage, however, is a lay term; in medicine the term abortion is used also for expulsion of the fetus before it becomes viable, before the end of the ____________ month of pregnancy.
7th	**10–94.** Thus the three stages of pregnancy are: the period of the (1) ____________, about the first (2) ____________ days after the conception; the period of the (3) ____________, up to the end of the (4) ____________ month; and the period of the (5) ____________.
1. ovum 2. 10 3. embryo 4. third 5. fetus	**10–95.** The term abortion refers to the expulsion of either the embryo (embryonic abortion) or the fetus so long as it is nonviable, that is, before it becomes ____________ ____________ ____________. (Your own words.)
capable of surviving expulsion from the uterus (or equivalent statement)	**10–96.** With the delivery of a viable or live child, birth—the most important business of the reproductive process—is completed. The female reproductive organs will go through a short period of rest and readjustment to be ready to begin the process again. Before you begin the section on the urinary system, you may take a break.

urine

urinary

is

The Urinary System

10–97. As the organs of the reproductive system perform the life funciton of reproduction, the organs of the urinary system perform the life function of excretion. The urinary organs, as their name indicates, excrete the waste materials of metabolism from the body in the form of ________________.

10–98. Approximately 75% of the excretory function of the body is performed by the organs of the ________________ system.

10–99. The waste products of metabolism can be harmful and even fatal if allowed to accumulate in the body. Thus, the proper functioning of the urinary system, which plays the largest role in removing these materials, (is/is not) ________________ essential to the health of the body.

Urinary System

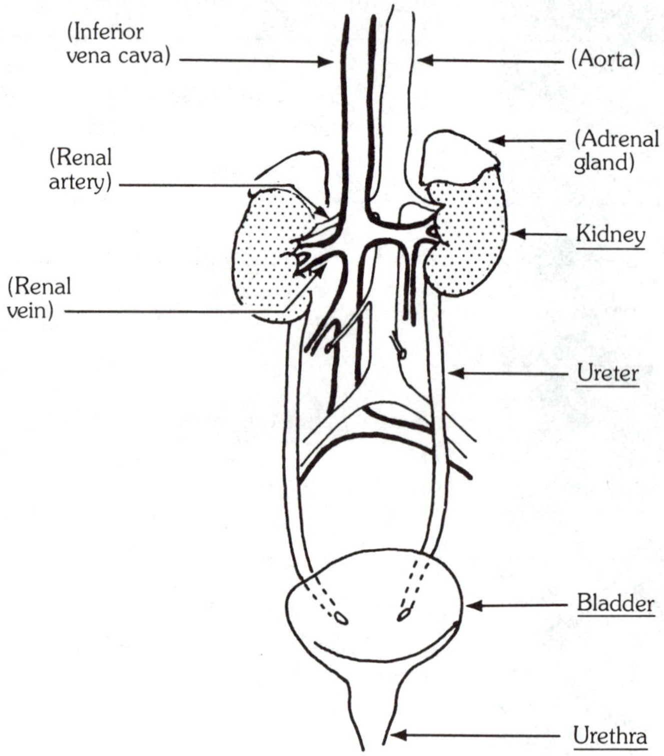

10–100. Look at the drawing of the urinary system, in which the names of the principal urinary organs are underlined.

The most important of these organs are the two KIDNEYS, which are responsible for the formation of urine. Urine is formed by the (1) ________________ and passes into two narrow tubes called (2) ________________.

1. kidneys 2. URETERS (YOOR—et—ers)	**10–101.** Look at the drawing. Urine passes from the kidneys, where it is formed, through the tubes called (1) ________ and into the (2) ________, where it gradually accumulates.
1. ureters 2. bladder	**10–102.** As the bladder fills with urine, its muscular walls distend until the point is reached—usually with about a pint of urine—when the sensory nerves in the walls alert the brain that the ________ is full.
bladder	**10–103.** Look at the drawing again. When the bladder is full, urine is discharged from the body through the ________.
urethra (yoo—REE—thra)	**10–104.** In the male, both urine and semen leave the body by the same route: through the ________ of the penis.
urethra	**10–105.** In the female body, on the other hand, the opening of the urethra is anterior to that of the vagina and quite separate from it; the reproductive and the urinary systems of the female (do/do not) ________ share a common pathway.

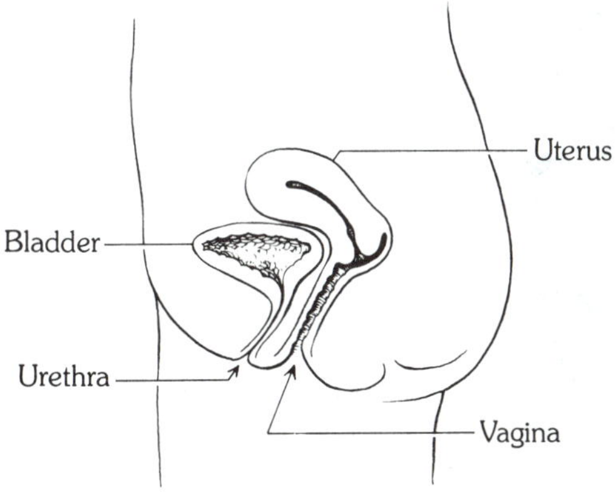

do not	**10–106.** Look at the drawing of the urinary system again. As you can see, the kidneys are in very close touch with the largest blood vessels in the body. The kidneys produce urine initially from the waste materials in the ________ that circulates through them.
blood	**10–107.** A single RENAL ARTERY normally supplies all the blood to a single kidney. Once inside the kidney, however, a vast system of arterioles and capillaries branch and rebranch from this one ________ artery.
renal	**10–108.** The arterioles and capillaries within the kidney form into duct systems called NEPHRONS (NEF—rons), which are the functional units of the kidneys. The kidney's function of producing urine is performed by the ________.

nephrons

10–109. Each kidney contains about one million of these functional units or ________________.

nephrons

10–110. The capillaries of the nephron form together into a cluster called the ________________.

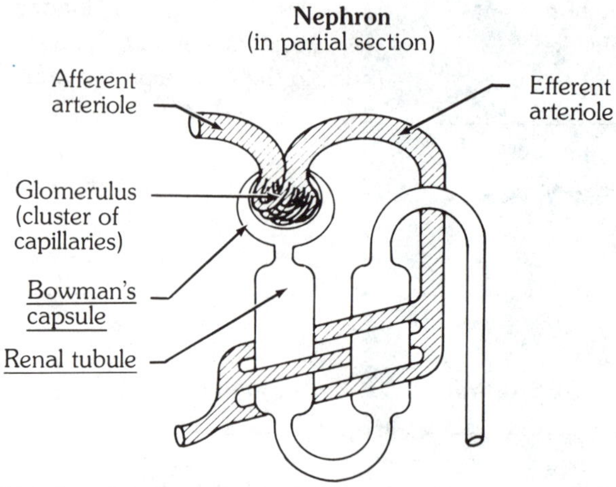

GLOMERULUS (glo—MARE—you—lus)

10–111. Blood is carried into the glomerulus through an afferent arteriole. Blood is carried out of the (1) ________________ through an (2) ________________ arteriole.

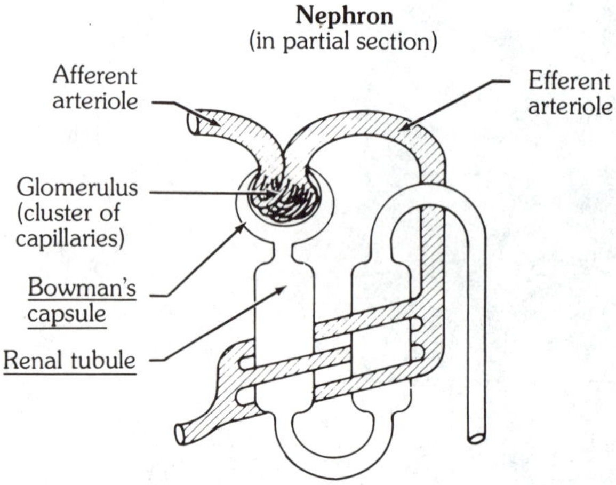

1. glomerulus
2. efferent

10–112. Each glomerulus is enclosed by a funnellike structure called the ________________ ________________.

Bowman's capsule

10–113. A tubule called the (1) ______________ tubule leads out of the (2) ______________ ______________, which encloses the glomerulus.

Nephron
(in partial section)

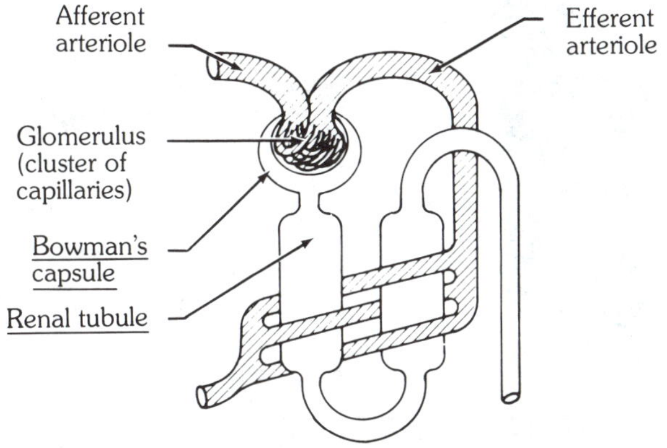

1. renal
2. Bowman's capsule

10–114. Fill in the blanks in the drawing below.

Nephron
(in partial section)

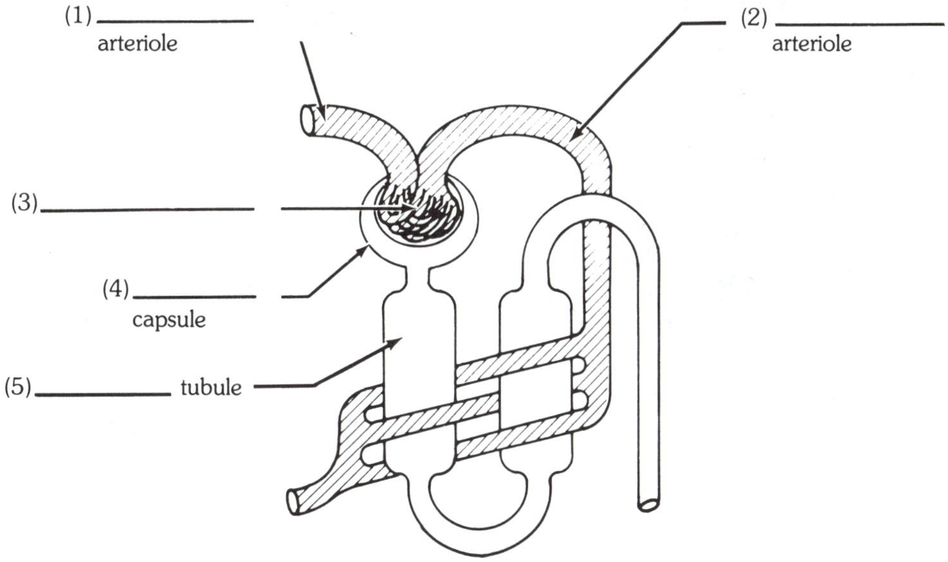

1. afferent
2. efferent
3. glomerulus
4. Bowman's
5. renal

10–115. The efferent arteriole, which carries the blood out of the (1) ________________, branches again into capillaries that are elaborately intertwined with the renal tubule that leaves the (2) ________________ ________________.

Nephron
(in partial section)

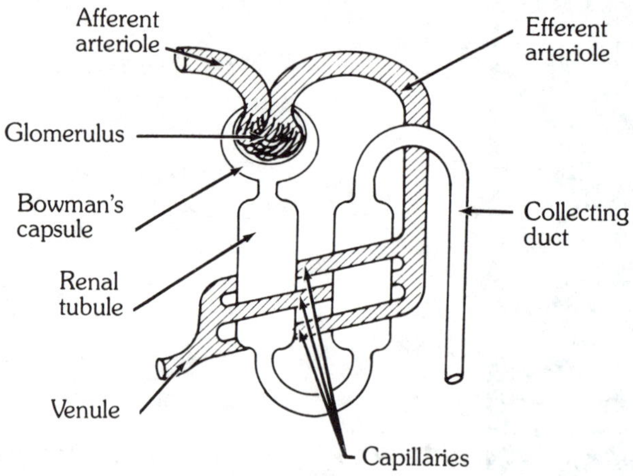

1. glomerulus
2. Bowman's capsule

10–116. The capillaries from the efferent arteriole that twine around the renal tubule empty ultimately into a single ________________.

Nephron
(in partial section)

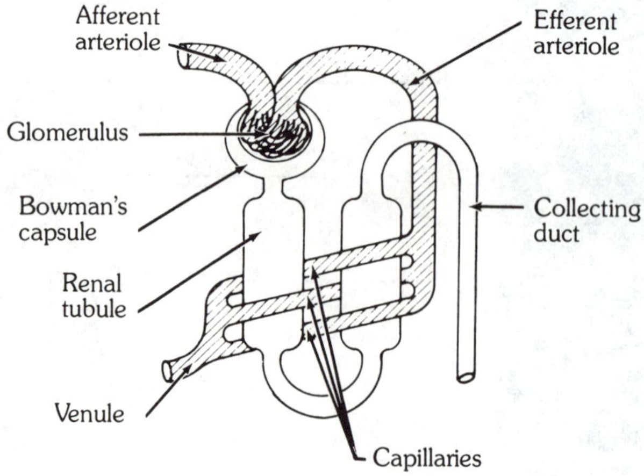

venule

10–117. The blood that entered the glomerulus from the arterial system through the afferent arteriole is returned to the venous system through the ________________.

venule

10–118. Urine is formed in the renal tubule, which leads into the ________________ duct.

collecting

10–119. Urine is carried ultimately to the ureters through the (1) ______ duct, which receives the urine from the (2) ______ tubule in which it was formed.

1. collecting
2. renal

10–120. Fill in the blanks in the drawing below.

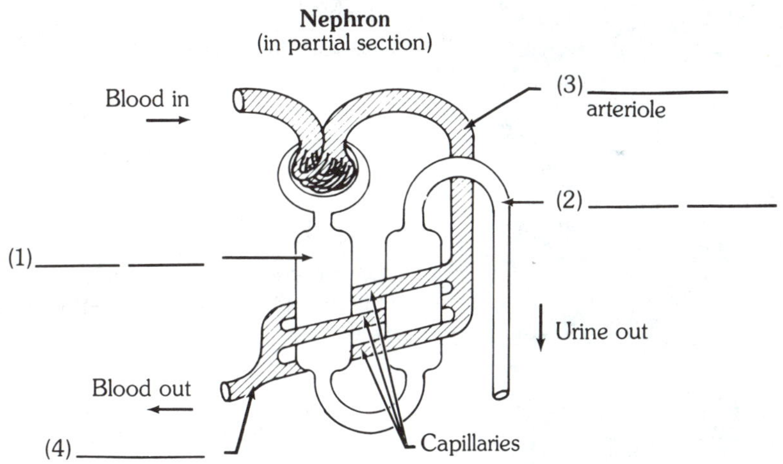

1. renal tubule
2. collecting duct
3. efferent
4. venule

10–121. The functional units of the kidney are the (1) ______, of which each kidney contains about (2) ______ ______ (number).

1. nephrons
2. one million

10–122. Urine is formed within the nephron, in the renal (1) ______, which begins at a funnellike structure called the (2) ______ ______ and leads out of the nephron into a (3) ______ ______.

1. tubule
2. Bowman's capsule
3. collecting duct

10–123. Blood enters the glomerulus from the arterial system through an afferent (1) ______, which carries it into the cluster of capillaries called the (2) ______. Blood passes out of this cluster again into an (3) ______ ______, which branches again into capillaries that intertwine with the renal tubule. Blood leaves the nephron for return to the venous system through a (4) ______.

1. arteriole
2. glomerulus
3. efferent arteriole
4. venule

10–124. The following frames will cover the processes involved in the production of urine.

(No answer required.)

10–125. The first stage of urine production is the filtration of blood by the glomerulus. As blood flows through the capillaries of the glomerulus, about 20% of it is (1) ______ through the walls of the capillaries and into the Bowman's (2) ______.

1. filtered
2. capsule

10–126. Two of the normal constituents of whole blood are too large to pass through the walls of the capillaries: particles of serum protein and the blood cells. The substance that is filtered into the Bowman's capsule, then, is simply blood minus its particles of serum (1) ______________ and its (2) ______________.

---

1. protein
2. cells

10–127. The substance filtered into the Bowman's capsule by the glomerulus is called the glomerular filtrate. The glomerular filtrate represents about 20% of the volume of blood passing through the glomerulus, and consists of blood minus (1) ______________ ______________ and blood (2) ______________.

---

1. serum protein
2. cells

10–128. The force required to produce the glomerular filtrate is provided by the blood pressure.

More or less than 20% of the blood in the glomerulus may be filtered through the capillary walls according to variations in the blood ______________ of the individual.

---

pressure

10–129. The glomerular filtrate is funneled from the Bowman's capsule into the renal ______________.

---

tubule

10–130. The glomerular filtrate is not urine. Whole blood, as you know, contains many substances needed by the body, and all of these (except protein and blood cells) are filtered indiscriminately into the renal tubule. It is essential to the health of the body that these substances (be/not be) ______________ allowed to leave the body in urine.

---

not be

10–131. The process by which substances essential to the body are transferred out of the glomerular filtrate and back into the blood stream is called reabsorption. Electrolytes, glucose, and water are the most important of the substances returned to the blood by the process of ______________.

---

reabsorption

10–132.

Reabsorption

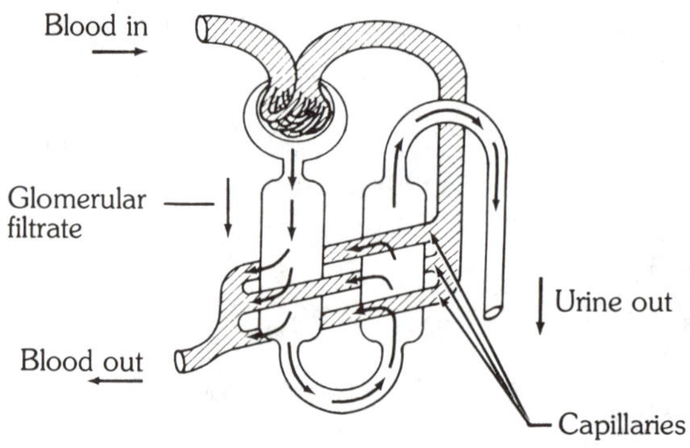

(No response required.)

	10–133. "Reabsorption" means that as the glomerular filtrate flows through the renal tubule, the cells in the walls of the tubule itself reabsorb whatever substances are needed and pass them out into the ________ that are twined around the tubule.
capillaries	**10–134.** Over 99% of the glomerular filtrate is normally (1) ________ into the walls of the tubule and returned to the (2) ________ in the surrounding capillaries.
1. reabsorbed **2. blood**	**10–135.** The renal tubule, however, is very selective about what it reabsorbs. In a manner not yet fully understood, the tubule is able to ________ for reabsorption just the amount of any one substance actually needed by the blood.
select	**10–136.** Thus the glomerular filtrate contains the charged ions that make up the electrolyte content of blood. The renal tubule is able to determine whether there is a healthy balance between the positively and negatively charged ________ in the filtrate.
ions	**10–137.** If there are too many ions of either charge, the renal tubule will not reabsorb the excess ions, leaving them to be excreted in the urine. By thus returning a balanced proportion of electrolytes to the blood stream, the renal tubule plays a vital role in maintaining the ________ balance of the whole body.
electrolyte	**10–138.** You recall from Chapter 1 that acid-base balance is an important chemical characteristic of body fluids and that buffer systems are regulatory mechanisms to achieve this ________.
balance	**10–139.** Because metabolic waste products contain a high concentration of acids, some of the acid substances in the blood are excreted when urine is formed. Conversely, the bases, available to buffer the (1) ________ are reabsorbed. Reabsorption of bases is important to (2) ________ the acids formed during the ongoing metabolic processes.
1. acids **2. buffer**	**10–140.** Besides electrolytes, another substance selectively reabsorbed by the tubule is glucose or sugar. When the glucose level in the filtrate is within normal limits, the walls of the tubule reabsorb all of it; in other words, no ________ or sugar is excreted.
glucose	**10–141.** Glucose is excreted in urine only when the level of glucose in the filtrate rises above healthy limits, and the excess (is/is not) ________ reabsorbed by the tubule.

is not	**10–142.** Thus the renal tubule regulates not only the (1) ____________ balance but also the level of (2) ____________ ____________ in the blood.
1. electrolyte 2. glucose (sugar)	**10–143.** A third substance reabsorbed by the renal tubule is water. Of the total volume of water in the filtrate, the tubule normally returns about 85% to the capillaries in a passive way and exercises its selective faculty over the remaining ____________ %.
15	**10–144.** How much of this remaining 15% will be reabsorbed by the tubule is determined by a posterior pituitary hormone called the ANTIDIURETIC (an—ti—die—you—RET—ik) HORMONE. The posterior pituitary is stimulated to produce ____________ hormone by changes in the body's water balance.
antidiuretic	**10–145.** "Diuretic" means "tending to increase the volume of urine excreted." The volume of urine is made up almost entirely of water. Thus the antidiuretic hormone would act to (increase/decrease) The amount of water left to be excreted in urine.
decrease	**10–146.** When the level of water in the body falls below normal—for instance, through excessive sweating—the renal tubule is stimulated to reabsorb whatever amount of water is needed by an increased production of ____________ hormone from the posterior pituitary.
antidiuretic	**10–147.** Similarly, when the body takes on more than the usual amount of water, the pituitary will produce (1) ____________ (more/less) antidiuretic hormone, and the renal tubule will reabsorb (2) ____________ (more/less) water for return to the blood.
1. less 2. less	**10–148.** Antidiuretic hormone is commonly referred to by its initials, ADH. You should remember that ADH acts to increase reabsorption of only the ____________ % of water in the filtrate not normally reabsorbed by the tubule as a matter of course.
15	**10–149.** Another factor besides ADH that influences water reabsorption is the RENAL SOLUTE LOAD: the load of ____________ or substances dissolved in the renal tubule.
solute(s)	**10–150.** The load or concentration of solutes filtered into the renal tubule influences water reabsorption through its effect on the process of osmosis, which you learned in an earlier chapter is the tendency of your body fluids to pass through semipermeable membranes from the side with the (1) ____________ (lower/higher) concentration of solutes to the side with the (2) ____________ (lower/higher) concentration.

1. lower 2. higher	**10–151.** The passive reabsorption of the first 85% of water takes place by osmosis; that is, as the solutes (glucose, ions, etc.) initially filtered into the renal tubule are selectively reabsorbed, their concentration rises inside the ________ surrounding the renal tubule, and the water simply follows them.
capillaries	**10–152.** However, if there is an abnormally high renal solute load—many more solutes in the filtrate than are reabsorbed—the concentration in the tubule may be such that an excessive amount of water (passes out of/remains in) ________ the tubule and is subsequently secreted.
remains in	**10–153.** Thus the effect of an abnormally high renal (1) ________ load is typically diuretic: it tends to (2) ________ the amount of urine excreted.
1. solute 2. increase	**10–154.** The diuretic effect of an abnormally high (1) ________ ________ ________ may be harmful if it results in the loss of (2) ________ that should otherwise have been retained in the body.
1. renal solute load 2. water	**10–155.** The third part of urine production is <u>secretion</u>: substances are transferred out of the (1) ________ in the capillaries and into the cells of the tubule, which then (2) ________ these substances into the filtrate.
1. blood 2. secrete	**10–156.** The substances secreted into the filtrate are foreign elements in the blood that were not filtered into the Bowman's capsule by the glomerulus. These ________ substances are excreted from the body in urine.
foreign	**10–157.** The formation of urine is now completed. Blood has been (1) ________ through the glomerulus into the Bowman's capsule, which passes it into the renal tubule. Substances essential to the body have been (2) ________ by the tubule and returned to the capillaries entwined around the tubule. Foreign elements have been (3) ________ into the filtrate.
1. filtered 2. reabsorbed 3. secreted	**10–158.** The water volume of the glomerular filtrate has been reduced first by passive (osmotic) reabsorption and then by the action of ADH until the end product is highly concentrated; this end product is the urine, which passes through and eventually leaves the nephron, the kidney, the bladder, and finally the ________.
body	**10–159.** This is the end of the chapter on the Genitourinary system.

11 The Special Senses

organs

11–1. The subject of this chapter is special sense ability. The capacities to touch, smell, taste, see, and hear are due to the functional integration of several different structural types of tissue. The study of special sense abilities requires the study of special sense ____________________.

1. muscle
2. nervous

11–2. The function of the special sense organs is a result of the integrated action of particular tissues to receive a stimulus, interpret it, and respond to it. The tissue types involved, therefore, are connective, (1) ____________________ and (2) ____________________ tissues.

receptor

11–3. A review of the basic pathways of the nervous system is in order. The particular part of the body that receives a stimulus is a ____________________.

11–4. Each of the sense (1) ____________________ is, therefore, a (2) ____________________ organ.

1. organs 2. receptor	**11–5.** After an impulse leaves a receptor organ it travels by (1) ________ pathways of the peripheral nervous system to the CNS. The parts of the CNS, to which the impulse can be transmitted are the (2) ________ or (3) ________ ________.
1. sensory (afferent) 2. brain 3. spinal cord	**11–6.** If an impulse is directed to the dorsal spinal column, there are four possible paths for the impulse to take: (1) to the ________ column on the same side. (2) cross through the ________ gray matter and travel to the ________ column on the opposite side. (3) to travel upward to the ________. (4) to travel down the spinal column.
1. ventral 2. horns (commissure) 3. brain	**11–7.** When the impulse takes routes (1) or (2) in Frame 6, a (1) ________ response occurs almost immediately because the efferent pathway stimulates (2) ________ muscle. This type of pathway is called a (3) ________ ________.
1. motor 2. skeletal 3. reflex arc	**11–8.** Although the reflex arc will protect you from danger, it <u>will not</u> help you learn about the stimuli in your environment. To learn about stimuli, the impulses must travel up the spinal column by (1) ________ tracts to the (2) ________ ________ and (3) ________. ## Touch
1. ascending (spinothalamic) 2. brain stem 3. brain	**11–9.** When you touch a flame, your body protects itself by a reflex act of pulling your hand away. However, so that you may <u>LEARN</u> that flames are dangerous, the impulse travels also by spinothalamic tracts to the (1) ________. From the (2) ________ in the thalamus, the impulse is transmitted to the (3) ________ (motor/sensory) area of the cerebral cortex.
1. brain 2. nuclei 3. sensory	**11–10.** Cutaneous, or (1) ________, sensations can be felt when a stimulus exerts an effect on (2) ________ sites, just below the skin surface, and the impulse is transmitted by way of the sensory pathways of spinal nerves to the CNS.
1. skin 2. receptor	**11–11.** Cutaneous sensations include pressure, touch, cold, and ________. ## Smell
heat	**11–12.** A second special sense is the ability to detect and interpret odors, or to ________. The organ you think of when considering the ability to smell is the nose.

smell	**11–13.** The receptor sites for odor-stimuli are located in the lining of the upper part of the nose. From these sensory cells impulses travel along afferent pathways to the (1) ________________. The first pair of cranial nerves serves as the (2) ________________ pathway.
1. brain 2. afferent (sensory)	**11–14.** You should recall that the cranial nerves are comprised either of (1) ________________ fibers or (2) ________________ fibers or both.
1. sensory 2. motor	**11–15.** The OLFACTORY (ol—FAC—to—ry), or first cranial, nerve is composed of ________________ fibers, only.
sensory	**11–16.** When an odor stimulates the (1) ________________ sites in the lining of the nostrils, impulses are sent to the (2) ________________ by the sensory fibers of the olfactory nerve. **Taste**
1. receptor 2. brain	**11–17.** The next special sense you will learn about is taste. The receptor areas for taste stimuli are on the tongue. These ________________ sites are called TASTE BUDS.
receptor	**11–18.** Taste buds are distributed in several areas of the tongue. If you look at your tongue in a mirror you can see tiny, rounded elevations of tissue. These tiny mounds of tissue are called PAPILLAE (pa—PIL—ay), and the ________________ ________________ (receptors) are on the papillary surface.
taste buds	**11–19.** The afferent, or (1) ________________, pathways from the receptors, or taste buds, to the CNS are along the seventh and ninth cranial nerves. The part of the CNS responsible for interpretation of taste stimuli is the (2) ________________ ________________, the area responsible for hunger and the securing of food, necessary to survival.
1. sensory 2. limbic system	**11–20.** When you take a bite of a candy bar, you taste a sweet flavor because the receptor sites, (1) ________________ ________________, on your tongue are stimulated. The nerve impulses are transmitted to your brain on the fibers of the seventh (2) ________________ ________________.
1. taste buds 2. cranial nerve	**11–21.** The sense of taste is related, closely, to the sense of smell. When you have a "stuffy" nose, both your sense of (1) ________________ and your sense of (2) ________________ are changed.

Answers	Frames
1. smell 2. taste	**11–22.** Since several pairs of cranial nerves, the limbic system, and receptors located in the (1) ____________, all operate together to help you taste food, the sense of taste is a product of (2) ____________ ____________ tissue.

Vision

Answers	Frames
1. tongue 2. functionally integrated	**11–23.** The sensory capacity to see involves even more complex functional relationships among various tissues and fluids. The sense organ responsible for vision is the ____________.
eye	**11–24.** The eye is a complex structure that has the abilities to (1) allow certain amounts of light to ____________ through the PUPIL; (2) bend the ____________ rays so that an accurate, 3-dimensional object will be seen; (3) undergo chemical changes when light stimulates the ____________ areas; and, (4) send messages to the occipital lobe of the ____________ by way of the afferent pathway of the third cranial nerve.
1. enter 2. light 3. receptor 4. brain	**11–25.** In order to "see" an object, your eye must allow light rays to enter through the ____________, the dark rounded area in the center of your eye. If you look in a mirror that is dimly lit, you will see your pupil get larger (DILATE). Then, if you turn on a bright light, your pupil will quickly become smaller (CONSTRICT).
pupil	**11–26.** When the brightness of the light around you changes, special muscles work to make the ____________ larger or smaller.
pupil	**11–27.** The IRIS, the shaded (blue, gray, brown, or green) part of your eye consists of two layers of smooth muscle. The (1) ____________ functions to dilate or constrict your pupil, depending on how much (2) ____________ is available.
1. iris 2. light	**11–28.** The structures of the eye that give it the ability to allow specific amounts of light to enter are the (1) ____________ and (2) ____________.
1. iris 2. pupil	**11–29.** Light must enter the eye in certain amounts. Also the light rays must be bent into particular angles so that the (1) ____________ you look at can be seen in three- (2) ____________ form.

1. objects 2. dimensional	**11–30.** The ________________ of light rays is a second important ability of the eye. Bending of light rays is called REFRACTION. Specialized structures are needed to perform this function.
bending	**11–31.** Structures that lie in front of and behind the pupil are responsible for light ray refractions, or ________________. These structures are the CORNEA (KOR—ne—a) and the LENS.
bending	**11–32.** The illustration in frame 32 shows light ray refraction. The cornea is composed of fibrous tissue and bends light rays so they can pass through the ________________. 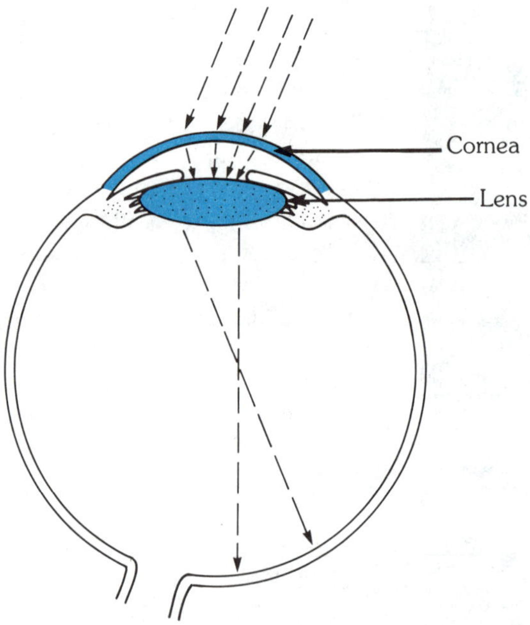
pupil	**11–33.** The lens is also composed of fibrous tissue. After light rays have passed through the pupil, the lens ________________ or bends them. In the spaces between the cornea and lens and in the larger area behind the lens are special fluids that help to maintain the shape of the eyeball.
refracts	**11–34.** The ability of the eye to refract light rays is due to the bending actions of the (1) ________________ and the (2) ________________.
1. cornea 2. lens	**11–35.** Refer to the illustration in frame 36. You see a thin layer of tissue covering the eyeball. This covering is the SCLERA (SKLER—a) and is composed of fibrous connective tissue. The purpose of the ________________ is to protect and support the eyeball.

sclera

11–36. Once the light rays have passed through the lens, they CONVERGE to a particular area on the inner lining of the eyeball.

The inner lining of the eyeball is called the ____________.

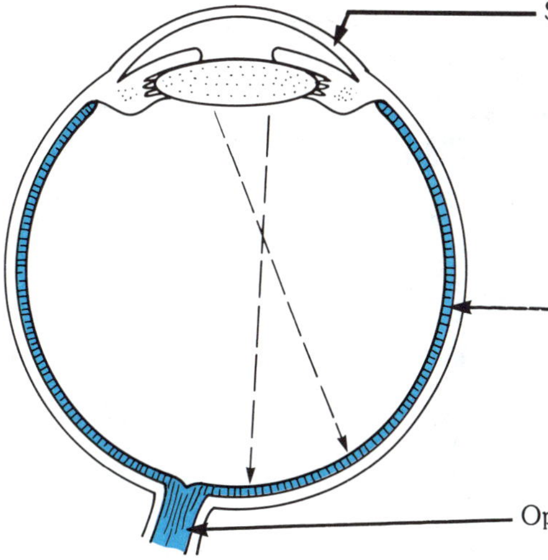

retina

11–37. The retina is the part of the eyeball on which visual stimuli arrive to be transmitted to the CNS. The neurons of the retina are called RODS and CONES. The rods and cones are the ____________ sites for visual stimuli.

receptor

11–38. When the receptor sites, the (1) ____________ and (2) ____________, are stimulated, their chemical composition undergoes changes.

1. rods
2. cones

11–39. The chemical products that result from the changes brought about by light stimulate the optic nerve (cranial nerve 3). It is the (1) ____________ fibers of the (2) ____________ ____________ that transmit impulses from the retina to the CNS.

1. sensory (afferent)
2. optic nerve

11–40. The optic nerve transmits the impulses to the area of the CNS concerned with vision, the ____________ lobe of the cerebral cortex.

occipital

11–41. Functioning of the eye and its specialized structures to receive light rays and transmit impulses to the CNS requires the functional integration of several dissimilar tissue types. The eye is a special sense ____________.

organ

11–42. The sclera is a thin outer layer of fibrous connective tissue that acts to (1) ____________ and (2) ____________ the eyeball.

1. support 2. protect	**11–43.** The (1) ________________ and lens are made of fibrous tissue and perform the action of (2) ________________ light rays.
1. cornea 2. refracting or bending	**11–44.** Muscle tissue is necessary to control the amount of light that enters the eye. The ________________ is the structure composed of smooth muscle tissue.
iris	**11–45.** After light rays have been conducted into the eyeball and converge, nervous tissue is necessary for the transmission of the stimuli to the brain. The retina acts by chemical changes in the rods and ________________ to produce impulses.
cones	**11–46.** The sensory impulses are collected on the fibers of the third (1) ________________ nerve and are transmitted to the (2) ________________ lobe, the area of the brain involved in vision.

Hearing

1. cranial 2. occipital	**11–47.** The last special sense you will learn about is the ability to hear. The organ that is involved in your capacity to hear is the ________________. Just as you discovered when studying the eye, learning about the structure and function of the ear is not as simple as it may appear.
ear	**11–48.** Like the eye, the ear is a complex sense <u>organ</u> that has specialized structures with the abilities to (1) permit <u>SOUND WAVES</u> to ________________ through the external ear; (2) conduct the ________________ waves to the internal structures of the ear; (3) receive the sound stimuli at ________________ sites; and, (4) send messages to the temporal lobe of the ________________ on the afferent pathway of the eighth cranial nerve.
1. enter 2. sound 3. receptor 4. brain	**11–49.** The ear is an organ. It is an organization of tissue that is dissimilar in <u>structure</u>, but is integrated in ________________.
function	**11–50.** The function of the ear is to collect (1) ________________ waves, conduct them to internal structures and to change the sound waves to nerve (2) ________________, and transmit them to the brain.

1. sound 2. impulses	**11–51.** The histologically different types of tissue that make up the ear sense organ are (1) cartilaginous ________________ tissue of the external ear that makes it flexible; (2) bony ________________ tissue of the middle ear that helps to conduct sound waves; (3) ________________ tissue in the inner ear that responds to stimuli and conducts them to the CNS.
1. connective 2. connective 3. nervous	**11–52.** The anatomy of the ear is divided into three divisions, the external ear, the (1) ________________ ear and the (2) ________________ ear.
1. middle 2. inner	**11–53.** The external ear is composed of ________________ connective tissue, covered by skin. The structures of the external ear are illustrated in frame 54.
cartilaginous	**11–54.** 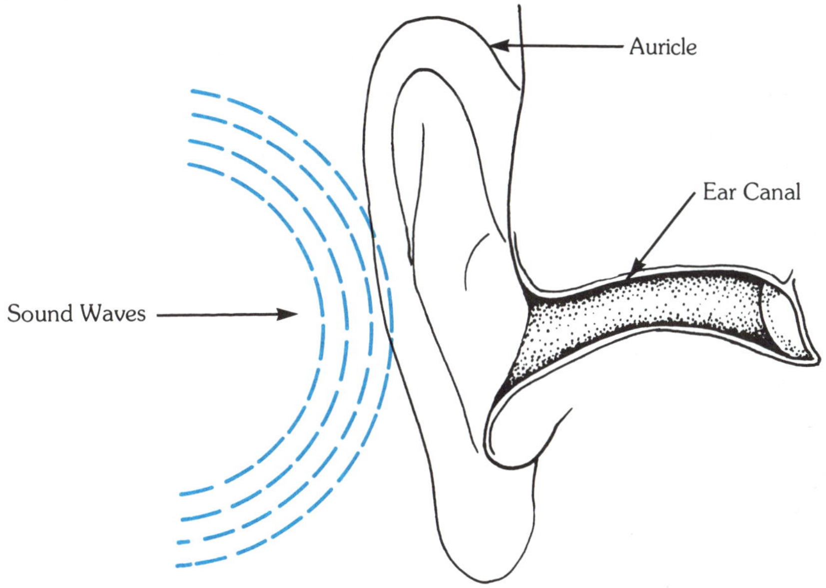 The outer, flexible portion of the external ear is called the ________________.
auricle	**11–55.** Sound waves arrive at the auricle and move through the ear ________________.
canal	**11–56.** In the tissue of the ear canal are glands that produce wax, or cerumen. These glands are called ________________ glands.
ceruminous	**11–57.** The purpose of ear wax, or ________________, is to lubricate the ear canal to protect the tissue from drying.

cerumen

11-58. Sound waves that have been received through the external ear are conducted into the middle ear by the structures composed of ________________ connective tissue. Frame 59 illustrates the structures of the middle ear.

bony

11-59.

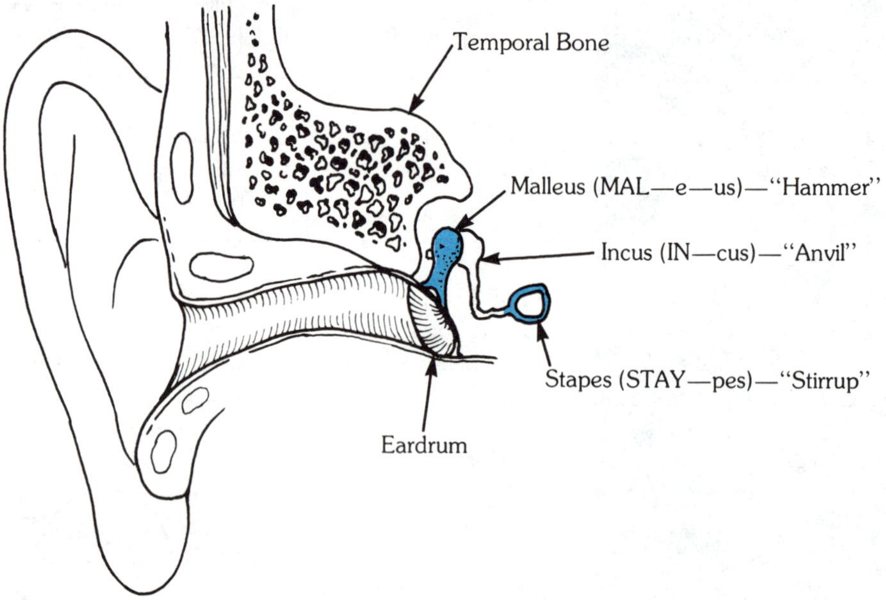

The bony structures of the middle ear are located in a cavity of the ________________ bone.

temporal

11-60. At the end of the ear canal is a tissuelike membrane called the (1) ________________. The (2) ________________ or "hammer" is the bone in the middle ear that is attached to the membrane.

1. eardrum
2. malleus

11-61. The "head of the hammer," or (1) ________________ bone, comes into contact with the (2) ________________ bone or "anvil."
(Refer to frame 59.)

1. malleus
2. incus

11-62. The last of the small bones in the middle ear comes into contact with the incus at one end. The bone is shaped like a stirrup and is called the ________________.

stapes

11-63. When sound waves move through the ear canal they first come in contact with the membrane, or ________________, at the end of the canal.

eardrum

11-64. When the eardrum is stimulated by (1) ________________ waves, it begins to vibrate. The vibration causes the (2) ________________ bone to move against the incus, or anvil. Movement of the incus causes the (3) ________________, or stirrup, to move back and forth.

1. sound
2. malleus
3. stapes

11–65. The movements of the three bones in the (1) ________________ ear conduct sound waves from the external ear, through the air-filled middle ear, to the sensory receptor sites in the (2) ________________ ear.

1. middle
2. inner

11–66. You recall that the cornea and lens of the eye are responsible for bending light rays to enable them to converge on the retina. Likewise, the malleus, incus, and stapes of the middle ear transmit ________________ waves to the receptor areas in the inner ear.

sound

11–67. As the stapes is moved, owing to the vibration caused by the conduction of sound waves, it is pulled away from or pushed toward the inner ear. Between the middle ear and ________________ ear is an opening called the OVAL WINDOW.

(Refer to the illustration in frame 68.)

inner

11–68.

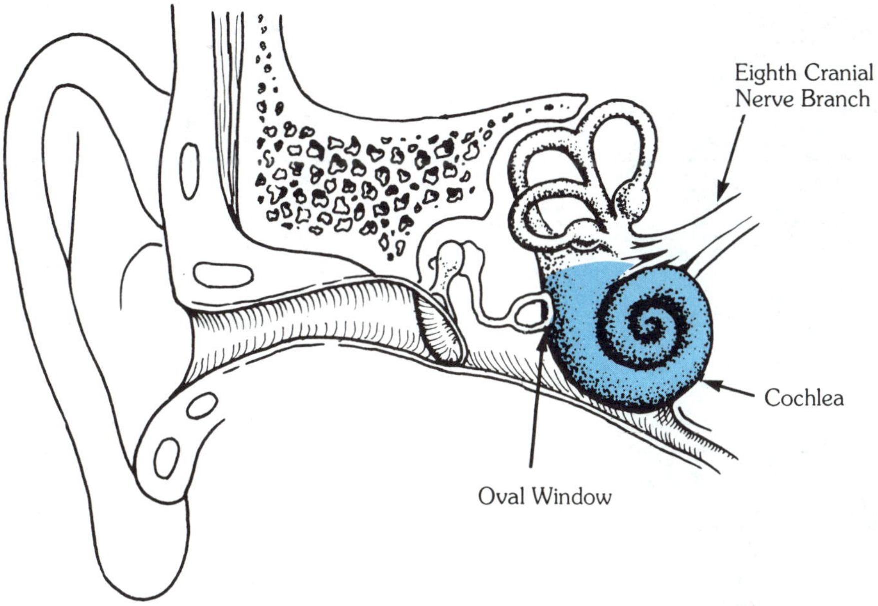

The inner ear structures are a bony-covered maze of canals. The largest portion of the bony tissue is called the ________________.

cochlea

11–69. Inside the cochlea are ducts filled with fluids. As the sound waves are conducted through the middle ear, the ________________ bone is pushed against the oval window.

stapes

11–70. The pushing of the stapes exerts a force on the fluid-filled ducts. This force causes the pressure of the fluid to (increase/decrease) ________________.

increase

11–71. On the lining of the cochlear ducts is the organ of CORTI (KOR—ti), made up of a long strand of cells with many tiny hairs extending from each cell.

When the fluid pressure in the cochlear duct increases, the organ of corti vibrates.

(No response required.)

11–72. When the organ of Corti vibrates, the tiny hair cells move and initiate nerve impulses. The receptor site for hearing is called the organ of ________________.

Corti

11–73. The nerve impulses are collected by dendrites that transmit the impulse to the cell bodies. The axons of these cell bodies form a major branch of the eighth ________________ nerve.

---

cranial

11–74. From the inner ear by way of the eighth cranial nerve, sound stimuli are transmitted to the temporal lobe of the ________________.

---

brain

11–75. A short review of the organization of the special senses is in order. The special sense abilities are due to the function of special sense ________________.

---

organs

11–76. Each of the special sense organs is a structure composed of histologically (1) ________________ (different/same) tissues that are integrated to (2) ________________ for one purpose.

---

1. different
2. function

11–77. Each sense organ is composed of ________________ sites that transmit the stimuli as nerve impulses.

---

receptor

11–78. The afferent (sensory) pathways for the senses of smell, taste, vision, and hearing are along the branches of particular ________________ nerves.

---

cranial

11–79. The afferent (sensory) pathways for cutaneous sensations are along fibers of the spinal ________________.

---

nerves

11–80. Following peripheral nerve conduction by cranial or spinal nerves, the sensory impulses arrive at specialized areas of the ______ ______ ______.

CNS

11–81. The actions of the brain are important so that the sensory impulses can be ________________.

---

identified or interpreted

11–82. It is through the special senses that we take in information about our environment.

You have reached the end of Chapter 11.

12 The Digestive System

12–1. The DIGESTIVE SYSTEM is essentially a long tube extending from the opening of the mouth at one end to the opening of the rectum—the anus—at the other end.

The major parts of the digestive system are the mouth, the pharynx, the esophagus, the stomach, the intestines, and the associated organs: the liver, gall bladder and pancreas.

12–2. The function of the digestive system is to break down foodstuffs—carbohydrates, fats, and proteins—into molecules that can be absorbed by the blood stream and carried to the cells of the body.

You will investigate the digestive system by tracing the breakdown of foodstuffs in their course from the mouth to their eventual entry into the
(1) ________________ stream and hence their arrival at the
(2) ________________ of the body.

1. **blood**
2. **cells**

12–3. The foodstuffs broken down by the digestive system can be divided into two general classes: (1) substances that serve as direct sources of energy and (2) substances that are necessary for life activities but that do not serve as direct sources of ________________.

energy

12–4. Vitamins, minerals, and water are necessary for life activities but do not provide energy directly. Carbohydrates, fats, and proteins, however, do serve as immediate sources of ______________.

energy

12–5. The foodstuffs that provide energy when broken down by the digestive system are:
a. carbohydrates, fats, and proteins.
b. vitamins, minerals, and water.

a.

12–6. The foodstuffs that support life activities but do not provide energy directly when broken down by the digestive system are:
a. carbohydrates, fats, and proteins.
b. vitamins, minerals, and water.

b.

12–7. The process of digestion involves the breakdown of foodstuffs—both energy-providing and nonenergy-providing— into molecules that are absorbed into the (1) ______________ stream and are carried to all the (2) ______________ of the body.

The Mouth, Pharynx, and Esophagus

1. blood
2. cells

12–8. Before digestion begins, food is ingested—taken into the mouth—and the tongue, which is an accessory organ, moves it into a position in that it is ground into smaller pieces by the ______________, another accessory organ.

teeth

12–9. Chewing helps to prepare the food for swallowing but is not an integral or necessary part of the digestive process.

The digestive process starts when the SALIVARY GLANDS, another accessory organ, secrete ______________, which moistens the food.

saliva

12–10. Saliva, by moistening the food, facilitates swallowing. In addition, saliva begins the process of digestion—food breakdown—by attacking the carbohydrates.

The first of the foodstuffs to be broken down, the ______________, are in the energy-producing class of foods.

carbohydrates

12–11. The breakdown of the carbohydrates is started in the mouth by PTYALIN (TI—a—lin). One of the components of saliva is an enzyme (an organic substance that accelerates chemical transformation) that attacks carbohydrates. This enzyme is called ______________.

ptyalin

12–12. Saliva facilitates swallowing and begins the digestion of the carbohydrates in the food. Molecules of starch (carbohydrates) are split into smaller molecules of sugar (carbohydrates) by the enzyme called ______________.

The Digestive System

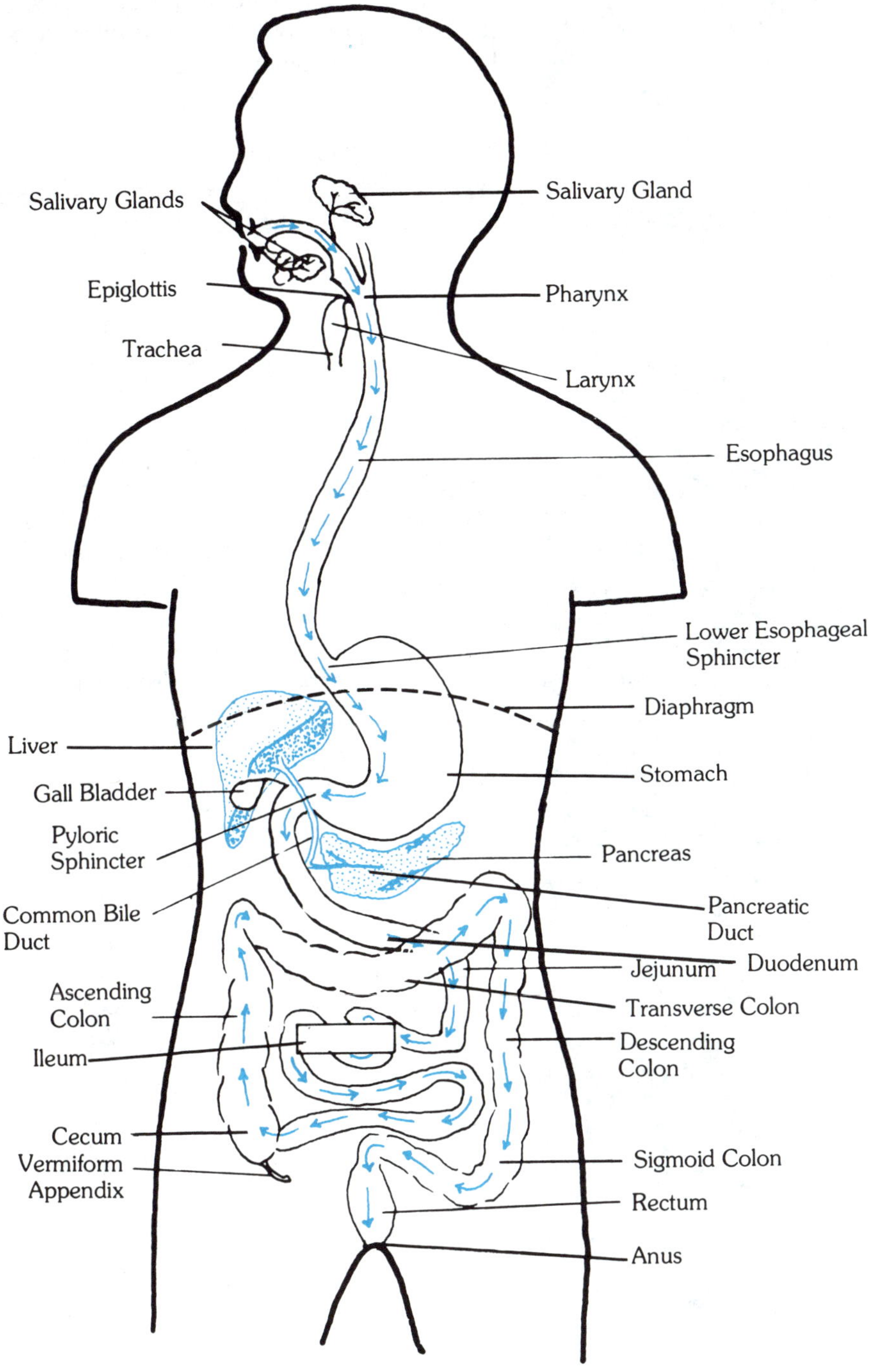

ptyalin

12–13. Refer to the diagram of the digestive system *(above)* as needed.

A pulpy mass of food that has been mixed with saliva is called a BOLUS (BO—lus). The next step in the digestive process is the movement of the bolus by the tongue into the ____________________, which is a common passageway for both food and air.

pharynx	**12–14.** While food remains in the mouth, the act of swallowing the food is voluntary; one can either begin to swallow or not, as one chooses. When food reaches the pharynx, however, swallowing ceases to be a voluntary act and becomes an automatic, or ________________ act or reflex.
involuntary	**12–15.** Swallowing a piece of food is a voluntary act while the food is in the (1) ________________. Swallowing becomes an involuntary, or automatic, act when the food reaches the (2) ________________.
1. mouth **2. pharynx**	**12–16.** The soft palate is moved up and back, closing off the respiratory pharynx. This action prevents the food and liquids from coming out of the ________________.
nose	**12–17.** The tongue moves the bolus (the pulpy mass of food and saliva) from the mouth to the pharynx. Then skeletal muscles in the pharynx propel the bolus into the tube leading to the stomach, namely the ________________.
<u>ESOPHAGUS</u> (e—SOF—a—gus)	**12–18.** As you will recall from the unit on respiration, the pharynx opens not only into the esophagus, but also into the ________________, or voice box.
larynx	**12–19.** You will also recall that food is prevented from entering the larynx by a reflex closure of the ________________, which is located at the entrance to the larynx.
epiglottis	**12–20.** The action of the soft palate prevents food from entering the nasal passage, and the action of the epiglottis prevents food from entering the larynx. These two actions ensure that the food will pass directly from the mouth to the (1) ________________ and then to the (2) ________________.
1. pharynx **2. esophagus**	**12–21.** Alternating waves of contraction and relaxation force the bolus down the tube that leads from the pharynx to the stomach. This tube is called the ________________.

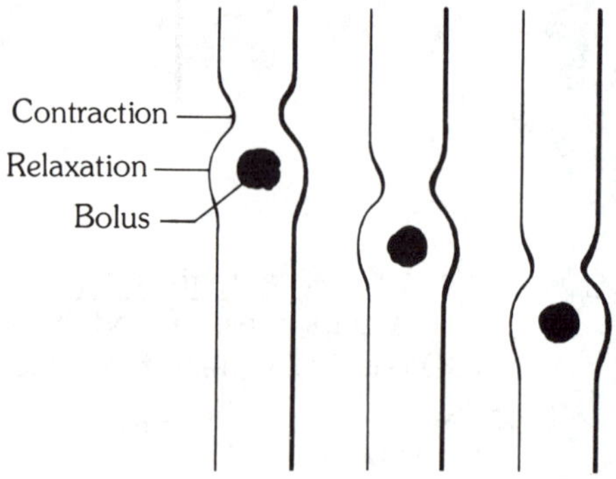

esophagus

12–22. The esophagus is about ten inches long. Food is forced down the length of the esophagus by alternating waves of (1) ________________ and (2) ________________ and the force of gravity.

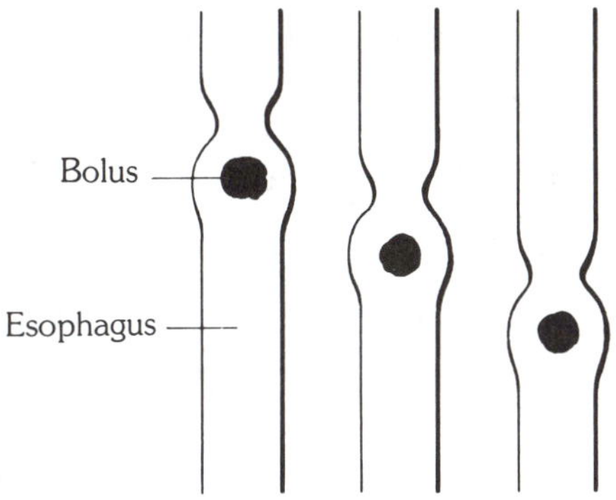

---

1. contraction
2. relaxation
(either order)

12–23. This process of alternating contraction and relaxation by which food is pushed down the esophagus is called PERISTALSIS (per—i—STAL—sis) or PERISTALTIC ACTION. In lower portions of the digestive tract, the small intestine for example, food is also moved by a similar ________________ action.

---

peristaltic

12–24. The process by which food is pushed down through the esophagus is called (1) ________________ and involves the alternating (2) ________________ and (3) ________________ of the muscles of the esophagus.

---

1. peristalsis (peristaltic action)
2. contraction
3. relaxation
[(2) and (3) in either order]

12–25. In the process of swallowing, food is moved from the mouth to the (1) ________________. An involuntary action then projects the food into the (2) ________________. At that point (3) ________________ ________________ takes over to move the food down into the stomach.

---

1. pharynx
2. esophagus
3. peristaltic action

12–26. Turn again to the drawing of the digestive system. As you can see, the lower end of the esophagus is made up of an area of muscle called the LOWER ESOPHAGEAL SPHINCTER (SFINCK—ter).

Just before the food reaches the stomach, it passes through a region called the ________________ ________________ ________________.

---

lower esophageal sphincter

12–27. Food is prevented from refluxing (returning) from the stomach to the esophagus by a region at the lower end of the esophagus. This area is called the ________________ ________________ sphincter.

---

lower esophageal

12–28. The lower esophageal sphincter is an area of smooth muscle in the wall of the esophagus that serves to ________________________________ ________________________________.

(Complete the sentence in your own words.)

prevent the reflux of food from the stomach to the esophagus

12–29. You have traced food from the mouth to the stomach where the first major contributions to the process of digestion take place.

Before continuing, let's review the steps in the digestive process that you have studied so far.

(No response required—move to frame 30.)

12–30. Upon being placed in the mouth, food is chewed. Chewing aids the swallowing process but (1) ________________ (does/does not) otherwise affect digestion.

The first contribution to the digestive process is made by the (2) ________________, which starts to break down the (3) ________________ in the food.

---

1. does not
2. saliva
3. carbohydrates

12–31. The salivary enzyme, (1) ________________, starts to split complex starches into simple sugars.

Once moistened with saliva, the food, now called a bolus, is moved by the tongue to the (2) ________________.

---

1. ptyalin
2. pharynx

12–32. Once in the pharynx, the bolus is involuntarily moved into the next part of the digestive tract, the (1) ________________.

The food is then moved downward by a process of alternating contractions and relaxations called (2) ________________.

---

1. esophagus
2. peristalsis

12–33. Just before entering the stomach, the bolus passes through an area of the esophagus that prevents food from refluxing from the stomach to the esophagus. This region is the ____________ ____________ ____________.

---

lower esophageal sphincter

12–34. The food has now reached the stomach, an organ that makes a significant contribution to the process of digestion.

(No response required—move to frame 35.)

The Stomach

12–35. **The Stomach**

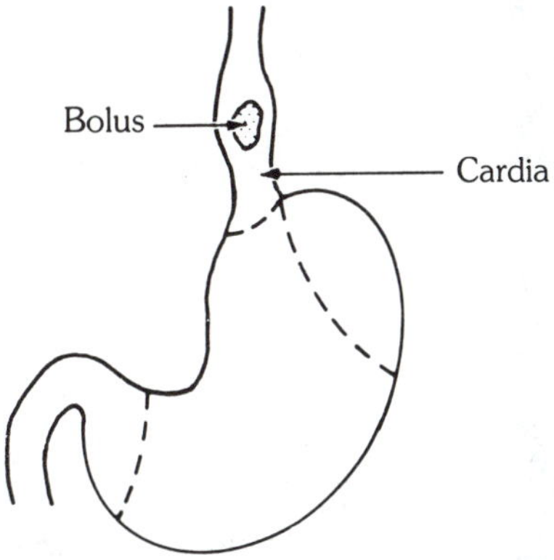

The stomach can be divided into four areas. As you can see from the diagram, food first reaches the area called the ________________.

CARDIA
(KAR—di—a)

12–36. The cardia surrounds the opening of the esophagus. To the left in the body (but right in the drawing) is a pouchlike portion called the ________________.

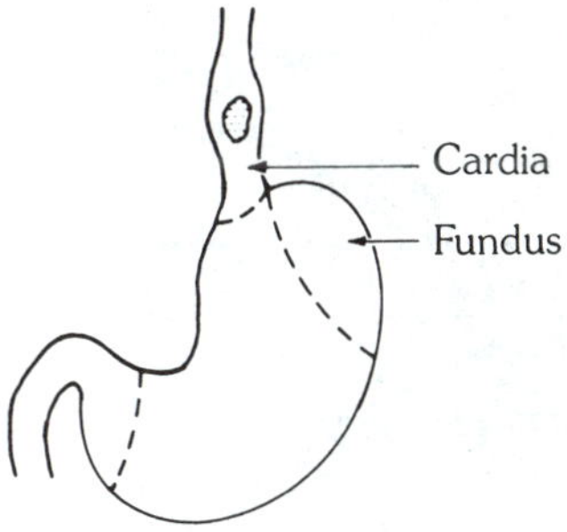

The Stomach

FUNDUS
(FUN—dus)

12–37. The area of the stomach that surrounds the opening of the esophagus is called the (1) ________________. The pouchlike portion to the left is called the (2) ________________.

1. cardia
2. fundus

12–38. The largest portion of the stomach is called the ________________.

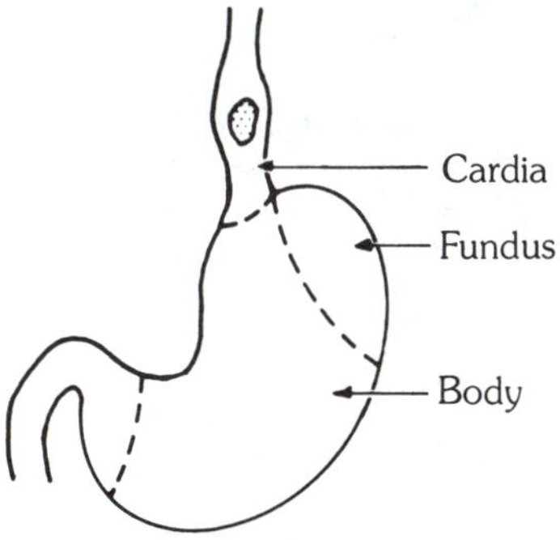

The Stomach

BODY

12–39. The fourth area of the stomach is the region leading into the intestine. This area is called the ______________.

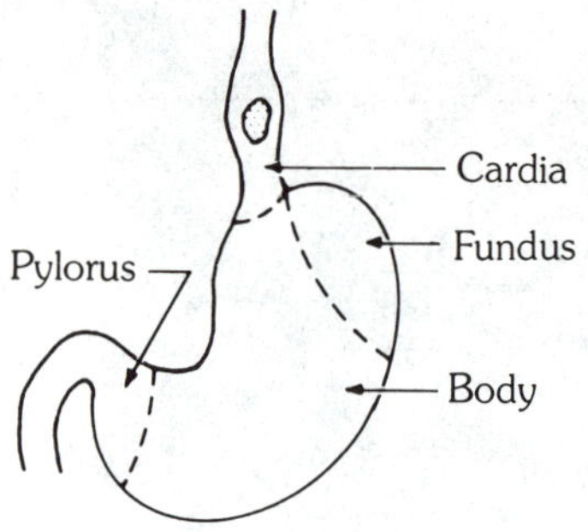

The Stomach

PYLORUS
(pi—LOR—us)

12–40. The area leading into the body of the stomach from the esophagus is the (1) ______________.
The area leading out of the stomach is the (2) ______________.

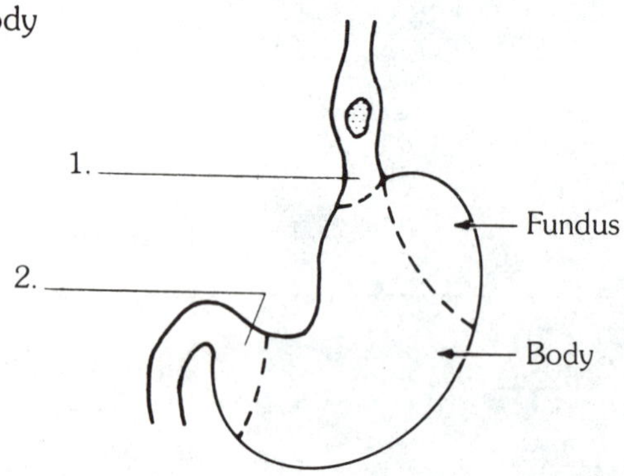

The Stomach

1. **cardia**
2. **pylorus**

12–41. The remaining two areas are the pouchlike distention to the left, called the (1) ______________, and the main portion of the stomach, called the (2) ______________.

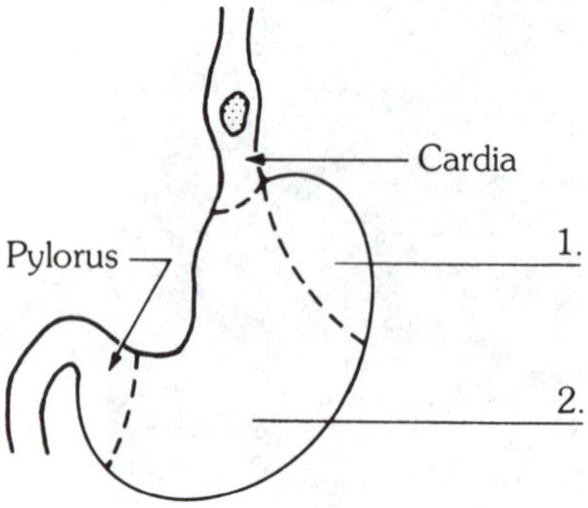

The Stomach

1. **fundus**
2. **body**

12–42. Match the names on the left with the descriptions on the right.

1. cardia	a. main portion of stomach
2. body	b. pouchlike area to left
3. pylorus	c. area immediately surrounding the esophagus
4. fundus	d. area leading to intestines

1. c
2. a
3. d
4. b

12–43. Now let's see how each of these areas of the stomach contributes to the digestive process.

Approximately 15 million GASTRIC GLANDS are spread through all of the areas of the stomach. Different types of cells make up these glands, and consequently the (1) ______________ ______________ differ in various parts of the stomach.

gastric glands

12–44. In the cardia and in the pylorus, the cells are predominantly of the MUCOUS (MU—kus) types. Thus, the secretion of the ______________ ______________ in these two areas is mainly mucus—a thick, slimy, clear, protective, semi-fluid substance.

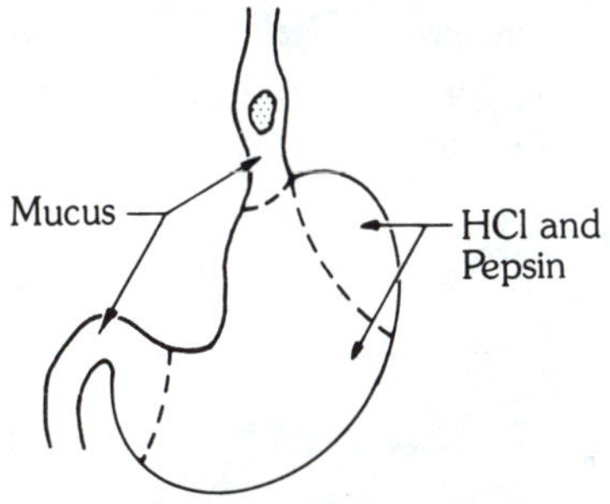

The Stomach

gastric glands

12–45. Mucus is predominantly secreted by the gastric glands in the areas that begin and end the stomach, namely in the (1) ______________ and in the (2) ______________. These glands are sometimes referred to as cardiac and pyloric glands, respectively.

1. cardia
2. pylorus

12–46. The glands in the fundus and in the body of the stomach are composed mainly of PARIETAL (pa—RI—e—tal) cells and CHIEF cells—in addition to some mucus cells. The parietal cells are responsible for the formation of hydrochloric acid, and the chief cells for the production of the enzyme PEPSIN (PEP—sin).

The two main substances secreted by the glands of the fundus and body of the stomach are (1) ______________ and (2) ______________.

1. hydrochloric acid
2. pepsin
(either order)

12–47. In the fundus and body of the stomach, the gastric glands mainly secrete hydrochloric acid and pepsin, whereas in the cardia and pylorus the primary secretion is ______________.

mucus

12–48. Match each of the four areas of the stomach with the correct secretion.

1. cardia
2. fundus
3. body
4. pylorus

a. pepsin
b. hydrochloric acid
c. mucus

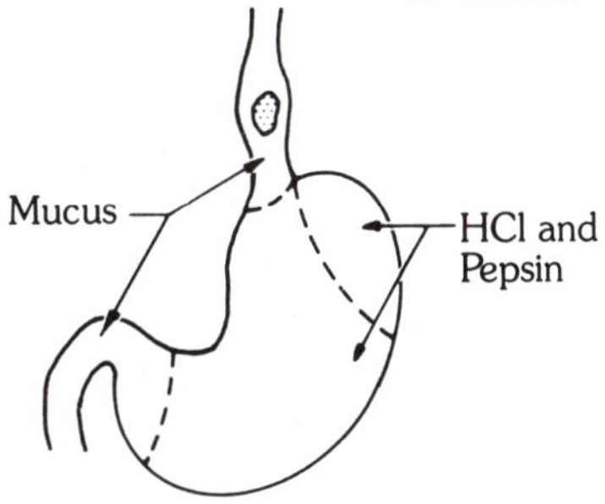

The Stomach

1. c 2. a and b 3. a and b 4. c	**12–49.** In the course of a 24-hour period, the gastric glands secrete approximately 1000 to 1500 ml of GASTRIC JUICE. This gastric juice is composed of the three secretions just mentioned: (1) ________________, (2) ________________ ________________ and (3) ________________ plus water and inorganic substances such as sodium, potassium, calcium, sulfate, and phosphate.
1. mucus 2. pepsin 3. hydrochloric acid (any order)	**12–50.** The muscular walls of the stomach contract and relax, churning the food and mixing it with the ________________ ________________ that is secreted.
	Functions of the Gastric Juice
gastric juice	**12–51.** Hydrochloric acid performs an important function—it releases the enzyme pepsin. Pepsin is secreted as PEPSINOGEN (pep—SIN—o—jen). Pepsinogen is not an active enzyme until it is converted to pepsin. This conversion takes place in the presence of ________________ ________________.
hydrochloric acid	**12–52.** Once pepsin has been formed, it is capable of converting more pepsinogen to pepsin. The whole process is started, however, when pepsinogen is converted to (1) ________________ by (2) ________________ ________________.
1. pepsin 2. hydrochloric acid	**12–53.** Thus, the important function of hydrochloric acid is to (1) ________________ pepsinogen by converting it to (2) ________________.
1. activate 2. pepsin	**12–54.** Pepsin is a digestive enzyme. Once it is released by hydrochloric acid (HCl), pepsin acts upon proteins, breaking them down into smaller molecules and AMINO ACIDS (a—me—no), preparing them for absorption into the bloodstream. Thus, the digestion of ________________ is begun in the stomach.
proteins	**12–55.** The enzyme ptyalin, which is secreted by the salivary glands, acts on carbohydrates. The enzyme pepsin, which is secreted in the fundus and body of the stomach, breaks (1) ________________ down into smaller molecules and (2) ________________ acids.
1. proteins 2. amino	**12–56.** The third important component of gastric juice is secreted mainly in the cardia and pylorus, although it is secreted to a small degree in all parts of the stomach. This component is called ________________.

mucus

12–57. If the stomach lining were not protected, it would be attacked by the hydrochloric acid (HCl) and the pepsin in the stomach. Fortunately, the secretion of the glands in the cardia and pylorus is a thick, slimy, protective, semi-fluid substance. This substance, called (1) ________________, serves to (2) ________________ the stomach from being damaged by the acid and enzymes present.

1. mucus
2. protect

12–58. To summarize the functions of the various components of gastric juice, first label each of the four sections of the stomach with respect to the substance secreted.

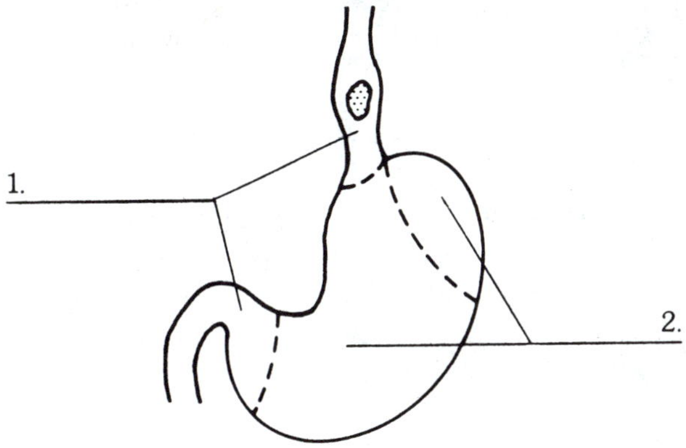

The Stomach

1. mucus
2. HCl pepsin

12–59. Now, match each of the secretions listed below with the appropriate function.

1. mucus
2. hydrochloric acid
3. pepsin

a. break proteins into smaller molecules and amino acids
b. line and protect the stomach
c. activate pepsinogen

1. b
2. c
3. a (and c)

12–60. Before going on to the next stage in the digestive process, let's discuss acidity in the stomach.

As you would suspect from its composition, gastric juice is ________________.

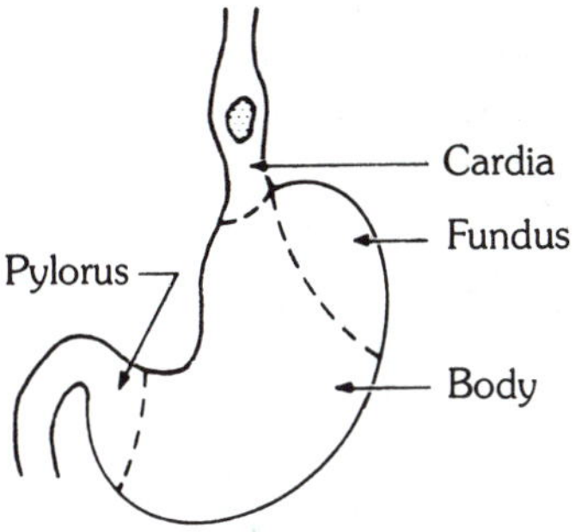

The Stomach

acidic

12–61. Recall the pH scale from Chapter 1 (repeated below).

← increasingly acidic							neutral ↑						increasingly basic →	
0	1	2	3	4	5	6	7	8	9	10	11	12	13	14

The pH of gastric juice is approximately 1.4.

From this scale you can determine that gastric juice is

a. highly basic.
b. somewhat basic.
c. somewhat acidic.
d. highly acidic.

d. highly acidic

12–62. Despite the fact that the gastric juice is highly acidic, under normal circumstances this acid does not attack the stomach wall itself. Under normal conditions, the hydrochloric acid production is increased only when food is present or on the way. In addition, the acid is diluted by water and mucus. These factors tend to reduce the ________________ (increase the pH) of the gastric juice.

Phases of Gastric Secretion

acidity

12–63. The final topic to be discussed before concluding the stomach's role in digestion is the phases of gastric secretion.

Gastric secretion increases in response to certain stimuli. Typically, there are three phases of gastric secretion, representing responses to three different types of ________________.

stimuli

12–64. The first phase of gastric secretion is the CEPHALIC (se—FAL—ik) PHASE. During this phase, conscious stimuli such as the desire for food, or the sight, smell, or taste of food, can ________________ gastric secretion.

increase or stimulate

12–65. Impulses carried from the brain to the stomach by way of the vagus nerve can cause an increase in the secretion of gastric juice. This phase of secretion, called the cephalic phase, is a response to such conscious stimuli as ________________
____
____.

(List as many as you can.)

desire for, or taste of, or sight of, or smell of food.

12–66. The gastric juice that is secreted in response to the desire for food or the sight or smell or taste of food during the ________________ phase is high **in HCl and pepsin content.**

cephalic

12–67. The cephalic phase is the first phase of gastric secretion. The second phase is the GASTRIC PHASE. The contact of ingested food with the glands in the stomach stimulates the second, or ________________ phase.

gastric

12–68. During the gastric phase, food reaching the stomach stimulates the secretion of the hormone called GASTRIN. This hormone is released by the pyloric glands, travels in the bloodstream, and stimulates the gastric glands to secrete gastric juice. Thus, in the gastric phase the sequence of stimuli that causes an increased secretion of gastric juice are:

1. food reaching the (1) ____________________;
2. the secretion of the hormone (2) ____________________;
3. the hormone stimulation of the (3) ____________________ ____________________.

---

1. stomach
2. gastrin
3. gastric glands

12–69. A gastric juice that is high in HCl and pepsin content is secreted during the first or (1) ____________________ phase.

A gastric juice that is lower in pepsin content than that secreted during the first phase is secreted during the second, or (2) ____________________ phase.

---

1. cephalic
2. gastric

12–70. The third phase of gastric secretion is called the INTESTINAL PHASE. The stimulus during this phase is the entrance of certain foods, such as water, proteins, milk, and alcohol, into the small intestine. In repsonse to these foods, the cells of the ____________________ stimulate the cells of the stomach to secrete gastric juice.

---

intestine

12–71. When certain foods reach the small intestine, the cells there are stimulated and, in turn, the gastric glands are stimulated to secrete gastric juice. This third phase of gastric secretion is called the ____________________ phase.

---

intestinal

12–72. To summarize—match each of the phases of gastric secretion with the appropriate stimulus (or stimuli).

1. gastric
2. cephalic
3. intestinal

a. smell or taste of food
b. food reaching the stomach
c. food reaching the intestine
d. secretion of the hormone gastrin
e. desire for or sight of food

---

1. b and d
2. a and e
3. c

12–73. To summarize the stomach's role in digestion:

Food enters the stomach from the esophagus after passing through the lower esophageal sphincter. The stomach can be divided into four distinct areas—the cardia, fundus, body, and pylorus. In the cardia and pylorus, the gastric secretion is primarily (1) ____________________, whereas in the fundus and body of the stomach the secretion is made up primarily of (2) ____________________ and (3) ____________________.

---

1. mucus
2. hydrochloric acid
3. pepsin
(2. and 3. in either order)

12–74. The HCl functions by converting pepsinogen to (1) ____________________.

The function of the enzyme that is activated by HCl is to break (2) ____________________ down into smaller molecules and amino acids.

The secretion of mucus serves to (3) ____________________ the stomach lining.

1. pepsin
2. proteins
3. protect

12–75. The gastric juice of the stomach has a pH of approximately 1.4 and is thus highly (1) ________________.

The secretion of HCl is increased in the response to food and other stimuli. Damage to the stomach lining is avoided by the presence of water and (2) ________________.

1. acidic
2. mucus

12–76. There are three phases of gastric secretion: the first, or (1) ________________ phase, which is a response to stimuli when food is seen, smelled, or tasted; the second, or (2) ________________ phase, which is a response to the secretion of gastrin in the pylorus; and the third, or (3) ________________ phase, which is a response to certain foods in the small intestine.

The Small Intestine

1. cephalic
2. gastric
3. intestinal

12–77. Food entering the stomach has been chewed and mixed with saliva. In this state it is called, as you recall, a bolus.

By the time food leaves the stomach it has been changed into a semi-fluid called CHYME (kyme). It is in this state that food moves from the stomach to the beginning of the small ________________.

intestine

12–78. You will recall that a bolus is prevented from refluxing from the stomach to the esophagus by the action of the lower esophageal sphincter. At the other end of the stomach, food, now called (1) ________________, is prevented from escaping from the stomach into the small intestine before it has adequately mixed with HCl and pepsin by the action of the pyloric (2) ________________.

(Use the diagram of the digestive system as needed.)

1. chyme
2. sphincter

12–79. As the drawing on p. 273 indicates, chyme, the product of the digestive action of the stomach, passes through the ________________ ________________ into the DUODENUM (du—o—DE—num), the first portion of the small intestine.

pyloric sphincter

12–80. At the opening between the esophagus and the stomach is the (1) ________________ sphincter. At the opening between the stomach and the duodenum is the (2) ________________ sphincter.

1. lower esophageal
2. pyloric

12–81. Chyme from the stomach passes through the pyloric sphincter into the first portion of the small intestine, which is called the ________________.

duodenum

12–82. Refer to the diagram again.

Chyme passes from the duodenum into the second and third portions of the small intestine, which are called, respectively, the (1) ________________ and the (2) ________________.

1. JEJUNUM (je—JOO—num)
2. ILEUM (IL—e—um)

12–83. The duodenum of an average adult is somewhat less than a foot long, whereas the jejunum averages 9 feet in length, and the ileum averages 13 feet. Thus, the average total length of the small intestine of an adult is about ________________ feet.

23

12–84. The shortest portion of the small intestine is the duodenum, which is less than one foot long. The second portion, called the (1) ________________ is about 9 feet long, and the third and longest portion, the (2) ________________, is about 13 feet long.

---

1. jejunum
2. ileum

12–85. The large intestine, into which indigested food passes after it leaves the small intestine, is approximately 5 feet long but has a greater diameter than the small intestine. In other words, the small intestine is (1) ________________ (longer/shorter) than the large intestine but has a (2) ________________ (larger/smaller) diameter.

---

1. longer
2. smaller

12–86. Because of its great length, the small intestine is arranged in irregular coils within the abdominal cavity. The diagram is not an exact representation of the way these coils look but is a simplified drawing, designed to show the direction in which the digested food or chyme moves through the intestines.

(No response required—move to frame 87.)

12–87. List the three parts of the small intestine in the order in which chyme moves through them after it leaves the stomach.

(1) ________________
(2) ________________
(3) ________________

---

1. duodenum
2. jejunum
3. ileum

12–88. The small intestine contains intestinal glands that secrete intestinal juice. In the duodenum, chyme from the stomach is mixed with ________________ juice produced by the small intestine.

---

intestinal

12–89. Chyme in the duodenum is also mixed with secretions from the two organs drawn in red in the diagram, the (1) ________________ and the (2) ________________.

---

1. liver
2. pancreas
(either order)

12–90. The pancreas, located beneath the liver and the stomach, is both an endocrine and an exocrine organ. The endocrine portion of the pancreas has been discussed in Chapter 9 of this program. The pancreatic juices that enter the duodenum are secreted by the other portion of the pancreas, the (endocrine/exocrine) ________________ portion.

exocrine	**12–91.** When chyme enters the duodenum, it stimulates the duodenum to produce hormones, SECRETIN (SEC—re—tin) and CHOLECYSTOKININ (COLE—ee—cys—to—KI—nin). These hormones are then carried to the exocrine portion of the ________ where they stimulate the secretion of pancreatic juice.
pancreas	**12–92.** The pancreas, after being stimulated by the hormones (1) ________ and (2) ________, produces pancreatic juice.
1. secretin 2. cholecystokinin (either order)	**12–93.** The pancreatic juices are capable of digesting all three classes of food: carbohydrates, fats, and proteins. These juices are mixed with chyme when the chyme is in the ________, the first portion of the small intestine.
duodenum	**12–94.** The pancreatic juice contains an enzyme that splits complex sugars into simple sugars. This enzyme is called PANCREATIC AMYLASE (AM—i—lace). The action of the salivary enzyme, (1) ________, on carbohydrates is to split complex sugars. Complex sugars are also split into simple sugars by the pancreatic enzyme called (2) ________ ________.
1. ptyalin 2. pancreatic amylase	**12–95.** The pancreatic enzyme responsible for splitting fat molecules is LIPASE (LI—pace). The fact that fats are also known as LIPIDS (LIP—idz) may help you to remember that the pancreatic enzyme called (1) ________ splits (2) ________ molecules. Small lipid molecules are called FATTY ACIDS.
1. lipase 2. fat (lipid)	**12–96.** The pancreatic enzyme responsible for splitting small protein molecules is TRYPSIN (TRIP—sin). The enzyme ________ breaks down small proteins into amino acids.
trypsin	**12–97.** Match the items in these two columns. 1. pancreatic amylase — a. salivary enzyme, splits complex sugars into simple sugars 2. lipase — b. pancreatic enzyme, splits complex sugars into simple sugars. 3. trypsin — c. pancreatic enzyme, splits fat molecules 4. ptyalin — d. pancreatic enzyme, splits protein molecules into amino acids
1. b 2. c 3. d 4. a	**12–98.** Pancreatic juices are mixed with chyme while the chyme is in the duodenum. BILE, secreted by the liver, is also mixed with the chyme while it is in the ________ portion of the small intestine.

duodenum	**12–99.** The liver secretes a fluid called ________________, which acts upon the fats in chyme.
bile	**12–100.** In the duodenum, the fatty elements of chyme are digested partly by the pancreatic enzyme, (1) ________________, and partly by the (2) ________________ secreted by the liver.
1. lipase 2. bile	**12–101.** Bile and pancreatic juice are both reservoirs of alkali. As a result, they help to neutralize the ________________ in the chyme that is produced in the stomach.
acids	**12–102.** Alkaline secretions, such as bile and pancreatic juice, mix with the highly acidic chyme to produce intestinal juices with a pH of approximately 6.0 to 7.0. Duodenal juice is (more basic/more acidic) ________________ than chyme.
more basic	**12–103.** Thus, we find that bile fills two important roles: it acts upon (1) ________________, and in combination with (2) ________________ ________________ it helps to (3) ________________ the acids.
1. fats 2. pancreatic juice 3. neutralize	**12–104.** Bile is secreted by the liver but is stored in the GALL BLADDER, which, as you can see from the diagram, is located (above/below) ________________ the liver.
below	**12–105.** Bile may enter the duodenum directly from the (1) ________________ that secretes it, or it may (more commonly) enter from the (2) ________________ ________________ where it is stored and concentrated.
1. liver 2. gall bladder	**12–106.** Refer to the diagram of the digestive system. Chyme, mixed with the secretions of the liver, pancreas, and small intestine is forced from the duodenum through the (1) ________________ and (2) ________________ portions of the small intestine.
1. jejunum 2. ileum	**12–107.** The alternating contraction and relaxation of the muscles of the intestines moves chyme through them. In the intestines, as in the esophagus and stomach, the chyme is pushed along by ________________ action.
peristaltic	**12–108.** The interior of the small intestine is covered with microscopic fingerlike projections called VILLI (VIL—li). Water and the final products of digestion are absorbed into blood and lymph vessels through the projections called (1) ________________ in the (2) ________________ intestine.

1. villi
2. small

12–109. The function of the villi in the small intestine is to ________________ water and the final products of digestion into the blood and lymph vessels.

absorb

12–110. The carbohydrates, fats, and proteins have been broken down into simple (1) ________________, simple protein molecules and (2) ________________ ________________, and fatty (3) ________________.

1. sugars
2. amino acids
3. acids

12–111. Simple sugars, simple protein molecules and amino acids, and fatty acids are now absorbed by the many small projections of the small intestine, the ________________.

villi

12–112. **Villus**

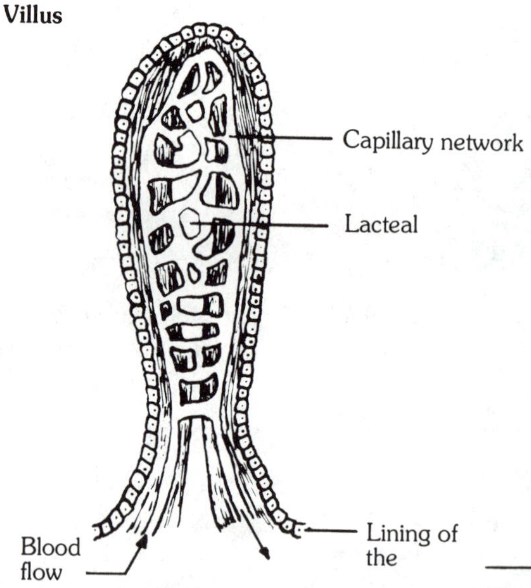

________________ ________________.

(Fill in the blanks.)

small intestine

12–113. The products of digestion are absorbed by passing through the lining of the small intestine into either the (1) ________________ ________________ or the (2) ________________.

1. capillary networks
2. lacteals
(either order)

12–114. The fatty acids are absorbed by the lacteals, and the simple protein molecules, amino acids, and simple sugars are absorbed by the ________________ ________________.

capillary networks

12–115. The digestive process is completed when the food is broken down and absorbed by the villi. The simple sugars enter the (1) ________________ ________________ of the villi, and the fatty acids enter the (2) ________________ of the villi.

1. capillary networks 2. lacteals	**12–116.** Little or no absorption takes place in the esophagus or the stomach. Absorption does not begin in earnest until food reaches the ______________ ______________.
small intestine	**12–117.** Products of the digestive process from the small intestine then pass into the first portion of the ______________ ______________ or colon.

The Large Intestine

large intestine	**12–118.** The large intestine's main function is to absorb water from the intestinal contents and then transfer the waste products with little or no ______________ value from the body.
nutrient	**12–119.** The first part of the large intestine is called the ______________. (Refer to the diagram of the digestive system.)
CECUM (SEE—come)	**12–120.** Peristalsis in the large intestine, or colon, pushes the waste products up from the cecum into the (1) ______________ colon, across the (2) ______________ colon and down through the (3) ______________ colon, and the (4) ______________ colon. (Refer to the diagram.)
1. ascending 2. transverse 3. descending 4. sigmoid	**12–121.** The first portion of the large intestine is called the (1) ______________. The second portion consists of the ascending, transverse, descending, and sigmoid (2) ______________.
1. cecum 2. colons	**12–122.** Waste products from the colon then pass into the last portion of the large intestine, called the (1) ______________, and are finally excreted through the (2) ______________.
1. rectum 2. anus	**12–123.** The waste product of digestion is called FECES (FE—seez). Normally, this waste product, (1) ______________, is stored only in the sigmoid (S-shaped) portion of the (2) ______________.
1. feces 2. colon	**12–124.** Usually feces are stored in (1) ______________ section of the (2) ______________.
1. sigmoid 2. colon	**12–125.** Normally the rectum is empty. The desire to defecate waste occurs when the rectum becomes distended with feces. When food, water, or even a cigarette stimulate the stomach, a reflex is initiated called the GASTROCOLIC reflex. The sigmoid contracts and forces feces into the rectum. This action produces the desire to ______________.

defecate	**12–126.** Stimuli such as food or water induce the stomach to initiate a (1) ________________ reflex which in turn induces the (2) ________________ to contract.
1. gastrocolic 2. sigmoid	**12–127.** In response to the gastrocolic reflex, peristalsis of the sigmoid colon forces feces into the last portion of the large intestine, namely, the ________________.
rectum	**12–128.** In response to the gastrocolic reflex, feces enter the rectum and distend it. Distention of the rectum stimulates peristalsis of the entire colon, which then pushes the feces out through the ________________.
anus	**12–129.** Guarding the opening of the anus are two circular sphincters. The internal sphincter is composed of smooth muscle, and the external sphincter is composed of skeletal muscle. Thus, the internal sphincter is under involuntary control, and the external one is under ________________ control.
voluntary	**12–130.** At the opening of the anus, the internal sphincter is under (1) ________________ (involuntary/voluntary) control, whereas the external sphincter is under (2) ________________ (involuntary/voluntary) control.
1. involuntary 2. voluntary	**12–131.** The internal sphincter relaxes during the reflex stimulation of the colon, thus allowing the ________________ to escape.
feces	**12–132.** Defecation may be blocked by the voluntary contraction of the (internal/external) ________________ sphincter composed of skeletal muscle.
external	**12–133.** In summary, it was mentioned that the main function of the large intestine or (1) ________________ is to absorb (2) ________________ from the intestinal contents and to then move (3) ________________ products out of the body.
1. colon 2. water 3. waste	**12–134.** The waste products of digestion that are composed of non-nutrient materials and bacteria are called (1) ________________. One reflex that brings about the desire to remove the waste products from the body is called the (2) ________________ reflex.
1. feces 2. gastrocolic	**12–135.** To review the entire process of digestion: The entire digestive canal is referred to as the ALIMENTARY (al—i—MEN—ter—e) CANAL. The passages and spaces through which food passes between its ingestion through the mouth and its excretion through the anus make up the ________________ canal.

alimentary

12–136. Food enters the mouth, where it is ground and crushed by the teeth. The ______ pushes the food against the teeth and also propels it toward the pharynx where it can be swallowed.

---

tongue

12–137. In the mouth, food is mixed with ______, which moistens it and prepares it for swallowing.

---

saliva

12–138. Saliva contains an enzyme called (1) ______, which splits complex (2) ______ into simple (3) ______.

---

1. ptyalin
2. starches
3. sugar

12–139. After leaving the mouth, food passes into the (1) ______, which is a common passageway for food and air. When food reaches the pharynx, the act of swallowing the food becomes (2) ______ (voluntary/involuntary).

---

1. pharynx
2. involuntary

12–140. Food then passes from the pharynx into the (1) ______. Closure of the glottis by the (2) ______ at this point prevents food from entering the (3) ______ or voice box.

---

1. esophagus
2. epiglottis
3. larynx

12–141. The bolus of food moves through the esophagus by means of a process called (1) ______ or (2) ______ action, which involves the alternating (3) ______ and (4) ______ of the esophagus.

---

1. peristalsis
2. peristaltic
3. contraction
4. relaxation
   (3 and 4 either order)

12–142. The peristaltic action of the esophagus forces the food through the (1) ______ ______ sphincter into the (2) ______ where gastric juices mix with it to form a semi-liquid called (3) ______.

---

1. lower esophageal
2. stomach
3. chyme

12–143. In the stomach:
(1) ______ ______ activates pepsinogen.
(2) ______ breaks down proteins into smaller molecules and amino acids. (3) ______ is secreted to protect the stomach lining.

---

1. hydrochloric acid
2. pepsin
3. mucus

12–144. Chyme passes from the stomach into the (large/small) ______ intestine.

---

small

12–145. Chyme from the stomach passes through the (1) ______ sphincter into the first portion of the small intestine, which is called the (2) ______.

1. pyloric
2. duodenum

12–146. After leaving the duodenum, chyme passes first through the (1) ______________, or second portion of the small intestine, and then through the (2) ______________, or third portion of the small intestine.

---

1. jejunum
2. ileum

12–147. In the duodenum, chyme is mixed with intestinal juice, with (1) ______________ juice and with (2) ______________ from the liver.

---

1. pancreatic
2. bile

12–148. Pancreatic juice contains (1) ______________, which digests fat, pancreatic (2) ______________, which splits complex sugars into simple sugars, and (3) ______________, which breaks down small protein molecules into amino acids.

---

1. lipase
2. amylase
3. trypsin

12–149. Bile, secreted by the (1) ______________, acts upon the (2) ______________ present in chyme.

Bile and pancreatic juice also help to neutralize the (3) ______________ in the chyme.

---

1. liver
2. fats
3. acid(s)

12–150. Bile is stored in the ______________ ______________ where it is concentrated.

---

gall bladder

12–151. Absorption of the nutrients and digested food takes place mainly in the (large/small) ______________ intestine.

---

small

12–152. The interior of the small intestine is covered with microscopic projections called (1) ______________. The function of these microscopic projections is (2) ______________________________

______________________________.

(Use your own words.)

---

1. villi
2. to absorb nutrients and water (or any synonymous statement)

12–153. The fatty acids are absorbed in the (1) ______________ of the villi.

The amino acids and simple sugars are absorbed in the (2) ______________ ______________ of the villi.

---

1. lacteals
2. capillary network

12–154. The remaining products of digestion from the small intestine pass into the large intestine. These products pass first into the (1) ______________. The products then pass through ascending, transverse, descending, and sigmoid portions of the (2) ______________.

---

1. cecum
2. colon

12–155. Waste products from the colon pass finally into the last portion of the large intestine, which is called the (1) ______________, and are excreted through the (2) ______________.

1. rectum 2. anus	**12–156.** Under normal circumstances, feces is stored only in the (1) ________________ portion of the (2) ________________.
1. sigmoid 2. colon	**12–157.** One reflex that brings about the desire to defecate is called the ________________ reflex.
gastrocolic	**12–158.** At the opening of the anus are two sphincters. The internal sphincter is under (1) ________________ (voluntary/involuntary) control, and the external sphincter is under (2) ________________ (voluntary/involuntary) control.

The Liver

1. involuntary 2. voluntary	**12–159.** In this last section, some of the functions of the liver will be considered. The liver is essential to life. Its vital functions include secretion, the formation and storage of essential body substances, and the detoxification of blood. First, as you have already learned, the liver secretes ________________, which is essential for the emulsification of fats.
bile	**12–160.** In addition to secreting bile, the liver stores and regulates the release of nutrients that are carried to the body cells through the ________________ stream.
blood	**12–161.** Nutrients are absorbed through the villi of the small intestine and pass into capillaries or lacteals. The capillaries, by way of tributaries, empty into a single vein, the PORTAL VEIN, which leads to the liver. The nutrient-laden blood from the small intestine is then carried to the liver through this ________________ vein.
portal	**12–162.** The liver stores nutrients that are mainly composed of GLYCOGEN (GLY—co—jen) and releases them into the bloodstream as they are needed by the body. The nutrients processed from the digestion of food are not released immediately but are stored by the ________________ until the body's output of energy requires their release.
liver	**12–163.** Another function of the liver, then, is to store and regulate the amount of nutrients, mainly ________________, which are released into the bloodstream.
glycogen	**12–164.** In addition to glycogen, the liver also stores HEPARIN (HEP—a—rin), an anticoagulant that maintains the fluidity of the blood. This anticoagulant, called (1) ________________, is both produced and stored in the (2) ________________.

1. heparin 2. liver	**12–165.** The liver also produces and stores fibrinogen and prothrombin, which serve as coagulants by aiding in the formation of blood clots at injured body sites. Thus, the liver produces and stores both an (1) ________________ (heparin) and two (2) ________________ (prothrombin and fibrinogen).
1. anticoagulant 2. coagulants	**12–166.** The liver also serves as a storehouse of VITAMINS, including vitamins A, B_{12}, and K. Vitamin B_{12}, which is stored in the ________________, is essential to the formation of red blood cells.
liver	**12–167.** After birth and during the later stages in the development of the embryo, red blood cells are produced by the red marrow of bones. In the early stages of embryonic development, before the red marrow is able to assume this function, the liver stores vitamin B_{12}, which is essential for the production of ________________ cells.
red blood	**12–168.** The liver, by storing vitamin B_{12}, produces red blood cells in the (mature adult/developing embryo) ________________.
developing embryo	**12–169.** Vitamin K, which is also stored in the liver, contributes to the production of prothrombin. Prothrombin, as you learned, is formed and stored in the liver and serves to assist in the ________________ of blood.
coagulation (clotting)	**12–170.** Iron and copper are also stored in the liver. These two elements, (1) ________________ and (2) ________________, are important in the production of hemoglobin.
1. iron 2. copper	**12–171.** The liver stores (and in some cases also forms) many essential body substances and releases them into the bloodstream as needed. Select the substances stored by the liver from the list below. a. heparin — f. mucus b. copper — g. prothrombin c. pepsin — h. fibrinogen d. iron — i. ptyalin e. vitamins — j. glycogen
a, b, d, e, g, h, and j	**12–172.** In addition to the secretion of bile and the formation and storage of many essential body substances, the liver also serves to remove toxic, or ________________, substances from the blood.
harmful	**12–173.** Many substances absorbed through the intestines into the bloodstream are toxic and potentially dangerous to the body. The blood from the intestines passes through the portal vein into the liver, where it is detoxified. Thus the liver removes harmful or ________________ substances from the blood.

toxic

12–174. In summary—select the functions of the liver from the list below.

a. produces red blood cells in the embryo
b. secretes bile
c. regulates the storage and release of nutrients
d. secretes ptyalin
e. stores glycogen, vitamins, copper, and iron
f. produces coagulants
g. produces gastrogen
h. produces anticoagulants
i. detoxifies blood

a, b, c, e, f, h, and i